Topics in Applied Physics   Volume 4

Topics in Applied Physics   Founded by Helmut K. V. Lotsch

Volume 1   **Dye Lasers**   Editor: F. P. Schäfer

Volume 2   **Laser Spectroscopy** of Atoms and Molecules
            Editor: H. Walther

Volume 3   **Numerical and Asymptotic Techniques in Electromagnetics**
            Editor: R. Mittra

Volume 4   **Interactions on Metal Surfaces**   Editor: R. Gomer

Volume 5   **Mössbauer Spectroscopy**   Editor: U. Gonser

Volume 6   **Picture Processing** and **Digital Filtering**
            Editor: T. S. Huang

Volume 7   **Integrated Optics**   Editor: T. Tamir

Volume 8   **Light Scattering in Solids**   Editor: M. Cardona

Interactions on Metal Surfaces

Edited by R. Gomer

With Contributions by
E. Bauer   M. Boudart   R. Gomer   S. K. Lyo
D. Menzel   E. W. Plummer   L. D. Schmidt
J. R. Smith

With 112 Figures

Springer-Verlag Berlin Heidelberg GmbH 1975

Professor ROBERT GOMER

The James Franck Institute, The University of Chicago, 5640 Ellis Avenue,
Chicago, IL 60637, USA

ISBN 978-3-662-30842-4   ISBN 978-3-540-37393-3 (eBook)
DOI 10.1007/978-3-540-37393-3

Library of Congress Cataloging in Publication Data. Gomer, Robert. Interactions on metal surfaces. (Topics in applied physics; v. 4). Includes bibliographies and index. 1. Surface chemistry. 2. Metallic surfaces. I. Bauer, Ernst. II. Title. QD506.G63.   541'.3453.   75-1281.

© by Springer-Verlag Berlin Heidelberg 1975
Originally published by Springer-Verlag Berlin Heidelberg New York in 1975
Softcover reprint of the hardcover 1st edition 1975

Preface

Surface phenomena encompass an important but enormously vast field. Any attempt to treat all or even a major portion of them in a book of this size is clearly impossible. I have selected metal surfaces and interactions on them, principally chemisorption, as the main topics of this book because they are of fundamental importance to almost all aspects of surface science and because progress, both theoretical and experimental is now very rapid in this field. Although catalysis goes considerably beyond the confines of the rest of the book a discussion of it has been included (Chapter 7) because of its intrinsic importance, its close relation to other subjects discussed here, and finally to indicate that there is more to interactions on surfaces than the adsorption of hydrogen on the (100) plane of tungsten.

To keep within the limits of this monograph-like series choices had to be made even within these restrictions. The intent of the book is to acquaint the reader with the theoretical underpinnings, the most important techniques currently being used to investigate metal surfaces and chemisorption, and finally to present some of the results of modern research in this area. It was necessary to omit a number of important topics, for instance field emission and field ion microscopy. In part, these omissions were based on the availability of reference books and review articles, and in part on the editor's sense of priorities for this particular book.

Chapter 1 discusses the electronic properties of surfaces, with emphasis on clean surfaces, largely from the point of view of W. KOHN and his school, and indicates how the linear response approach can be extended to treat chemisorption. Chapter 2 devotes itself specifically to the theory of chemisorption with emphasis on LCAO-MO methods, but includes some discussion of other approaches. Chapter 3 gives a very brief summary of the principal techniques used in chemisorption research but is mainly devoted to discussing what is known about a number of actual systems. Chapter 4 deals with various desorption techniques and the theory of such processes. Chapter 5 discusses the theory of electron spectroscopies, specifically field and photoemission, which provide insight into the electronic structure of surfaces and adsorption complexes, and presents a number of quite recent results obtained by these techniques. Chapter 6 is devoted to a discussion of low energy electron diffraction and Auger phenomena, two of the most important techniques for characterizing the geometry of clean and adsorbate covered metal and semiconductor surfaces, and for determining chemical composition on the atomic scale. Chapter 7, as already pointed out, is devoted to a discussion of catalysis.

Discussions of specific adsorption systems occur not only in Chapter 3 but figure importantly also in Chapters 4 and 5. In some cases there is redundancy in subject matter, but generally not in point of view. This comes about because there is as yet a great deal we do not understand about chemisorption. It is rather recent in fact that there is essentially universal agreement on the experimental *facts* for a given system. It is hoped that the occasional disagreements in interpretation, honestly presented, will not confuse the reader and will serve to give him the true flavor of the current status of this field.

Chicago, January 1975   ROBERT GOMER

Contents

1. Theory of Electronic Properties of Surfaces.
By J. R. SMITH (With 21 Figures)

1.1. A Simple Example: Surface States, Continuum States,
and Local Orbitals 2
1.2. Electron Work Function 5
1.3. Impurity Screening—Static Dielectric Response 16
1.4. Surface States and Surface Plasmons 23
    1.4.1. Surface States 23
    1.4.2. Surface Plasmons 28
1.5. Local Density of States 32
1.6. Status—Current Challenges 34
References 36

2. Theory of Chemisorption. By S. K. LYO and R. GOMER
(With 7 Figures)

2.1. Qualitative Discussion 41
2.2. Newns-Anderson Model 43
2.3. Reformulation of the Theory Using a Complete Basis Set . 51
2.4. Adsorbate-Adsorbate Interaction 57
2.5. Valence-Bond (Schrieffer-Paulson-Gomer) Approach . . 58
2.6. Linear Response (Kohn-Smith-Ying) Method 60
2.7. Concluding Remarks 60
References 61

3. Chemisorption: Aspects of the Experimental Situation.
By L. D. SCHMIDT (With 13 Figures)

3.1. Structures of Clean Surfaces 64
    3.1.1. Body Centered Cubic Transition Metals 65
    3.1.2. Face Centered Cubic Metals 67
    3.1.3. Semiconductors 67
    3.1.4. Insulators 68
3.2. Methods of Adsorbate Characterization 68
    3.2.1. Diffraction 69
    3.2.2. Kinetics 70

3.2.3. Auger Electron Spectroscopy (AES) 71
3.2.4. Miscellaneous Techniques 71
3.3. Crystallographic Anisotropies 72
   3.3.1. Stepped Surfaces 74
3.4. Binding States 77
   3.4.1. Hydrogen on Tungsten 77
   3.4.2. Carbon Monoxide on Tungsten 84
3.5. Adsorbate-Adsorbate Interactions 89
   3.5.1. Repulsive Interactions 89
   3.5.2. Attractive Interactions and Ordered Structures . . . 91
   3.5.3. Two-Dimensional Phase Transitions 93
3.6. Adsorption on Similar Metals 95
   3.6.1. H_2, N_2, and CO on W, Mo, and Ta 95
   3.6.2. CO on Ni, Pd, and Cu 97
   3.6.3. Comparison between fcc and bcc Substrates . . . 98
3.7. Summary 98
References 99

4. Desorption Phenomena. By D. MENZEL (With 12 Figures)
4.1. Thermal Desorption 102
   4.1.1. Desorption Mechanisms and Rate Parameters . . . 102
   4.1.2. Experimental Methods of Thermal Desorption . . 105
   4.1.3. Some Results of Thermal Desorption Measurements 115
   4.1.4. Theories of Thermal Desorption 120
4.2. Electron Impact Desorption 124
   4.2.1. Experimental Methods of EID and Evaluation of
       Data 125
   4.2.2. Experimental Results 128
   4.2.3. Theory of EID 131
   4.2.4. Practical Importance of EID 134
4.3. Photodesorption 135
4.4. Ion Impact Desorption 136
4.5. Field Desorption 136
4.6. Conclusion 138
References 138

5. Photoemission and Field Emission Spectroscopy.
By E. W. PLUMMER (With 31 Figures)
5.1. Preliminary Discussion of Field and Photoemission . . . 144
   5.1.1. Field Emission 145
   5.1.2. Photoemission 148

5.2. The Measurement Process 150
    5.2.1. Field Emission 150
    5.2.2. Clean Surfaces 151
5.3. Adsorbate Covered Surfaces 157
    5.3.1. Tunneling Resonance 157
    5.3.2. Inelastic Tunneling 160
    5.3.3. Many-Body Effects and Photoassisted Field Emission 162
5.4. Photoemission 163
    5.4.1. Photoexcitation; Primarily Atoms and Molecules . 163
    5.4.2. Cross Section for Photoemission 166
    5.4.3. Angular Dependence of Emission 171
    5.4.4. Relaxation Effects 174
    5.4.5. Bulk vs. Surface Emission 182
    5.4.6. Angular Resolved Surface Emission 189
5.5. Experimental Results 193
    5.5.1. Clean Surfaces 193
    5.5.2. Adsorption Studies 200
        Hydrogen Adsorption 200
        CO Adsorption—Multiple Binding States—
            Dissociation 206
        Hydrocarbon Decomposition 214
References 219

6. Low Energy Electron Diffraction (LEED) and Auger Methods.
   By E. BAUER (With 28 Figures)
6.1. Interaction of Slow Electrons with Condensed Matter . . 227
    6.1.1. Elastic Scattering 227
    6.1.2. Inelastic Scattering 230
    6.1.3. Quasielastic Scattering 233
    6.1.4. Consequences of the Scattering Processes 234
6.2. Low Energy Electron Diffraction (LEED) 235
    6.2.1. Experimental Methods 235
    6.2.2. Theoretical Methods 236
    6.2.3. Results 244
        Clean Surfaces 244
        Adsorption Layers 247
    6.2.4. Special Topics 251
        Inelastic Low Energy Electron Diffraction (ILEED) 251
        Polarized Electrons 252
        Low Energy Electron Diffraction Microscopy . . . 252
6.3. Auger Electron Methods 253
    6.3.1. Physical Principles 253

Free Atoms 253
Condensed Matter 257
6.3.2. Experimental Methods 259
6.3.3. Results 261
6.3.4. Special Topics 263
Details of the Auger Emission Process 263
In-Depth Auger Analysis and Auger Electron
Microscopy 266
6.4. LEED Structure Nomenclature and Superstructures . . . 268
6.5. Appendix: Recent Reviews of LEED and Auger Phenomena 270
References 271

7. Concepts in Heterogeneous Catalysis. By M. BOUDART

7.1. Definitions 275
7.2. Affinity, Reactivity and Catalytic Activity 277
7.3. Non-Ideal Catalytic Surfaces: the Ammonia Synthesis . . 280
7.4. Non-Ideal Catalytic Surfaces: the Water Gas Shift Reaction 283
7.5. Structural or Geometric Factors 285
7.6. Electronic or Ligand Factors 288
7.7. Promoters and Poisons 291
7.8. Active Centers 293
References 297

Additional References with Titles 299
Author Index 303
Subject Index 307

Contributors

BAUER, ERNST

Physikalisches Institut, Technische Universität Clausthal,
D-3392 Clausthal-Zellerfeld, Fed. Rep. of Germany

BOUDART, MICHEL

Department of Chemical Engineering, Stanford University,
Stanford, CA 94305, USA

GOMER, ROBERT

The James Franck Institute, The University of Chicago,
Chicago, IL 60637, USA

LYO, SUNG K.

The James Franck Institute, The University of Chicago,
Chicago, IL 60637, USA

MENZEL, DIETRICH

Physik-Department, Technische Universität München,
D-8046 Garching bei München, Fed. Rep. of Germany

PLUMMER, E. WARD

Department of Physics, University of Pennsylvania,
Philadelphia, PA 19104, USA

SCHMIDT, LANNY D.

Department of Chemical Engineering and Materials Science,
University of Minnesota, Minneapolis, MN 55455, USA

SMITH, JOHN R.

Research Laboratories, General Motors Technical Center,
Warren, MI 48090, USA

1. Theory of Electronic Properties of Surfaces

J. R. Smith

With 21 Figures

Surface or interface electronic structure has been generally recognized as being of pivotal importance in many technologies: solid state and gaseous electronics, catalysis, adhesion, and corrosion—to name a few. Despite its practical importance, our fundamental understanding of surface electronic properties has been greatly overshadowed by progress in understanding the bulk.

There is good reason for this situation. Experimentally, microscopically clean surfaces have been difficult to obtain. Theoretically, the loss of symmetry in the direction perpendicular to the surface greatly complicates calculations.

Recently, there has been rapid progress in our understanding of surfaces, however. The commercial availability of vacuum systems capable of 10^{-10} Torr as well as a number of analytical tools (see Chapters 3–6), have allowed experimentalists to obtain well characterized, clean surfaces. The well developed methods of bulk theory are being applied with encouraging success to the computation of many surface electronic properties.

In this chapter, it is hoped that the reader will obtain some feeling for the basic physical principles involved in the electronic properties of surfaces. It is not meant to be a review, however. Because of space limitations, we were not able to discuss surface energy determinations, but fortunately there is a recent review by LANG [1.1] (see also [1.2]). The reader interested in the companion field of adhesion is referred to the review of KRUPP [1.3] and the more recent theoretical efforts of FERRANTE and SMITH [1.4]. Small particle properties are related to those of solid surfaces, and for an introduction to the former we suggest the recently successful work of JOHNSON et al. [1.5]. Surface phonons are indirectly related to surface electronic properties, and we refer the interested reader to the review of WALLIS [1.6].

In Section 1.1, the subjects of surface states, continuum states, and local orbitals in the surface region are introduced via a simple example. Section 1.2 is devoted to the electron work function with results for a wide range of metals, including the effects of crystallinity and of fractional monolayers of adsorbed gases. Dielectric response,

including impurity screening, is discussed in Section 1.3 with application
to chemisorbed hydrogen. Resonance levels, electron scattering cross
sections, binding energies, vibrational modes, and electronic charge
distributions are considered. In Subsection 1.4.1 the subject of surface
states is treated, including recent theoretical results for the (111) plane
of Si. In Subsection 1.4.2, surface plasmons are discussed. Experimental
and theoretical results are given for Al, as examples. The nature of the
charge fluctuation associated with the surface plasmon is investigated.
The subject of the local density of states in the surface region is
discussed in Section 1.5. Results for a Ni d-band are presented.
Finally, Section 1.6 is devoted to a short description of the current
status of surface calculations.

1.1. A Simple Example: Surface States, Continuum States, and Local Orbitals

As an introduction to surface electronic structure, consider the simple
model surface [1.7] specified by the one-dimensional potential of
Fig. 1.1. The potential repeats periodically throughout the bulk and
then rises to the vacuum level, forming a surface barrier. This surface
barrier may be considered the wall of the box containing the electrons
within the solid. Within this surface barrier there is a well whose
depth is less than that of the bulk wells. This is reminiscent of a

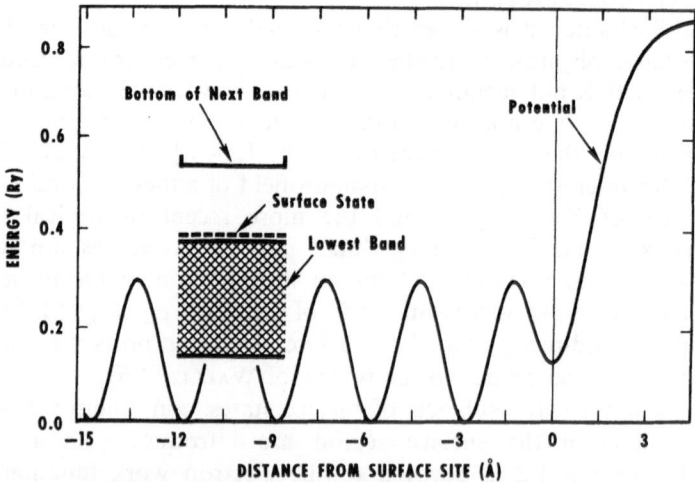

Fig. 1.1. Surface potential plot. The inset shows the relevant part of the energy spectrum
of the potential [1.7]

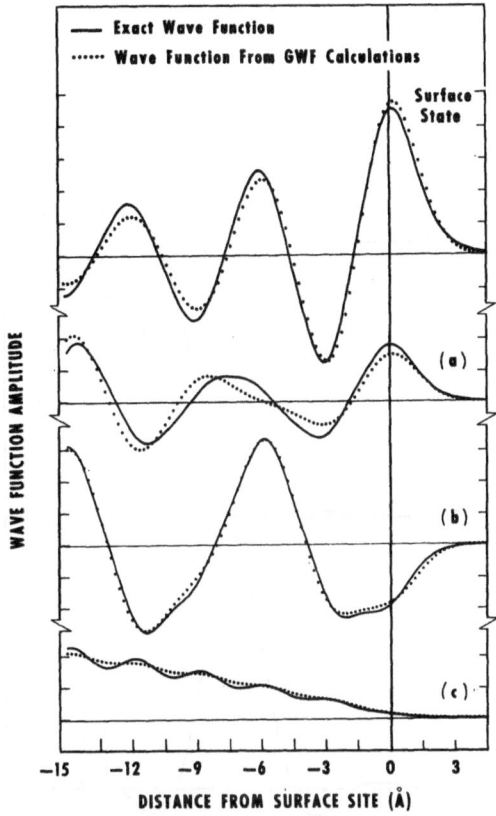

Fig. 1.2a-c. Top curves pertain to the surface state wave function, followed by examples of continuum states near to the top (a), middle (b), and bottom (c) of the lowest band [1.7]

chemisorbed layer, or an atomic layer of foreign particles which is chemically bonded to the surface.

There are two observations that can be made by inspection. The imposition of the surface barrier eliminates the periodicity in the direction perpendicular to the surface. Further, it is a quite strong and yet local perturbation.

The bulk bandwidth of the lowest band and the first band gap are also shown in Fig. 1.1. There is a surface state in the band gap. A surface state is an electronic state bound to the surface region. The local defect produced by the surface barrier can, under certain conditions, bind electrons.

Some of the wave functions of interest are plotted in Fig. 1.2. Let us consider some of the qualitative features of the wave functions.

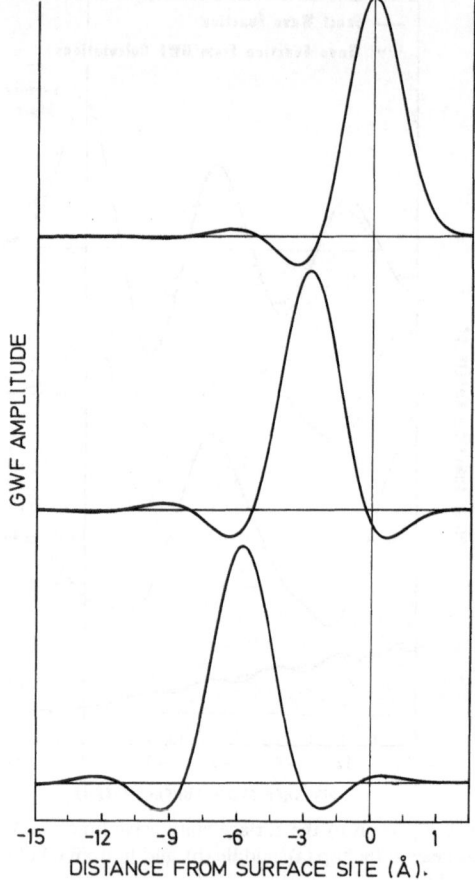

Fig. 1.3. Generalized Wannier functions belonging to the first three lattice sites [1.7]. The maxima of the GWF's lie very close to the minima of the successive potential wells shown in Fig. 1.1. By the third site in from the vacuum, the local function is essentially the bulk Wannier function

The surface state wave function is unique in that it decays in both the bulk and vacuum directions. Surface states will be discussed further in Subsection 1.4.1. The states within the band, or continuum states, decay only in the vacuum direction. They are, as it were, the tails of the bulk wave functions. This decay of the wave functions leads to a charge asymmetry or, in turn, an electrostatic barrier at the surface which contributes to the electron work function (Sect. 1.2).

This tailing effect and the appearance of surface states means that the local density of states [1.7], the density of states weighted by the square of the local-wave-function-magnitude, will vary in an important way as one proceeds from the bulk through the surface region. The breaking of bonds in forming the surface will generally lead to a narrowing of the local bands, but the appearance of bound or surface states in the band gap will tend to broaden the local density of states. These effects are seen experimentally in, e.g., ion-neutralization, field emission, and photoemission spectroscopies. A discussion of this can be found in Chapter 5 and in [1.8]. The effect of chemisorption on the local density of states will be discussed in Section 1.3, and the results of a calculation of the local density of states for a Ni d-band can be found in Section 1.5.

In Fig. 1.3 are plotted some of the generalized Wannier functions (GWF) for the lowest band [1.7]. The concept of a generalized Wannier function for nonperiodic systems has been quite recently introduced by KOHN and ONFFROY [1.9]. They are local, orthonormal functions, generally one for each lattice site. Total band energies, charge densities and local densities of states can be written in terms of these functions [1.7, 9]. The GWF were all determined via a single four parameter variational calculation. The dotted curves in Fig. 1.2 were obtained using the local functions of Fig. 1.3, showing a rather high accuracy.

Note that in the surface well (top graph of Fig. 1.3) there is considerable asymmetry in the local function. In the first plane in from the surface, the function is somewhat more symmetric. By the second plane in from the surface, the local function is essentially the bulk Wannier function. Thus these "local orbitals" differ from their bulk counterparts only in the first two or three planes in from the vacuum. This is characteristic of the width of the surface region that we deal with for many properties, i.e., 3–12 Å (see also Fig. 1.21).

1.2. Electron Work Function

The electron work function, as usually defined for metals [1.10–12] and for semiconductors [1.13, 14], is depicted schematically for 0 K as the quantity ϕ_e in Fig. 1.4. It is the energy difference between the Fermi level and the vacuum level just outside the surface in the case of both semiconductors and metals. Thus, if the work function of a reference solid surface is known, the work function of another solid surface can be determined from the contact potential difference. For metals, the photoelectric threshold is equal to the work function. This is not true in general for semiconductors [1.14]. For semiconductors

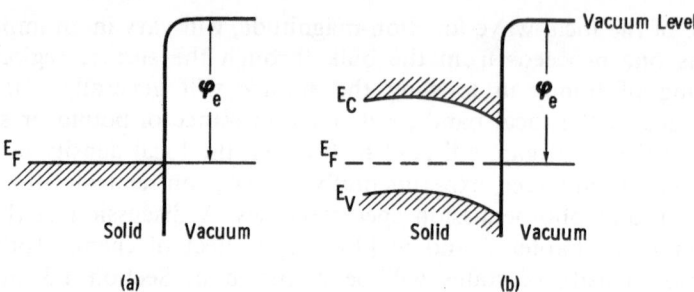

Fig. 1.4a and b. Schematic depicting the electron work function ϕ_e for a) a metal surface and b) a semiconductor surface. E_F is the Fermi energy, while E_c and E_v denote the bottom of the conduction band and the top of the valence band, respectively

the photoelectric threshold is equal to the energy difference between the vacuum level and the top of the valence band (sometimes called the ionization potential) when the bands are flat to the surface over the escape depth of emitted electrons and when surface state emission is negligible. There are many other ways of measuring the work function, and the interested reader is referred to RIVIÈRE's review [1.15] of experimental methods and results.

In the remainder of this section we will concentrate on theoretical calculations of the work function. Further, we will consider only metals, but will return to the semiconductor ionization potential for the case of Si in Subsection 1.4.1.

For a metal at 0 K,

$$\phi_e = -V(\infty) + E_{N-1} - E_N = -V(\infty) - \mu, \tag{1.1}$$

where $V(\infty)$ is the total electrostatic potential just outside the surface, E_N is the ground state energy of the metal with N electrons and μ is the chemical potential. In order to calculate μ, we must introduce a formalism which will be used extensively in later sections of this chapter.

HOHENBERG and KOHN (HK) [1.16] have shown that the ground-state energy E_v of a confined interacting electron gas can be written as a functional of the electron number density $n(r)$. Further, they have shown that $E_v[n]$ assumes a minimum value for the correct $n(r)$, if admissible density functions conserve the total number of electrons. Thus, $n(r)$ can be determined from [1.12]

$$(\delta/\delta n)\{E_v[n] - \mu N\} = 0, \tag{1.2}$$

where $N = \int n(r)\, dr$. HK write

$$E_v[n] = -\int V^{ex}(r) n(r)\, dr + 1/2 \iint \frac{n(r)\, n(r')}{|r - r'|}\, dr\, dr' + G[n] \qquad (1.3)$$

(atomic units are used throughout unless stated otherwise), where $V^{ex}(r)$ in the first term is a static external potential (that of the metal ion cores in our clean surface work function calculation); the second term is the ordinary electron-electron interaction energy; and $G[n]$ is the sum of the kinetic, exchange, and correlation energies of the electronic system.

Combining (1.3) and (1.2), we have

$$-V(r) + \delta G[n]/\delta n(r) = \mu, \qquad (1.4)$$

where the electrostatic potential is given by

$$V(r) = V^{ex}(r) - \int \frac{n(r')}{|r - r'|}\, dr'. \qquad (1.5)$$

The point at which the left-hand side of (1.4) is evaluated is arbitrary. It is instructive [1.10] to take the volume average of (1.4) and combine it with (1.1)

$$\phi_e = -[V(\infty) - \langle V \rangle] - \langle \delta G[n]/\delta n(r) \rangle. \qquad (1.6)$$

Equation (1.6) is the desired result because it exhibits the fact that the work function divides into two parts: $-[V(\infty) - \langle V \rangle]$ is the electrostatic barrier at the surface referred to in Section 1.1; $\langle \delta G[n]/\delta n(r) \rangle$ is a bulk term, the bulk chemical potential relative to the mean interior electrostatic potential. Because of the surface term, ϕ_e depends on crystalline plane exposed and impurity effects at the surface. Both of these effects will be discussed later in this section.

Because one can write ϕ_e in terms of volume averages (1.6), one might expect that reasonably accurate polycrystalline work functions of some metal surfaces could be obtained using a uniform-positive-background or jellium model depicted in Fig. 1.5. This is in fact the case, as was first shown for a wide range of metals by SMITH [1.12] (see also JONES and MARCH [1.11]).

The sum of the kinetic, exchange, and correlation energies was approximated by the local and first order terms of a gradient ex-

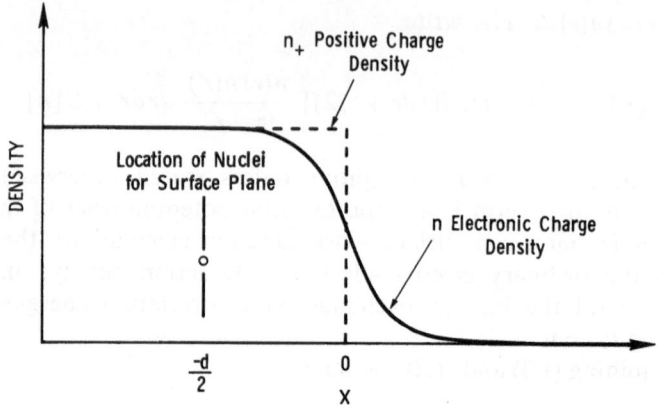

Fig. 1.5. Jellium model. (d is the distance between planes parallel to the surface)

pression [1.16]

$$G[n] = \frac{3}{10}(3\pi^2)^{2/3}\int n^{5/3}\,dr - \frac{3}{4}(3/\pi)^{1/3}\int n^{4/3}\,dr$$

$$-0.056\int \frac{n^{4/3}}{0.079 + n^{1/3}}\,dr + \frac{1}{72}\int \frac{(\nabla n)^2}{n}\,dr\,. \tag{1.7}$$

The integrands of the first three terms on the right-hand side of (1.7) represent, respectively, the kinetic, exchange, and correlation energy densities of a uniform electron gas of density n. The Wigner interpolation formula was used to represent the correlation energy of a homogeneous electron gas at metallic densities. The fourth term is the first of the inhomogeneity terms, i.e., those terms containing one or higher orders of the gradient operator acting on n.

The electrostatic barrier is given, in this model, by

$$-(V(\infty) - \langle V \rangle) = 4\pi \int_{-\infty}^{\infty} dx\, x[n(x) - n_+(x)], \tag{1.8}$$

where (Fig. 1.5)

$$n_+(x) = \begin{cases} n_+, & x \leq 0 \\ 0, & x > 0. \end{cases}$$

Correspondingly, the bulk contribution to ϕ_e is

$$\langle \delta G[n]/\delta n(r) \rangle = (1/2)\,(3\pi^2)^{2/3}\,n_+^{2/3} - (3/\pi)^{1/3}\,n_+^{1/3}$$

$$- \frac{0.056\,n_+^{2/3} + 0.0059\,n_+^{1/3}}{(0.079 + n_+^{1/3})^2}. \tag{1.9}$$

It remains to calculate $n(x)$ variationally, combining (1.2), (1.5), and (1.7). The inclusion of Poisson's equation (1.5) ensures self-consistency between the electronic charge density and the potential used to calculate it. Self-consistency is absolutely necessary in metal surface calculations, due to the spread of the electronic density into the vacuum, as shown in Fig. 1.5. To simplify the variational calculation, it was assumed [1.12] that the electron number density belongs to the following family of functions

$$n = n_+ - \frac{1}{2}\,n_+\,e^{\beta x}, \; x < 0$$

$$n = \frac{1}{2}\,n_+\,e^{-\beta x}, \; x \geq 0, \tag{1.10}$$

where β is a family parameter and n_+ is the bulk electron density (Fig. 1.5). The variational results then follow from quite simple analytical manipulations.

For the metals shown in Fig. 1.6, $0.40 \leq 1/\beta \leq 0.43$ Å. Since β is an inverse screening length, one can see that the decay of the electron gas into the vacuum is quite rapid. A comparison with experimental work function data is shown in Fig. 1.6. Rough agreement is obtained over a rather wide range of r_s ($r_s \equiv (1/4\pi\,n_+)^{1/3}$). For large r_s (e.g., C_s), the electrostatic barrier, $-[V(\infty) - \langle V \rangle]$ is only of the order of tenths of eV. Thus for large r_s the surface barrier is primarily due to the exchange and correlation potential part of $\langle \delta G[n]/\delta n(r) \rangle$. There is a gradual increase of the electrostatic contribution as r_s decreases, however. For example for Al, $-[V(\infty) - \langle V \rangle] = 6.0$ eV.

LANG [1.18] and LANG and KOHN (LK) [1.10] refined the jellium calculation of [1.12] by using a more accurate expression for the kinetic energy part of $G[n]$, following KOHN and SHAM [1.19]. In the second calculation, crystallinity was included. LK approximate $G[n]$ as

$$G[n] = T_s[n] - \frac{3}{4}(3/\pi)^{1/3}\int n^{4/3}\,dr - 0.056\int \frac{n^{4/3}}{0.079 + n^{1/3}}\,dr, \tag{1.11}$$

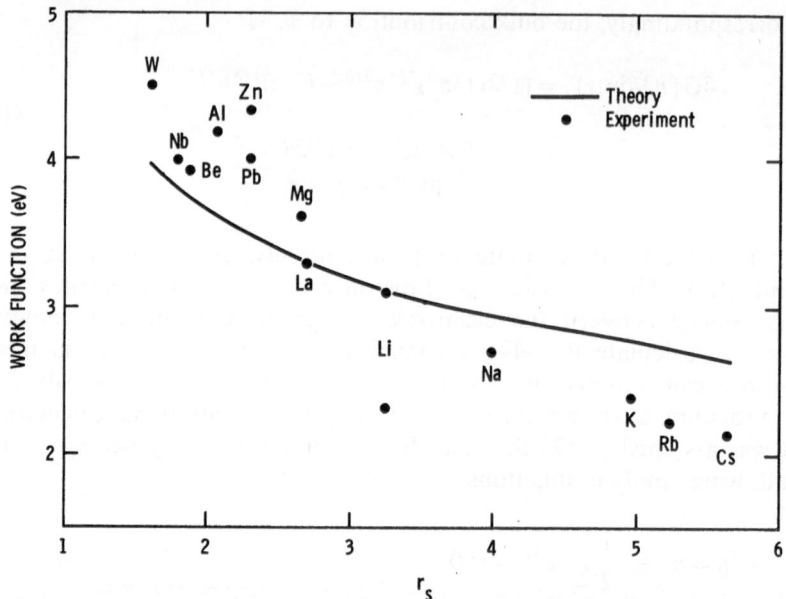

Fig. 1.6. Comparison of theoretical values [1.12] with the results of experiments on polycrystalline samples. For those metals listed in common with Fig. 1.7, the experimental values are those chosen in [1.10]. Otherwise, the data was taken from [1.17]

where $T_s[n]$ is the kinetic energy of a system of noninteracting electrons of density $n(r)$. Then the $n(r)$ which satisfies (1.2), (1.3), and (1.11) is found [1.13] by solving the equations

$$\left[-1/2 V^2 - V(r) - (3/\pi)^{1/3} n^{1/3}(r) \right.$$
$$\left. - \frac{0.056 n^{2/3}(r) + 0.0059 n^{1/3}(r)}{(0.079 + n^{1/3}(r))^2} \right] \psi_i(r) = \varepsilon_i \psi_i(r) \qquad (1.12)$$

and setting

$$n(r) = \sum_{i=1}^{N} |\psi_i(r)|^2 . \qquad (1.13)$$

$V(r)$ is given, as before, in (1.5), thus providing self-consistency. LK take $V^{ex}(r)$ to be the electrostatic potential of the jellium charge density n_+ (Fig. 1.5). Thus their solution for the electron number density n depends only on x. The dipole barrier and bulk contribution to ϕ_e are again given by (1.8) and (1.9) respectively.

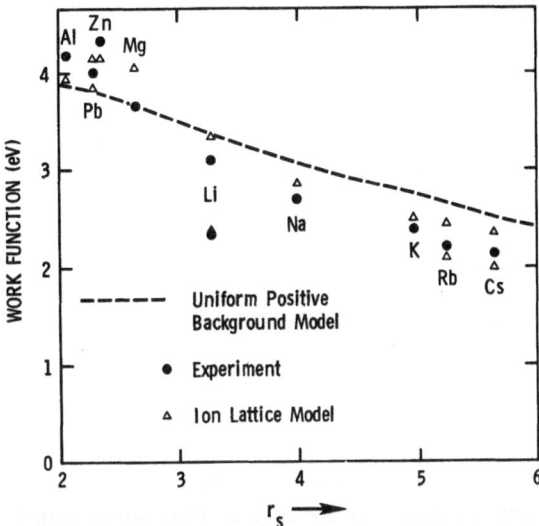

Fig. 1.7. Comparison of theoretical values [1.10] with the results of experiments on polycrystalline sample. The experimental values are those chosen in [1.10]. The ϕ_e values in the ion-lattice model were computed for the (110), (100), and (111) faces of the cubic metals and the (0001) face of the *hcp* metals (Zn and Mg). For qualitative purposes, the simple arithmetic average of these values for each metal is indicated by a triangle (two triangles are shown for those cases in which there are two possible pseudopotential radii r_c). The experimental and theoretical points for Zn should be at $r_s = 2.30$; they have been shifted slightly on the graph to avoid confusion with the data for Pb

Their results for the jellium model are shown as a dashed line in Fig. 1.7. They are quite close to those obtained in [1.12]. (See [1.18] for a direct comparison.)

Their most important contribution was to consider the effect of crystallinity on the work function, which will now be described [1.20]. LK introduce the discrete ion cores via first order perturbation theory, using pseudopotentials of the form employed by Ashcroft and Langreth [1.21]

$$V_{\text{PS}}^{\text{ex}}(r) = \begin{cases} 0, & r \leqq r_c, \\ z/r, & r > r_c. \end{cases} \tag{1.14}$$

The difference between an array of pseudopotentials of the form given in (1.14) and the potential of the positive charge density of the jellium provides a perturbing potential $\delta V^{\text{ex}}(r)$. Using the definition of the work function given in (1.1), the first order change in ϕ_e due to $\delta V^{\text{ex}}(r)$

is then, by standard perturbation theory,

$$\delta\phi_e = \int \delta V^{ex}(r)\, n_\sigma(x)\, dr,\tag{1.15}$$

where $n_\sigma(x)$ is the difference in the (ground state) electron number density distributions occurring when one electron is removed from the unperturbed system, the jellium metal. Since

$$\int n_\sigma(x)\, dr = -1,\tag{1.16}$$

$$\delta\phi_e = -\int \delta V^{ex}(x)\, n_\sigma(x)\, dx / \int n_\sigma(x)\, dx,\tag{1.17}$$

where

$$\delta V^{ex}(x) \equiv \int \delta V^{ex}(r)\, dy\, dz.\tag{1.18}$$

The $(N-1)$-electron ground state has, by Gauss's law, a weak electric field perpendicular to the surface at a point outside the surface. For a sufficiently weak electric field, $n_\sigma(x)$ will depend linearly on the field strength (see Section 1.3). Because, by (1.17), $\delta\phi_e$ is homogeneous in $n_\sigma(x)$, $n_\sigma(x)$ can now be taken to be the surface-charge density linearly induced by an *arbitrary* weak electric field perpendicular to the surface. The screening charge density for a weak applied field and $r_s = 2$ (corresponding approximately to Al), is given [1.22] in Fig. 1.8, using the coordinate system defined in Fig. 1.5. LK denote by x_0 in Fig. 1.8 the center of mass of the screening charge,

$$x_0 = \int_{-\infty}^{\infty} x n_\sigma(x)\, dx / \int_{-\infty}^{\infty} n_\sigma(x)\, dx.\tag{1.19}$$

They showed quite generally [1.22] that in fact x_0 is the location of the image plane. This important result was verified in a particular application by YING et al. [1.23]. A knowledge of the image plane location is useful in e.g., chemisorption, field emission, and small-gap condenser analyses [1.22].

Once $n_\sigma(x)$ is known, $\delta\phi_e$ can be calculated via (1.15). A simple arithmetic average of the values for the closest packed planes are listed as triangles in Fig. 1.7. Two results are listed for those elements for which ASHCROFT and LANGRETH gave two choices for the ion core radius, r_c. There is an improvement over the jellium results (dashed line) on comparison with the polycrystalline experimental results indicated by the dots.

More recently, APPELBAUM and HAMANN (AH) [1.24] have treated the (100) surface of Na self-consistently and beyond the perturbative

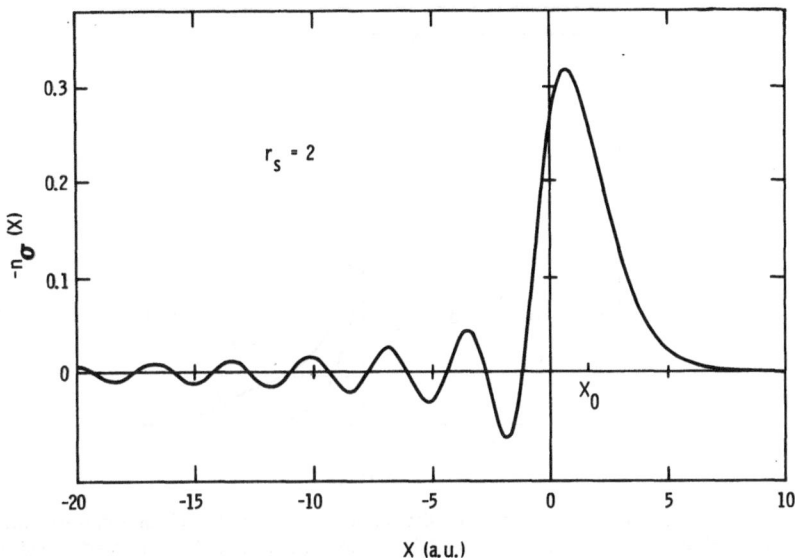

Fig. 1.8. Plot of the screening charge density $n_\sigma(x)$ for a weak applied field at $r_s = 2$, taken from [1.22]. x_0 locates the center of mass of the screening charge and the image plane as well. The edge of the uniform positive background (Fig. 1.5) is at $x = 0$

approximation. They solved (1.5), (1.12), and (1.13) to a good approximation, taking $V^{ex}(r)$ to be of the form given in (1.14). This means that the electron density solution n is three dimensional. They obtained a ϕ_e value for Na(100) of 2.71 eV, which is to be compared with 2.75 eV obtained by LK and 2.93 eV by SMITH. ALLDREDGE and KLEINMAN (AK) [1.25] have computed the three-dimensional charge density and the self-consistent energy band structure of a (001) film of Li. They also solved (1.5), (1.12), and (1.13) to a good approximation, using a non-local ion core pseudopotential $V^{ex}(r, r')$. They obtained $\phi_e = 3.70$ eV. This is 0.4 eV larger than the larger of LK's two results for Li(100) and 0.6 eV larger than SMITH's value. AK point out that the difference with LK may be due mainly to the repulsive (and non-local) part of their pseudopotential not being optimal. The methods of AH and AK will be described in more detail in Subsection 1.4.1.

It was mentioned earlier in this section that adsorbed particles on the solid surface can affect ϕ_e through the surface electrostatic barrier term $-[V(\infty) - \langle V \rangle]$. For example, a monolayer of hydrogen on W(100) increases the work function by ~ 0.9 eV [1.26]. On the other hand, it has been known for many years [1.27, 28] that a fractional

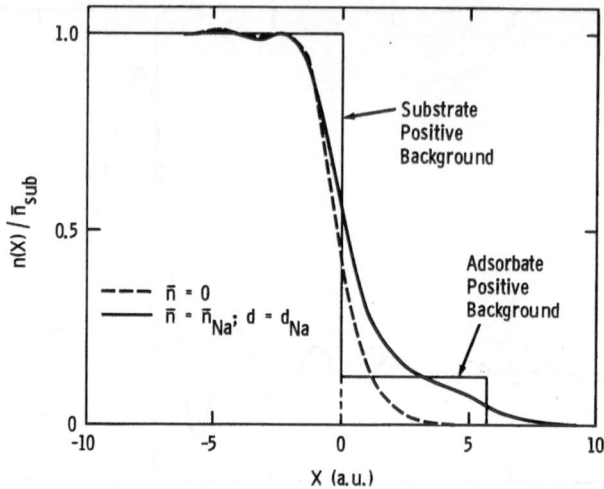

Fig. 1.9. Self-consistent electron density distributions $n(x)$ from [1.31] for a bare-substrate model ($r_s^{(sub)} = 2$, $\bar{n} = 0$), and for a model of substrate with full layer of adsorbed Na atoms ($\bar{n} = \bar{n}_{Na}$)

monolayer of an alkali metal like Cs can lower the work function of a refractory transition metal by up to 4 eV [1.29]. For a theoretical discussion relating to this comparison see [1.23].

This makes the alkalies quite an interesting theoretical problem. The remainder of this section will be devoted to LANG's [1.31] theory of work function changes induced by alkali adsorption.

Following the lead of the successful jellium ϕ_e calculations, LANG represented the alkali adsorbate layer by a jellium slab (see Fig. 1.9). The metal substrates for which most of the change in work function ($\Delta\phi_e$) vs alkali coverage data are taken are refractory transition metals. LANG argued, however, that many of the data were insensitive to the substrate element. He therefore took $r_s = 2.0$ for the substrate, which is roughly characteristic of the bulk density of Al. The thickness d of the adsorbate slab was taken to be equal to the bulk alkali spacing between closest packed lattice planes. The density $\bar{n}$ of the adsorbate slab was determined so that

$$\bar{n}d = N, \tag{1.20}$$

where N is the number of alkali atoms per unit area.

Since the surface is only a tiny fraction of the total volume of the crystal, $\langle \delta G[n]/\delta n(r) \rangle$ is changed only infinitesimally by a monolayer

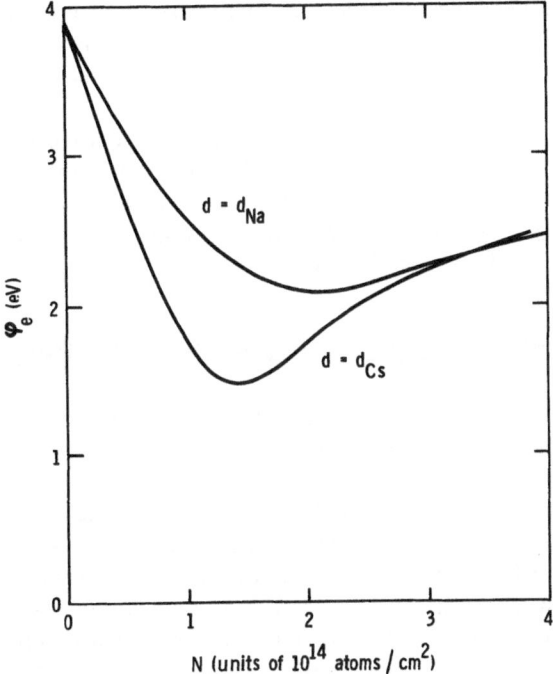

Fig. 1.10. Electron work function ϕ_e as a function of coverage of Na and Cs, taken from [1.31]. Values of the density at the work function minimum indicated here are substantially smaller than those determined for adsorption on a substrate such as W(110), whose zero-coverage ϕ_e is much larger than that of the model

of adsorbed alkali atoms. Thus, from (1.6),

$$\Delta\phi_e = -\Delta[V(\infty) - \langle V \rangle], \qquad (1.21)$$

or, the alkali influences the work function by altering the electrostatic barrier at the surface. To find the electrostatic barrier, one need only calculate $n(x)$ via (1.12), (1.13), and (1.5). $V^{ex}(r)$ is the sum of the electrostatic potentials of the substrate jellium ($r_s = 2.0$) and the adsorbate slab of density $\bar{n}$. In Fig. 1.9, $n(x)$ for zero coverage and monolayer coverage are plotted for Na adsorption. The results for ϕ_e for both Na and Cs as a function of adsorbate coverage are shown in Fig. 1.10. These curves exhibit a minimum at fractional monolayer coverage as does the experimental data.

The results of Fig. 1.10 also have a qualitative explanation in terms of an atomic orbital picture first proposed by GURNEY [1.32]. The adsorbed alkali atoms are partially ionized because their broadened valence level lies largely above E_F. The resulting electrical double layer lowers the electrostatic barrier at the surface, and hence the work function. As the coverage of the alkali atoms increases and hence the work function decreases, the broadened valence level is presumably lowered relative to E_F. This means the net positive charge of each alkali adatom is lowered, and hence the contribution per particle to the double layer is decreased. This can lead to the saturation or minimum shown in Fig. 1.10. For more recent employment of this picture see the references listed by LANG [1.31].

1.3. Impurity Screening — Static Dielectric Response

The static (zero frequency) screening of particles and defects, which plays an important role in the physics of bulk metals, is also of considerable interest in the physics of metal surfaces. Some typical applications are: the effect of the metal ions (i.e., crystallinity), on the electron density distribution near the surface; interactions of adatoms with metal surfaces; and mutual interactions of adatoms.

Several workers have dealt with static screening at a solid surface [1.33]. While considerable understanding of surface screening fundamentals has resulted from these treatments, they all deal with an electron gas bounded by an infinite potential step. Self-consistency is now known to be important in surface impurity screening [1.34], as it is also in "bare" or clean surface calculations [1.12, 10, 24, 25]. In the following we consider only a self-consistent response function formalism which will first be presented in quite general form. It will then be applied in approximate form to hydrogen chemisorption as an example [1.35, 23, 30]. Frequency dependent response will be discussed in Subsection 1.4.2.

The formalism of dielectric response follows succinctly from the density functional approach introduced in Section 1.2 [1.30, 34]. Consider first an unperturbed inhomogeneous electron gas in an external potential $V_0^{ex}(r)$ and with electron number density $n_0(r)$. For example, $V_0^{ex}(r)$ might correspond to a crystalline array of metal ion core potentials and $n_0(r)$ the electron number density distribution at a "clean" metal surface. Now introduce an additional small charge density $\varrho_1^{ex}(r)$ (such as an ion core of a chemisorbed particle), giving rise to a perturbing potential

$$V_1^{ex}(r) = \int \left[\varrho_1^{ex}(r')/|r - r'| \right] dr' . \tag{1.22}$$

What is the resulting change $n_1(r)$ of the electronic density?

Linearizing (1.4 and (1.5), i.e., writing $V^{ex} = V_0^{ex} + V_1^{ex}$, $n = n_0 + n_1$, etc., gives first the equations for the zeroth-order density

$$- V_0(r) + \left[\frac{\delta G[n]}{\delta n(r)} \right]_{n_0} = \mu_0 , \tag{1.23}$$

$$V_0(r) = V_0^{ex}(r) - \int \frac{n_0(r')}{|r - r'|} dr' . \tag{1.24}$$

The first order screening density is given by

$$- V_1(r) + \int n_1(r') \left[\frac{\delta^2 G[n]}{\delta n(r) \delta n(r')} \right]_{n_0} dr' = \mu_1 , \tag{1.25}$$

$$V_1(r) = V_1^{ex}(r) - \int \frac{n_1(r')}{|r - r'|} dr' . \tag{1.26}$$

A calculation of $n_1(r)$ then begins with a (self-consistent) solution of (1.23) and (1.24) for $n_0(r)$ and $V_0(r)$. These results can then be used in (1.25) and (1.26) to find $n_1(r)$ and $V_1(r)$.

One can define the density-potential response function $L(r, r')$ as

$$V_1(r) = \int L(r, r') \varrho_1^{ex}(r') dr' . \tag{1.27}$$

It follows from (1.27) that L is the induced potential V_1 due to a point perturbing charge $\varrho_1^{ex} = \delta(r - r')$. Once L is calculated for that system, one can use it through (1.27) to find the induced potential due to a general perturbing charge distribution. It can be shown that [1.23] reciprocity is obeyed:

$$L(r, r') = L(r', r) . \tag{1.28}$$

However, $L(r, r') = L(|r - r'|)$ only for homogeneous systems, and therefore unfortunately not in surface calculations.

Up to now, we have made no approximations other than linearization in our formalism. We would now like to apply it to chemisorption on a metal surface, and make the following approximations:

1) The unperturbed system (substrate) is represented by a jellium model (Fig. 1.5) with appropriate positive charge density.

2) The density functional $G[n]$ is given by (1.7). All of these approximations are discussed at length in [1.23] and [1.30], where it is concluded that they can provide a good zeroth order model for chemisorption on the low index planes of a refractory transition metal

like tungsten. This is a system for which there is considerable experimental data due to the possibility of obtaining clean refractory metal surfaces.

Because of the translational invariance parallel to the surface of the unperturbed (jellium) system, it is convenient to introduce the two-dimensional Fourier transform of the various quantities in the form

$$L(Q, x, x') = \int du\, e^{-i\boldsymbol{Q}\cdot\boldsymbol{u}}\, L(u, x; x') \qquad (1.29)$$

etc., when $\boldsymbol{Q} \equiv (0, Q_y, Q_z)$, $\boldsymbol{u} \equiv (0, y - y', z - z')$, $u \equiv |\boldsymbol{u}|$, and $Q \equiv |\boldsymbol{Q}|$. Then (1.25) and (1.26) become [1.30, 1.23]:

$$-\frac{1}{36 n_0(x)}\frac{d^2 n_1(Q, x)}{dx^2} + \frac{1}{36 n_0^2(x)}\frac{dn_0(x)}{dx}\frac{dn_1(Q, x)}{dx}$$
$$+ \left(X + \frac{Q^2}{36 n_0(x)}\right) n_1(Q, x) - V_1(Q, x) = \mu_1\, \delta(Q) \qquad (1.30a)$$

and

$$\frac{d^2 V_1(Q, x)}{dx^2} - Q^2 V_1(Q, x) = 4\pi[n_1(Q, x) - \delta(x - x')], \qquad (1.30b)$$

where

$$X = \frac{1}{3}(3\pi^2)^{2/3} n_0^{-1/3}(x) - \frac{1}{3}(3/\pi)^{1/3} n_0^{-2/3}(x)$$
$$- \frac{(\boldsymbol{\nabla} n_0(x))^2}{36 n_0^3(x)} + \frac{\boldsymbol{\nabla}^2 n_0(x)}{36 n_0^2(x)}. \qquad (1.31)$$

As discussed immediately following (1.27), we have taken $\varrho_1^{ex} = \delta(\boldsymbol{r} - \boldsymbol{r}')$ in Poisson's equation (1.30). Equations (1.30a) and (1.30b) can now be solved simultaneously (self-consistently) to yield the screening potential V_1 and density n_1 for a point perturbing charge. An immediate application of the results would be to hydrogen chemisorption (see Chapter 2 for a review of theoretical and experimental chemisorption). The proton immersed in the inhomogeneous electron gas of the metallic surface region is the point perturbing charge. We chose $r_s = 1.5$ for n_+ in an effort to approximate [1.12, 30] a close packed surface plane for a refractory transition metal like tungsten.

In the remainder of this section, results are given for a number of observables in the hydrogen-tungsten system.

First we determine the hydrogen ion-metal substrate interaction as a function of the coordinates of the proton $(u = 0, x')$. This is given,

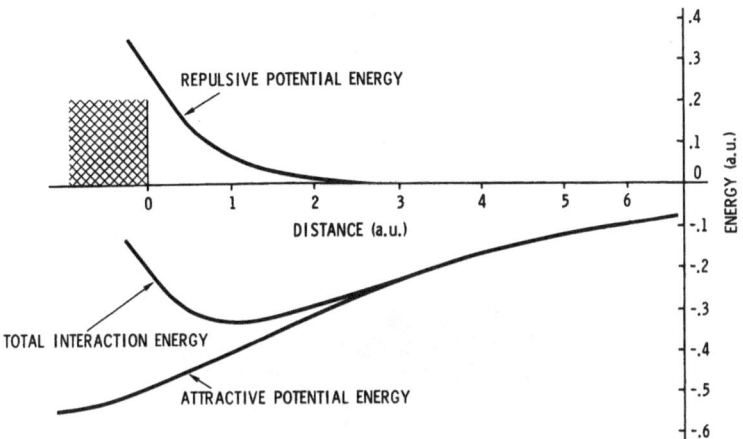

Fig. 1.11. Hydrogen ion-metal interaction energy versus separation distance [1.23, 30]

according to the Hellmann-Feynman theorem, by

$$W(x') = V_0(x') + 1/2\,\bar{V}_1(x = x', u = 0)\,, \qquad (1.32)$$

where $V_0(x')$ is the electrostatic potential of the "bare" surface and $\bar{V}_1(x = x', u = 0)$ is the potential of the screening charge, evaluated at the proton location. The results are shown in Fig. 1.11. There is competition between the repulsive term $V_0(x')$ and the attractive term $1/2\,\bar{V}_1(x = x', u = 0)$. This results in a minimum in $W(x')$ at $x' = 1.08$ a.u. The depth of the minimum is the ionic desorption energy, $E_I = 9$ eV. The experimental value [1.35, 36] for hydrogen on tungsten is $E_I = 11.3$ eV. This allows a measure of the accuracy of the calculation. For a first-principles calculation with no adjustable parameters, this sort of agreement is encouraging in light of the state-of-the-art in surface physics today. It should be noted that hydrogen is singular in that its ionic desorption energy E_I is much larger than that of other chemisorbed species.

The adsorbed hydrogen will exhibit vibrational modes in the potential well of Fig. 1.11. The excitation energy is given by the usual relation

$$\hbar\omega = \hbar\left(\frac{1}{m}\frac{d^2 W(x)}{dx^2}\right)^{1/2}_{x_m}, \qquad (1.33)$$

where $x_m = 1.08$ a.u. is the position of the proton at its energy minimum, and m is the proton mass. The theoretical value is 200 meV, while the

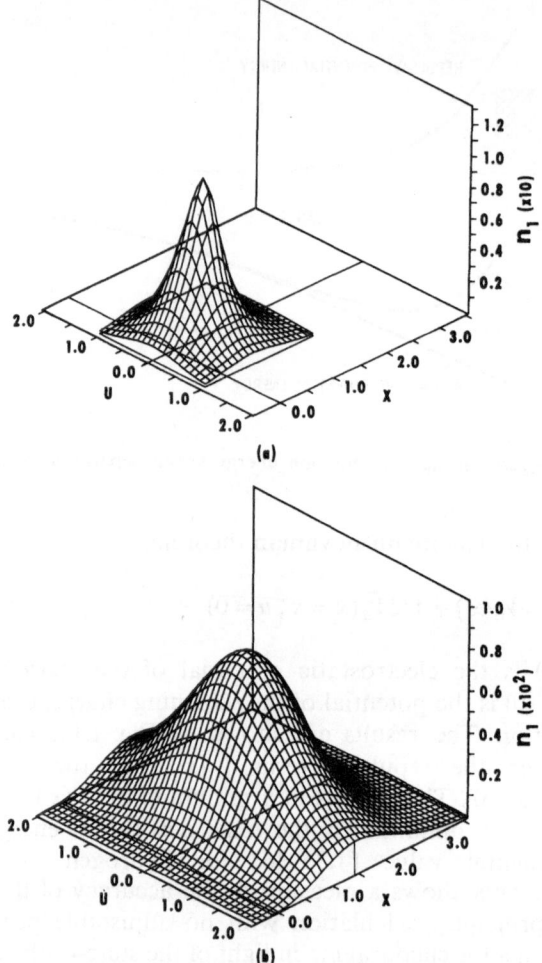

Fig. 1.12a and b. Screening charge density n_1 for two proton positions: a) at $u = 0$, $x = 0.571$ Å, the interaction energy minimum, i.e., the location for chemisorbed H; b) at $u = 0$, $x = 2.29$ Å, a small displacement in the vacuum direction from the adsorption site [1.23, 35]. u is the cylindrical coordinate in the plane parallel to the surface measured from an axis through the proton. Scaled units [1.33] are used for n_1. x and u are given in Å. The vertical line denotes the n_1 peak location. The peak density in a) is 13 times the peak density in b) (note change in scale for n_1)

experimental result is 140 meV [1.38, 39]. This fair agreement between the single-adsorbate theoretical result and experiment is evidence of dissociative adsorption [1.40]. This is consistent with computations [1.23, 30] of the short range (≤ 2 a.u.) interaction energy between hydrogen adatoms. This short range energy is repulsive, indicating dissociation. This supports the numerous experimental contentions (see, e.g., [1.41]), that hydrogen dissociates upon adsorption in the first adlayer on tungsten and certain other metals.

One might well ask, "what does the chemisorbed hydrogen look like?" The screening charge $n_1(r)$ is shown for two positions of the proton in Fig. 1.12, $x' = 0.571$ Å and $x' = 2.29$ Å. When $x' = 0.571$ Å, the hydrogen is located at the energy minimum (Fig. 1.11), and the screening charge is rather symmetric about the proton. However, when the proton moves to $x' = 2.29$ Å, the screening charge is quite asymmetric. Further, the screening charge spreads in the direction parallel to the surface, i.e., in the u direction (note the change in scale on the n_1 axis between Fig. 1.12a and b). As $x' \to \infty$, the spreading becomes complete, and n_1 becomes a function of x only.

With this rapid change in shape of the chemisorbed hydrogen as the proton moves about in the surface, one might expect that low energy electron scattering properties would depend on the location of the hydrogen. We will see later that this is in fact the case. When hydrogen is introduced onto a clean W(100) surface, a $c(2 \times 2)$ low energy electron diffraction (LEED—see Chapter 4) pattern is observed [1.42] which exhibits additional (half-order) beams. The intensity of these extra beams is comparable to the other (integral-order) beams of the pattern. The result is made more interesting by the fact that the isolated atom back-scattering cross sections for hydrogen are an order of magnitude smaller than those of the W atom [1.43] in the 50–100 eV energy range. Analogous data have been obtained for other light adsorbates on heavy substrates. Some authors (see, e.g., GERMER [1.44]) feel these data indicate reconstruction of the substrate. Others (see, e.g., BAUER [1.45a] and TRACY and BLAKELY [1.45b]) believe that reconstruction is not a necessary condition. More recently, JENNINGS and McRAE [1.46] have shown for H on W(100) that inter-layer multiple scattering can lead to fractional-order beams of comparable intensity to neighboring integral-order beams. They used isolated H atom phase shifts, however. In an attempt to obtain further information on the subject, we calculated [1.23, 35] differential scattering cross sections (DCS) for chemisorbed (Fig. 1.12a) and isolated hydrogen atoms. Figure 1.13 exhibits, for the first time, the effect of chemisorption on DCS. Note that the forward scattering DCS are considerably enhanced by chemisorption. Our results appear to support the picture of JENNINGS and McRAE, but one would have to do a multiple scattering calculation using the DCS (or, equivalently, our phase shifts) to be certain.

In Fig. 1.14, the total electronic potential in the vicinity of the chemisorbed hydrogen is plotted. This is the electrostatic potential of the proton (hence the "spike"), plus the electrostatic and (Kohn-Sham)

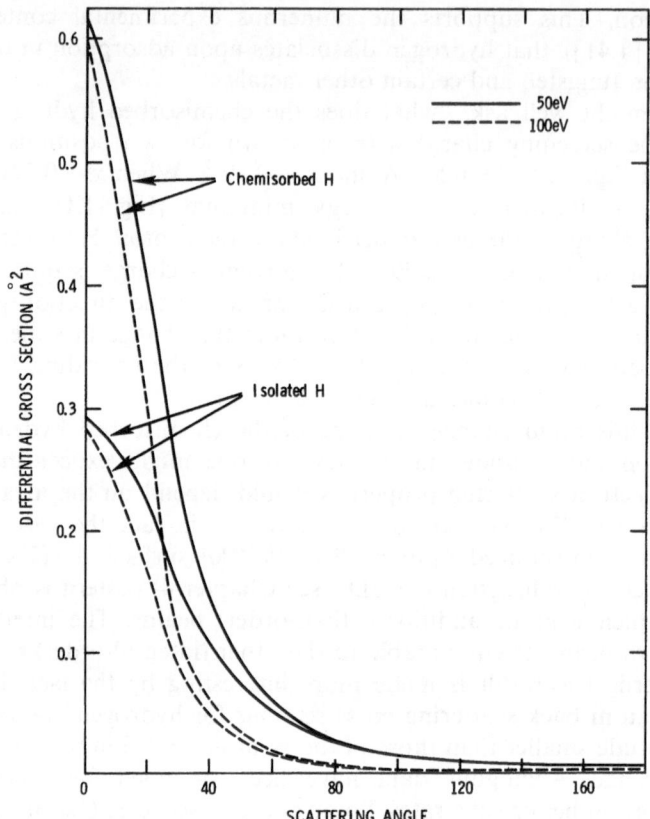

Fig. 1.13. Differential scattering cross sections for chemisorbed and isolated hydrogen atoms. The chemisorbed atom is located at the energy minimum of Fig. 1.11 [1.23, 35]

exchange potential of the screening charge. Notice that the potential dips below zero on the bulk side of the proton. This was much more striking in the earlier example, Fig. 1.1, where the peak at ≈ -1.5 Å is considerably lower than the vacuum level. One might well ask about the virtual energy levels associated with this potential well in Fig. 1.14. That is, it would be of considerable interest to determine the local density of states associated with the chemisorbed hydrogen. This interest is enhanced by the recent ability of experimentalists to detect local densities of states associated with adsorbed layers [1.47, 48]. We have found [1.23, 35] a resonance level associated with the potential well of Fig. 1.14 which lies 5.6 eV below the Fermi level. Plummer and Waclawski [1.48] find peaks in their photoemission energy spectra for hydrogen adsorption at 5.7 eV and 6.3 eV below the Fermi level for

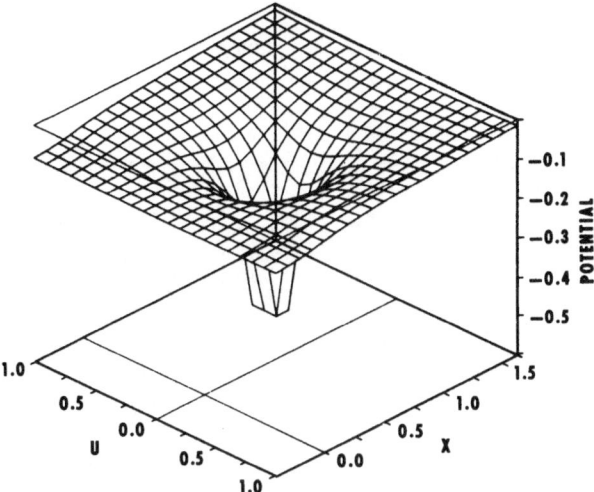

Fig. 1.14. Total electronic potential in the vicinity of the chemisorbed hydrogen. The potential is in a.u., while u and x are in Å [1.35]

W(100) and W(110) respectively. The agreement between theory and experiment indicates that these low-lying states are localized primarily on the adsorbate (compare with [1.49]). Our calculated level width is considerably larger than the observed value, however. We refer the interested reader to [1.23] and [1.35] for a discussion of this. This is the first *a priori* self-consistent calculation of a resonance level associated with an adsorbate potential well.

In Chapter 2, a LCAO-MO (linear combination of atomic orbitals-molecular orbital) treatment of H on W(100) is presented by Lyo and Gomer. A resonance at 5–6 eV below the Fermi level is also predicted.

Qualitative considerations [1.23] suggest that none of the approximations listed in the paragraph after (1.28) is unreasonable for the problem at hand. It is noteworthy that agreement between our theory and experiment is quite good. At the same time we consider it very important that improvements on all our approximations be carried out so that the theory can be put on a firmer and more quantitative basis.

1.4. Surface States and Surface Plasmons

1.4.1. Surface States

In Section 1.1 we saw an example of an electronic state which is bound to the surface region, i.e., it decays in both the vacuum and bulk directions. Tamm [1.50] first predicted that surface states could exist at

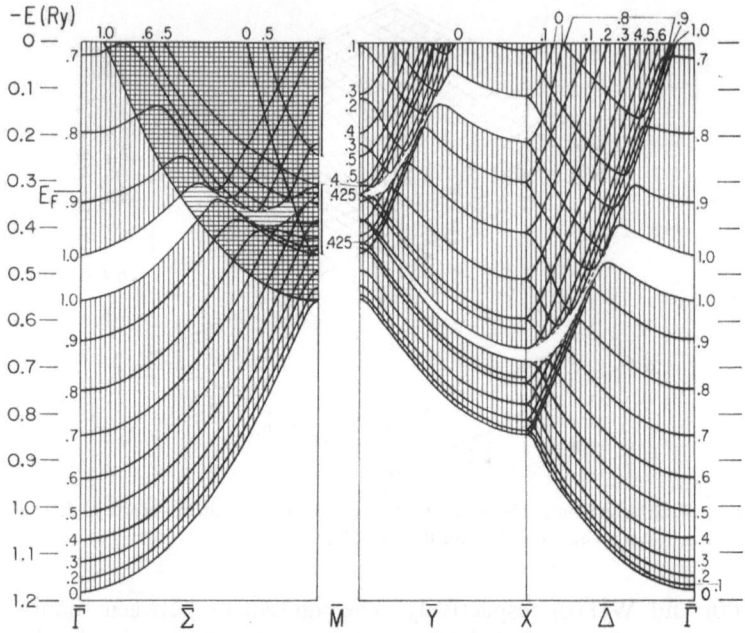

Fig. 1.15. Two-dimensional projection of the three-dimensional energy bands of aluminum [1.55]. The horizontal and vertical crosshatching represent continua of states with different symmetries. The numbers labeling the various bands represent values of k_x in units of $2\pi/a$, where a is the edge length of the fundamental cube

crystal interfaces. Many years later, the then budding semiconductor device industry learned that these states can strongly influence many properties of devices made from these materials. Surface states have also been reported experimentally at metal surfaces using field emission, photoemission, and Auger [1.51]. An extensive review (through 1970) of theory and experiment of surface states has been given by DAVISON and LEVINE [1.52] (see also [1.53]).

Let us consider a crystalline solid-vacuum interface. If there is to be a surface state at energy E with reduced wave vector parallel to the surface $k_\parallel$, then: a) the state must be a valid eigenfunction of Schrödinger's equation, b) the energy E must be less than the vacuum level (so that the electron is bound to the solid), c) there must be no continuum (band) states with the same E and $k_\parallel$ (so that electron cannot escape into the bulk). A thorough study of these requirements has been made by HEINE [1.54]. In Fig. 1.15 is a two-dimensional (001) projection of the three-dimensional energy bands of aluminum [1.54]. b) and c) require that surface states exist only within energy

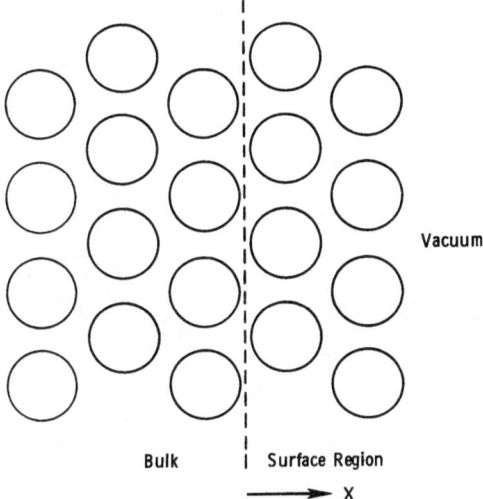

Vacuum

Bulk   Surface Region

⟶ X

Fig. 1.16. Schematic representation of a solid surface, indicating the mathematical separation into "surface" and "bulk" regions used in [1.58]

gaps of this two-dimensional projection. a) must then be satisfied in order for a surface state to exist in the gap. It is perhaps not surprising that the simultaneous satisfaction of these requirements depends rather sensitively on the potential used [1.56, 57] and hence on self-consistency [1.58, 25].

We will concentrate in the rest of this subsection on the two (very recent) self-consistent calculations of surface states [1.58, 25]. APPELBAUM and HAMANN (AH) have determined the ionization potential (c.f. Section 1.2), charge density, and surface state configuration for Si(111). In the following, their method is briefly summarized. For a review of earlier work on Si, see JONES [1.59]. The ion core potentials that make up $V^{ex}(r)$ [Eq. (1.3)] are represented by pseudopotentials [1.60]. Equations (1.12) and (1.13) are solved for two representative $k_{\parallel}$ points in the surface Brillouin zone, Γ at the zone center and J at the center of the zone edge (a two point variant of the Baldereschi summation scheme used for bulk semiconductors [1.61]). At both $k_{\parallel}$'s the equations were solved for all surface states below the Fermi energy and for ten continuum states per bulk valence band. The charge density was computed via these wave functions, which yielded a new potential [(1.5) and (1.12)]. This process was then iterated to self-consistency.

More particularly, AH divide the Si crystal into a "surface" and a "bulk" region by a mathematical plane (Fig. 1.16). These two

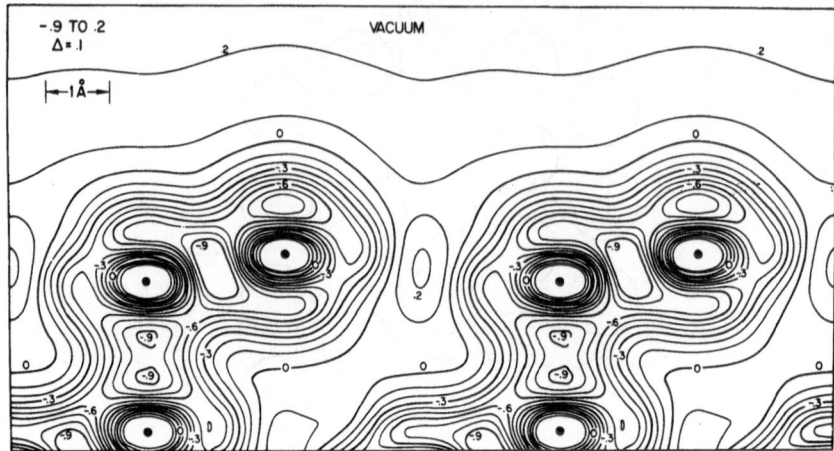

Fig. 1.17. Contours of constant surface potential plotted on a plane normal to the Si (111) surface and passing through a line connecting a surface atom with one of its nearest neighbors in the second plane of atoms [1.58]. The heavy black dots locate the centers of the Si atoms. The energy units are hartrees, the contours are spaced 0.1 hartree apart, and the scale is placed so the valence-band maximum falls at +0.06 hartree

regions are solved separately, and the wave functions joined at the plane. Since pseudopotentials are used, a plane wave basis set is manageable. In the bulk region, the potential is assumed to be fully periodic. Thus the solution for the evanescent and Block waves involves a matrix eigenvalue calculation. In the surface region, the wave function is expanded as

$$\psi(r) = \sum_{Q} \psi_{Q}(x)\, e^{i(k_{\parallel} - Q)\cdot r}, \tag{1.34}$$

where the Q are reciprocal lattice vectors parallel to the surface [see (1.29)]. In the basis of (1.34), (1.12) becomes a set of N coupled one dimensional equations, where N is the number of surface reciprocal lattice vectors used (~ 30). These equations can be numerically integrated to the matching plane. The Fermi energy lies in the gap surface-state band for Si, and is determined by filling the surface states until the surface region is neutral.

The resultant surface potential is plotted in Fig. 1.17. It is shown on a plane normal to the Si (111) surface and passing through a line connecting a surface atom with one of its nearest neighbors in the second plane of atoms. Notice the attractive wells between atoms indicative of covalent bonding. More importantly, above the surface atoms there are residual attractive potentials. AH attribute the

Table 1.1. For three different surface geometries, identified by giving the location of the second and first atomic layers relative to the third, the ionization potential (IP), electronic occupancy of the gap surface-state band in electrons per surface atom (n_{ss}) and the number of surface bands present are listed [1.58]. The geometric parameters corresponding to the ideal bulk lattice are 4.44 and 5.92 a.u.

Geometry [a.u.]	IP [eV]	n_{ss}	No. of surface-state bands
4.44, 5.26	5.44 ± 0.05	0.7016	3
4.44, 5.56	5.46 ± 0.004	0.5834	1
4.24, 5.06	5.51 ± 0.07	0.6350	3

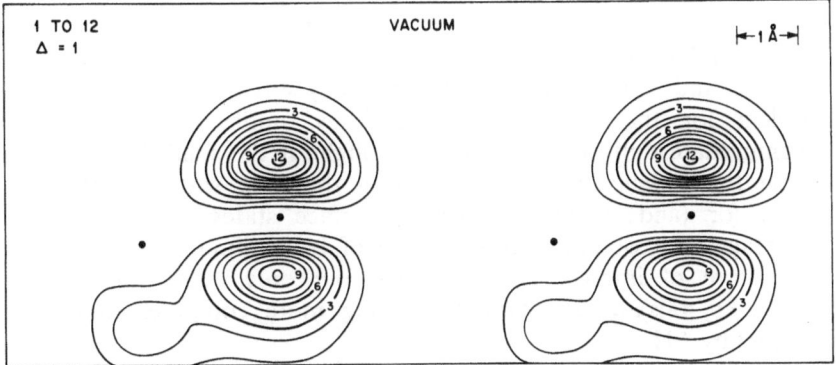

SILICON TOP SURFACE STATE CHARGE DENSITY

Fig. 1.18. "Dangling bond" surface state charge density [1.62] plotted on a plane normal to the Si(111) surface whose orientation is given in the caption of Fig. 1.17

localized "dangling-bond" surface states to these dangling potential wells. These surface states are only partially occupied, as indicated in Table 1.1, containing less than one electron per surface atom. The location of the top of the valence band with respect to the vacuum level (IP) is also given in Table 1.1. It is to be compared with an experimental value of 5.15 eV [1.14].

The charge density of one of the "dangling bond" surface states is shown in Fig. 1.18, again on a plane intersecting atoms in the first and second layers [1.62]. It is calculated for the geometry described by the first entry in Table 1.1. The band of surface states lies completely within the absolute valence band-contribution band gap. About 80% of the charge in the surface state lies in the area covered by the plot; only 20% tails off deeper into the bulk.

PANDEY and PHILLIPS (PP) [1.63] have formulated a simple semi-empirical tight-binding method which reproduces many of the AH results. They [1.64] disagree however, on the number of electrons per surface atom in the gap surface state band (n_{ss} in Table 1.1). PP find that $n_{ss} = 1$ for a relaxed surface (spacings 4.44, 5.26 a.u.—Table 1.1). KLEINMAN [1.65] has recently provided a rather general argument for the value of unity. PP note that it is possible that the discrepancy arises from the matching plane (Fig. 1.16) being too close to the vacuum. They [1.64] find a substantial density of gap surface states in the fourth layer.

We close this section with a brief discussion of a complementary method which has recently been developed to determine self-consistent surface states for a thirteen layer (001) film of lithium, due to ALLDREDGE and KLEINMAN (AK) [1.25] (see also [1.55]). AK also solve (1.5), (1.12) and (1.13), using a pseudopotential for the ion cores and a two-dimensional plane wave basis. AK do not divide the crystal artificially into "surface" and "bulk" regions, however. They solve a matrix eigenvalue problem for the entire film. Thus they, unlike AH, can easily accommodate non-local potentials throughout the solid. Further, AK obtain the *complete* band spectrum of the film, without having to search for (and possibly overlook) surface states. However, these advantages are obtained at the price of being limited to thin films (which are of interest in themselves, but only approximate semi-infinite crystals), and apparently taking more computer time than required by AH.

It will be interesting to follow these two approaches as they continue to develop.

1.4.2. Surface Plasmons

Throughout the chapter, we have considered only ground state properties of the solid surface. Our description of electronic properties would not be complete without a consideration of electronic excitation.

As an introductory example, Fig. 1.19 is an energy-loss profile in the (11) diffraction direction for a beam of electrons scattered from Al (100) in the LEED experiment (see Chapter 4) of BURKSTRAND [1.66]. The dominant peaks near 10 and 15 eV are attributed to surface and bulk plasmon excitations, respectively.

Surface and bulk plasmons have also been detected in transmission experiments [1.67] and by optical studies [1.68].

There have been several reviews recently of the subject of surface plasmons [1.69]. In the following, we will merely indicate the "nature of the beast".

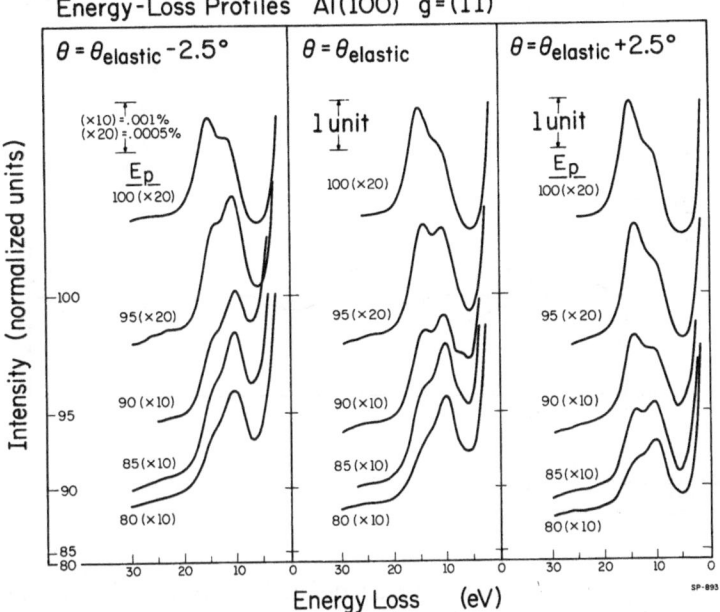

Fig. 1.19. Series of energy-loss profiles in the (11) diffraction beam for different primary energies E_p and collector angles θ. The ordinate marks correspond to the zero levels of each profile [1.66]

The concept [1.70] of volume or bulk plasmons in metals is well established. The plasma mode is a natural mode of oscillation of the electron gas. There are additional plasmon-like modes associated with surfaces and interfaces. This was first recognized by RITCHIE [1.71]. STERN and FERRELL [1.72] named these additional modes surface plasmons, and considered the effect of oxide coatings on them.

To obtain a more quantitative picture of the surface plasmon, one would have to generalize the response function of (1.27) for time dependent effects. This has been done formally by YING [1.73]. There is, in addition, a rather extensive literature on the subject of frequency dependent linear response functions [1.74].

A calculation which is as yet somewhat preliminary but is quite instructive as to the nature of surface plasmons is that of INGLESFELD and WIKBORG (IW) [1.75], who followed much of the earlier work of BECK and CELLI [1.76]. They obtain results for Al, the subject of BURKSTRAND'S experiment. IW dealt with a potential-density response function $P(k_\|, x, x'; \omega)$

$$n_1(k_\|, x, \omega) = \int_{-\infty}^{\infty} dx' P(k_\|, x, x'; \omega) V_1(k_\|, x'; \omega), \qquad (1.35)$$

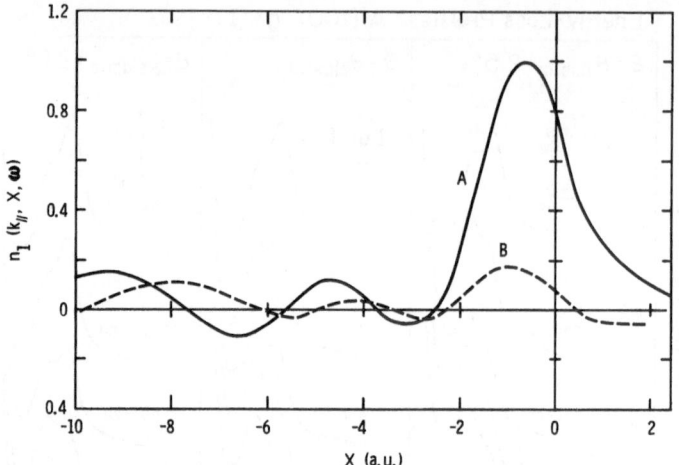

Fig. 1.20. A and B are $\mathrm{Re}\,\{n_1(k_\parallel, x, \omega)\}$ and $\mathrm{Im}\,\{n_1(k_\parallel, x, \omega)\}$ respectively, for Al: $k_\parallel = 0.05$ a.u. (from [1.75])

where n_1, V_1, x, x', and $k_\parallel$ are as defined previously in Section 1.3, and ω is the frequency of the time-varying perturbing potential V_1 (a one dimensional unperturbed system is assumed).

It is desired to compute the fluctuation in the static electron density which accompanies a surface plasmon of frequency ω and wave vector $k_\parallel$

$$n_1(\boldsymbol{r}, t) = n_1(\boldsymbol{k}_\parallel, x, \omega) \exp\left[i\left(\boldsymbol{k}_\parallel \cdot \boldsymbol{r} - \omega t\right)\right]. \tag{1.36}$$

Since the surface plasmon is a normal mode of the electronic plasma, the charge density fluctuation n_1 must produce its own perturbing potential through Poisson's equation

$$V_1(k_\parallel, x; \omega) = \frac{2\pi}{k_\parallel} \int\limits_{-\infty}^{\infty} dx' n_1(k_\parallel, x'; \omega) \exp\left(-k_\parallel |x - x'|\right). \tag{1.37}$$

Thus we encounter our old friend, the self-consistency requirement [c.f. discussion following (1.9)], which in this case is a necessary condition for the occurrence of a plasmon.

IW determined the response function $P(k_\parallel, x, x'; \omega)$ using the random-phase approximation [1.77]. They approximate the surface potential as a step-potential. Thus, while the fluctuations [(1.35) and (1.37)] are determined self-consistently, the response function is not. They chose $E_F = 11.6$ eV and step height $= 15.8$ eV in order to approximate Al. Their procedure is to search for values of ω and

$k_\parallel$ such that (1.35) and (1.37) are solved simultaneously for a fluctuation confined to the surface. Curves A and B in Fig. 1.20 show the real and imaginary parts respectively of the resulting charge fluctuation $n_1(k_\parallel, x, \omega)$ at $k_\parallel = 0.05$ a.u. Presuming that their Friedel-like oscillations eventually damp out (FEIBELMAN [1.78] has argued that the amplitude should fall off as x^{-2}) [1.75], Fig. 1.20 shows that the surface plasmon charge oscillation can be largely confined to within one atomic radius or so.

A bulk plasmon, however, is not confined to the surface region. Thus we see a parallel between the relation between surface and bulk plasmons and that between surface states and continuum states discussed in Section 1.1 and Subsection 1.4.1.

The resultant values of ω and $k_\parallel$ specify the dispersion relation which as FEIBELMAN [1.78] has shown, is unfortunately quite sensitive to the surface potential barrier shape and hence self-consistency. This dispersion relation can be obtained from analysis of LEED data, however. DUKE et al. [1.79] find that for Al(111),

$$\hbar\omega_s = 10.5\,(\pm0.1) + 2\,(\pm1)\,k_\parallel + 0\,(+2)\,k_\parallel^2$$
$$\Gamma_s = 1.85\,(\pm0.5) + 3\,(\pm2)\,k_\parallel\,,$$

(1.38)

for energies measured in eV and momenta in Å^{-1}, where $\hbar\omega_s$ is the real part of the energy of the surface plasmon and Γ_s is the imaginary part. The latter provides a measure of the lifetime of the surface plasmon. The $k_\parallel = 0$ limit of $10.5\,(\pm0.1)$ eV also agrees well with the high energy transmission experiments and with the values obtained by BURKSTRAND for $k_\parallel = 0$ (corresponding to the $\theta = \theta_{\text{elastic}}$ profiles in Fig. 1.17).

FEIBELMAN [1.80a] as well as HARRIS and GRIFFIN [1.80b] have shown that, for a jellium solid and within the RPA (see also [1.81]),

$$\hbar\omega\,(k_\parallel = 0) = (1/\sqrt{2})\,\omega_p\,,$$

(1.39)

independent of the shape of the barrier, where ω_p is the bulk plasma frequency $(4\pi e^2 n_+/m)^{1/2}$. For Al, (1.39) gives $\hbar\omega\,(k_\parallel = 0) = 11.2$ eV and $\omega_p = 15.8$ eV. Equation (1.39) is approximately consistent with (1.38) for $k_\parallel = 0$, except that the latter shows a nonzero imaginary part to the energy.

Finally we mention that surface plasmon dispersion and damping have been shown to be sensitive to adsorption. We refer the interested reader to [1.82].

1.5. Local Density of States

The density of states weighted by the amplitude squared of the wave function is denoted as the local density of states (LDS)

$$n(E, r) = \sum_i |\psi_i(r)|^2 \delta(E - E_i),$$ (1.40)

where the density of states is given by

$$n(E) = \int n(E, r) \, dr.$$ (1.41)

As we saw in Section 1.1, ψ_i decays into the vacuum at a solid surface. Surface states (see also Subsection 1.4.1) appear as well. Also resonance levels can be associated with chemisorbed particles, as described in Section 1.3. This means that $n(E, r)$ will have a positional dependence at the surface which is interestingly different from bulk behavior. These effects are being seen experimentally, and they are discussed at some length in Chapter 5.

Surface states in band gaps will tend to broaden the local density of states at a surface. There is also a narrowing effect [1.83] which is suggested by the following development. The second moment of the density of states is defined as (see CYROT-LACKMANN et al. [1.84, 85]):

$$\mu_2 = \int (E - E_0)^2 \, n(E) \, dE,$$ (1.42)

where E_0 is a suitably chosen reference energy. The quantity μ_2 is then a measure of the width of $n(E)$ about E_0. It may be expressed alternatively in terms of the wave functions $\psi_i(r)$ or in terms of local functions such as the generalized Wannier functions (GWF) discussed in Section 1.1 [1.9]

$$\mu_2 = N^{-1} \text{Tr} (\mathscr{H} - E_0)^2 = N^{-1} \sum_j \langle \psi_i | (\mathscr{H} - E_0)^2 | \psi_i \rangle$$

$$= N^{-1} \sum_n \langle a_n | (\mathscr{H} - E_0)^2 | a_n \rangle,$$ (1.43)

where an example of the Hamiltonian is given in (1.12), a_n is the GWF associated with the lattice site located by the lattice vector n, and the possibility of overlapping bands [1.7] has been ignored. Now let

$$E_0 = E_s \equiv \langle a_{n_s} | \mathscr{H} | a_{n_s} \rangle = \mathscr{H}_{n_s, n_s}$$ (1.44)

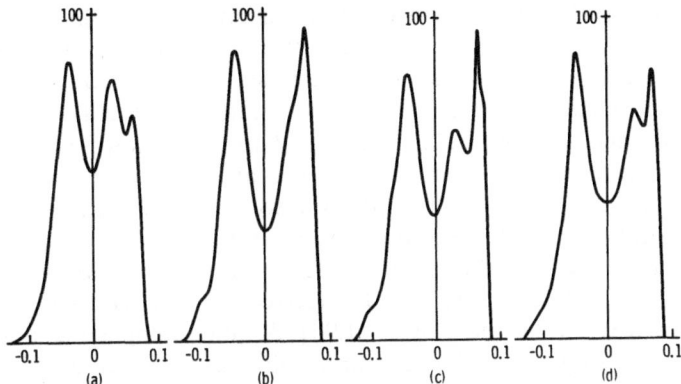

Fig. 1.21a-d. The local density of states of an atom in a) the (111) surface plane b) the first plane and c) the second plane below the surface, and d) in the bulk for a *fcc* d-band [1.88]

where a_{n_s} refers to one of the GWF in the surface plane. Then the contribution to the second moment from a_{n_s} is given by

$$|\mu_2|_s = \sum_{n \neq n_s} |\mathcal{H}_{n,n_s}|^2 . \tag{1.45}$$

Since the number of near neighbors is reduced at a surface, (1.45) suggests that $(\mu_2)_s < \mu_2$ and hence that the surface density of states is narrower than $n(E)$. This is only suggestive, because the surface orbitals are different from the bulk orbitals (see Fig. 1.3). Thus not only the number but the size of the matrix elements changes in going from the bulk to the surface. It has been shown by DAVENPORT et al. (DES) [1.83] that these self-consistency effects can be very important. Self-consistency has not been included in any "first principles" LDS calculations to date. There is a very interesting semi-empirical tight-binding treatment [1.62–64] by PANDEY and PHILLIPS in which the effects of relaxation of the surface plane are included parametrically (see Section 1.4.1).

If one rewrites (1.40) and (1.41) in terms of GWF, it is easily shown [1.7] that $n(E)$ can be written as a sum over lattice sites of terms of the form:

$$n_n(E) = -\pi^{-1} \sum_{\alpha=1}^{p} \lim_{\varepsilon \to 0} \text{Im} \{G_{\alpha,n}(E + i\varepsilon)\}, \tag{1.46}$$

where

$$G_{\alpha,n}(E) = \langle a_{\alpha,n}|(E - \mathcal{H})^{-1}|a_{\alpha,n} \rangle , \tag{1.47}$$

ε is a positive parameter, and there are p overlapping bands. HAYDOCK et al. [1.86–88] have recently provided a new continued fraction method for evaluating a term of the form given in (1.47). The d-band density of states in the surface layers was calculated using the Slater-Koster overlap parameters, as determined by Pettifor [1.89] for (bulk) Ni. Their resultant densities of states for the (111) surface, two sublayers, and the bulk of a *fcc* d-band are shown in Fig. 1.21. The narrowing effects mentioned earlier are readily apparent. Note also that by two layers in from the surface layer, the density of states is quite similar to that of the bulk (cf. Section 1.1).

For a somewhat different approach to the calculation of surface densities of states, the reader is referred to the Green's function theory of Kalkstein and Soven [1.90]. A quite complete solution for the case of a single s-band in a simple cubic lattice is presented.

1.6. Status — Current Challenges

The fundamental or *a priori* theory of those electronic properties of surfaces which we have considered in the preceding sections has often progressed in the following characteristic stages. The first stage is a noncrystalline (one dimensional) nonself-consistent treatment. Next, self-consistency [cf. discussion following (1.9)], is introduced into the noncrystalline calculation. Then the effects of crystallinity are taken into account approximately for nearly free electron metals via perturbation theory. Following that, crystallinity is treated beyond the perturbation approximation, but with a plane wave basis, allowing a wider range of materials to be considered. In the next stage, the array of ions in the surface region are to be allowed to relax from their bulk-like positions to new equilibrium positions. Finally, complex materials such as transition metals and their alloys are dealt with. Of course in all these stages many-electron effects are included in various approximations, and their relative importance depends on the observable being considered.

The aforementioned progression seems to be the most commonly followed, but there has been a most useful companion mode. A semi-empirical tight-binding approach (see, e.g. Section 1.5) allows the inclusion of full crystallinity in a simpler fashion, although full self-consistency is not attempted as yet.

In the following we will attempt to pinpoint the stage at which each of the subject areas alluded to in the preceding section lies. Lang [1] has pointed out that surface energy calculations are presently self-consistent and they include crystallinity in a perturbation approximation (this applies also to the theory of adhesion [1.4]).

Electron work function calculations are self-consistent and now include crystallinity beyond the perturbation approximation in a plane wave basis so that not only good metals [1.24, 25] but also small band gap semiconductors [1.58] are considered. Initial explorations of the effect of ionic relaxation are also available [1.25, 58].

Impurity screening in the surface region is self-consistent but such calculations do not as yet take into account crystallinity [1.23]. In that self-consistent category there is a single, rather crude but *a priori* calculation of the local density of states associated with an adsorbate [1.23].

Surface state theory and, in fact, surface band structure calculations [1.25] are at the same stage as electron work function calculations [1.25, 58].

The theory of surface plasmons as yet includes no crystallinity and self-consistent response functions are not used, although a self-consistent formalism has been completed [1.73]. The sensitivity of the surface plasmon dispersion relation to self-consistency has been revealed [1.77]. This area of surface physics is behind the other fields. That is to be expected, however, since electronic excitations are often inherently more difficult to deal with than ground state properties.

Surface density of states theories are not as yet self-consistent, although full crystallinity is taken into account [1.63, 64, 83–90]. There have also been efforts to include relaxation of the surface lattice (see, e.g., [1.62, 88]).

Of the ground state theories, there are none as yet which treat crystalline transition metals self-consistently [1.91]. This is particularly unfortunate because most of the surface experimental data is for refractory transition metals. There is good reason for this discrepancy in our knowledge, however. For transition metals, one must deal with a mixture of contracted and diffuse orbitals immersed in the inhomogeneous medium of the surface. This makes plane wave bases impractical in general. Perhaps the generalized Wannier function method [1.7, 9], which represents surface observables precisely in terms of local functions, or another local orbital method (see, e.g., [1.83]) will lead to advancement in this area.

There remains a long way to go in surface theory, but the enormous progress of the last three to four years indicates a productive future for the field.

Acknowledgements

The author is indebted to D. HAMANN, J. A. APPELBAUM, K. C. PANDEY, and J. C. PHILLIPS for sending unpublished data and to

L. KLEINMANN and G. ALLDREDGE for preprints of their work. It is a pleasure to acknowledge conversations with P. FEIBELMAN, S. C. YING, E. W. PLUMMER, J. C. TRACY, J. M. BURKSTRAND, and J. G. GAY.

References

1.1. N. D. LANG: Solid State Physics 28, 225 (1973).
    N. D. LANG, W. KOHN: Phys. Rev. B 1, 4555 (1970).
1.2. Surface energy considerations pertinent to surface composition of binary alloys have been implemented recently by F. L. WILLIAMS, D. NASON: Surface Sci. 45, 377 (1974); A recent investigation of correlation energy contributions to metal surface energies has been presented by J. HARRIS, R. O. JONES: Phys. Letters 46 A, 407 (1974).
1.3. H. KRUPP: Advan. Colloid Interface Sci. 1, 111 (1967).
1.4. J. FERRANTE, J. R. SMITH: Surface Sci. 38, 77 (1973).
1.5. See, e.g., K. H. JOHNSON, R. P. MESSMER: J. Vac. Sci. Technol. 9, 561 (1974). See also J. R. SMITH and J. G. GAY (to be published).
1.6. R. F. WALLIS: Progress in Surface Sci. 4, 233 (1973).
1.7. J. R. SMITH, J. G. GAY: Phys. Rev. Letters 32, 774 (1974); Phys. Rev. B 9, 4151 (1974); Phys. Rev. B 11, 4906 (1975). See also J. J. REHR, W. KOHN: Phys. Rev. B 10, 448 (1974), for an illuminating investigation of GWF asymptotic behavior.
1.8. H. D. HAGSTRUM, G. E. BECKER: Phys. Rev. B 8, 1580 (1973).
1.9. W. KOHN, J. ONFFROY: Phys. Rev. B 8, 2485 (1973).
1.10. N. D. LANG, W. KOHN: Phys. Rev. B 3, 1215 (1971); 6010 (1973).
1.11. W. JONES, N. H. MARCH: Theoretical Solid State Physics, Vol. 2 (Wiley-Interscience, New York, 1973), pp. 814, 1060—1064.
1.12. J. R. SMITH: Phys. Rev. 181, 522 (1969).
1.13. S. G. DAVISON, J. D. LEVINE: Solid State Phys. 25, 92 (1970).
1.14. F. G. ALLEN, G. W. GOBELI: Phys. Rev. 127, 141, 150 (1962).
1.15. J. C. RIVIÈRE: Solid State Surface Science, Vol. 1, ed. by M. GREEN (Dekker, New York, 1969), p. 179.
1.16. P. HOHENBERG, W. KOHN: Phys. Rev. 136, B 864 (1964).
1.17. V. S. FOMENKO: Handbook of Thermionic Properties (Plenum Press, Inc., New York, 1966).
1.18. N. D. LANG: Solid State Commun. 7, 1047 (1969).
1.19. W. KOHN, L. J. SHAM: Phys. Rev. 140, A 1133 (1965).
1.20. The first treatment of the work function of a single-crystal plane was due to J. R. SMITH: Phys. Rev. Letters 25, 1023 (1970). SMITH used what should now be considered a crude model, consisting of WIGNER-SEITZ polyhedra protruding into the vacuum.
1.21. N. W. ASHCROFT, D. C. LANGRETH: Phys. Rev. 155, 682 (1967); 159, 500 (1967).
    N. W. ASHCROFT: J. Phys. C 1, 232 (1968).
1.22. N. D. LANG, W. KOHN: Phys. Rev. B 7, 3541 (1973).
1.23. S. C. YING, J. R. SMITH, W. KOHN: Phys. Rev. B. Feb. 15, 1974. – L. KAHN, S. C. YING: Solid State Commun. (to be published).
1.24. J. A. APPELBAUM, D. R. HAMANN: Phys. Rev. B 6, 2166 (1972).
1.25. G. P. ALLDREDGE, L. KLEINMAN: Phys. Rev. B 10, 559 (1974).
1.26. E. W. PLUMMER, A. E. BELL: J. Vac. Sci. Technol. 9, 583 (1972).
1.27. H. E. IVES: Astrophys. J. 60, 209 (1924);
    I. LANGMUIR, K. H. KINGDON: Proc. Roy. Soc. (London) A 107, 61 (1925).

1.28. V. M. Gavrilyuk, A. G. Naumovets, A. G. Fedorus: Soviet Phys. JETP **24**, 899 (1967);
L. W. Swanson, R. W. Strayer: J. Chem. Phys. **48**, 2421 (1968).

1.29. Recently rather a large decrease in the work function (1.5 – 3 eV) has been found for hydrocarbon adsorption on Pt. See J. L. Gland, G. A. Somojai: Surface Sci. **38**, 157 (1973).

1.30. J. R. Smith, S. C. Ying, W. Kohn: Phys. Rev. Letters **30**, 610 (1973).

1.31. N. D. Lang: Phys. Rev. B **4**, 4234 (1971).

1.32. R. W. Gurney: Phys. Rev. **47**, 479 (1934).

1.33. J. Rudnick: Phys. Rev. B **5**, 2863 (1972);
D. M. Newns: Phys. Rev. B **1**, 3304 (1970);
D. E. Beck, V. Celli: Phys. Rev. B **2**, 2955 (1970);
J. W. Gadzuk: J. Phys. Chem. Solids **30**, 2307 (1969).

1.34. S. C. Ying, J. R. Smith, W. Kohn: J. Vac. Sci. Techn. **9**, 575 (1972).

1.35. J. R. Smith, S. C. Ying, W. Kohn: Solid State Commun. **15**, 1491 (1974).

1.36. The experimental E_I is obtained from the atomic desorption energy E_a the hydrogen ionization potential I the electron work function ϕ_e, and the Born-Haber cycle: $E_I = E_a + I - \phi_e$. E_a appears not to be very sensitive to the surface plane, and we used a representative value of 70 kcal/mole (see [1.37]). As mentioned earlier, the model is most appropriate to a close packed plane, and therefore we took $\phi_e = 5.3$ eV.

1.37. See, e.g.: T. E. Madey, J. T. Yates, Jr.: *Structure et Properties des Surface des Solides* (Editions du Centre National de la Recherche Scientifique, Paris, 1970), No. 187, p. 155;
T. W. Hickmott: J. Chem. Phys. **32**, 810 (1960).

1.38. E. W. Plummer, A. E. Bell: J. Vac. Sci. Technol. **9**, 583 (1972).

1.39. F. M. Propst, T. C. Piper: J. Vac. Sci. Technol. **4**, 53 (1967).

1.40. The energy of vibration of molecular hydrogen is 550 MeV, according to G. Herzberg: *Molecular Spectra and Molecular Structure. I. Spectra of Diatomic Molecules*, 2nd ed. (D. Van Nostrand, New York, 1963).

1.41. T. E. Madey: Surface Sci. **36**, 281 (1973).

1.42. K. Yonehara, L. D. Schmidt: Surface Sci. **25**, 238 (1971);
P. J. Estrup, J. Anderson: J. Chem. Phys. **45**, 2254 (1966).

1.43. M. Fink, A. C. Yates: Atomic Data (to be published).

1.44. L. H. Germer: Surface Sci. **5**, 147 (1966).

1.45. E. Bauer: Surface Sci. **5**, 152 (1966);
J. C. Tracy, J. M. Blakely: Surface Sci. **15**, 257 (1969).

1.46. P. J. Jennings, E. G. McRae: Surface Sci. **23**, 363 (1970).

1.47. J. C. Tracy, J. E. Rowe: In *Electron Spectroscopy*, ed. by D. A. Shirley (North-Holland Publishing Co., Amsterdam), p. 587;
H. D. Hagstrum, G. E. Becker: J. Chem. Phys. **54**, 1015 (1971).

1.48. E. W. Plummer, B. J. Waclwaski: Proc. Phys. Elec. Conf. (1973);
E. W. Plummer: (to be published);
B. Feuerbacher, B. Fitton: Phys. Rev. B **8**, 4890 (1973).

1.49. L. Anders, R. Hansen, L. Bartell: J. Chem. Phys. **59**, 5277 (1973). See also J. W. Gadzuk: In *Surface Physics of Crystalline Solids*, ed. by J. M. Blakely (Academic Press, New York, 1974).

1.50. I. Tamm: Z. Physik **76**, 849 (1932).

1.51. B. J. Waclawski, E. W. Plummer: Phys. Rev. Letters **29**, 783 (1972);
B. Feuerbacher, B. Fitton: Phys. Rev. Letters **29**, 786 (1972);
E. W. Plummer, J. W. Gadzuk: Phys. Rev. Letters **25**, 1493 (1970);
N. R. Avery: Phys. Rev. Letters **32**, 1248 (1974).

1.52. S. G. Davison, J. D. Levine: Solid State Phys. **25**, 1 (1970).

1.53. M. Henzler: Surface Sci. **25**, 650 (1971).

1.54. V. Heine: Proc. Phys. Soc. **81**, 300 (1963).

1.55. E. Carruthers, L. Kleinman, G. Alldredge: Phys. Rev. B **8**, 4570 (1973); **9**, 3325 (1974); **9**, 3330 (1974).

1.56. G. P. Alldredge, L. Kleinman: Phys. Rev. Letters **28**, 1264 (1972); Phys. Rev. B **10**, 1252 (1974).

1.57. J. A. Appelbaum, D. R. Hamann: Phys. Rev. Letters **31**, 106 (1973).

1.58. J. A. Appelbaum, D. R. Hamann: Phys. Rev. Letters **32**, 225 (1974).

1.59. R. O. Jones: J. Phys. C (Solid State Phys.) **5**, 1615 (1972).

1.60. J. A. Appelbaum, D. R. Hamann: Phys. Rev. B **8**, 1777 (1973).

1.61. A. Baldereschi: Phys. Rev. B **7**, 5212 (1973).

1.62. D. R. Hamann, J. A. Appelbaum: Private communication.

1.63. K. C. Pandey, J. C. Phillips: Phys. Rev. Letters **32**, 1433 (1974); Solid State Commun. **14**, 439 (1974).

1.64. K. C. Pandey, J. C. Phillips: To be published.

1.65. L. Kleinman: Phys. Rev. B **11**, 858 (1975).

1.66. J. M. Burkstrand: Phys. Rev. B **7**, 3443 (1973).

1.67. K. D. Sevier: *Low Energy Electron Spectrometer* (Interscience, New York, 1972), Chapter 8.

1.68. A. S. Barker, Jr.: Phys. Rev. B **8**, 5418 (1973).
J. G. Endriz, W. E. Spicer: Phys. Rev. B **4**, 4144 (1971).

1.69. E. N. Economou, K. L. Ngai: To be published;
R. H. Ritchie: Surface Sci. **34**, 1 (1973);
H. Raether: Surface Sci. **8**, 233 (1967).

1.70. D. Pines, D. Bohn: Phys. Rev. **85**, 874 (1952), **92**, 609, 626 (1953).

1.71. R. H. Ritchie: Phys. Rev. **106**, 874 (1957).

1.72. E. A. Stern, R. A. Ferrell: Phys. Rev. **120**, 130 (1960).

1.73. S. C. Ying: Proceedings of the Taormina Conference on Prospectives of Many-Electron Calculations on Solids; Nuovo Cimento (to be published.
See also S. C. Ying, J. J. Quinn, A. Eguiluz: To be published.

1.74. D. M. Newns: Phys. Rev. B **1**, 3304 (1970);
V. Peuckert: Z. Physik **241**, 191 (1971);
J. Harris, A. Griffin: Can. J. Phys. **48**, 2592 (1970);
D. E. Beck: Phys. Rev. B **4**, 1555 (1971);
P. A. Fedders: Phys. Rev. **153**, 438 (1967);
P. J. Feibelman, C. B. Duke, A. Bagchi: Phys. Rev. B **5**, 2436 (1972).

1.75. J. E. Inglesfield, E. Wikborg: Solid State Commun. **14**, 661 (1974); J. Phys. C (Solid State Phys.) **6**, 158 (1973). These authors feel that the increase of the Friedel oscillations into the bulk is due to a numerical approximation — the finite range over which the integral equation is evaluated (private communication).

1.76. D. E. Beck, V. Celli: Phys. Rev. Letters **28**, 1124 (1972).

1.77. L. Hedin, S. Lundquist: *Solid State Physics*, Vol. 24 (Academic Press, New York, 1969).

1.78. P. J. Feibelman: Phys. Rev. Letters **30**, 975 (1973); Phys. Rev. B **9**, 5077 (1974).

1.79. C. B. Duke, U. Landman: Phys. Rev. B **8**, 505 (1973);
C. B. Duke et al.: J. Vac. Sci. Technol. **10**, 183 (1973);
See also Bagchi and Duke: Phys. Rev. B **5**, 2784 (1972).

1.80. P. J. Feibelman: Phys. Rev. B **3**, 220 (1971);
J. Harris, A. Griffin: Phys. Letters **34**, 51 (1971).

1.81. Hydrodynamic theories have yielded bands of surfaces plasmons in addition to the "usual" band whose frequency is given by (1.39) in the infinite-wavelength limit.

See A. J. BENNETT: Phys. Rev. B 1, 203 (1970) and L. KLEINMAN: Phys. Rev. B 7, 2288 (1973).

1.82. P. J. FEIBELMAN: Surf. Sci. **40**, 102 (1973); Phys. Rev. B 9, 5077 (1974); D. M. NEWNS: Phys. Letters **38** A, 341 (1972).

1.83. J. W. DAVENPORT, T. L. EINSTEIN, J. R. SCHRIEFFER: Japanese J. Appl. Phys., Suppl. 2, Part 2, 691 (1974).

1.84. F. CYROT-LACKMANN, M. C. DESJONQUERES: Surface Sci. **40**, 423 (1973).

1.85. J. P. GASPARD, F. CYROT-LACKMANN: J. Phys. C (Solid State Phys.) **6**, 3077 (1973); **7**, 1829 (1974).

1.86. R. HAYDOCK, V. HEINE, M. J. KELLY: J. Phys. C (Solid State Phys.) **5**, 2845 (1972).

1.87. R. HAYDOCK, V. HEINE, M. J. KELLY, J. B. PENDRY: Phys. Rev. Letters **29**, 868 (1972).

1.88. R. HAYDOCK, M. J. KELLY: Surface Sci. **38**, 139 (1973).

1.89. D. G. PETTIFOR: J. Phys. C; Proc. Phys. Soc. London **2**, 1051 (1969); **3**, 367 (1970).

1.90. D. KALKSTEIN, P. SOVEN: Surface Sci. **26**, 85 (1971).

1.91. There is an interesting recent calculation which sizes some of the phenomena of interest at transition metal surfaces; P. FULDE, A. LUTHER, R. WATSON: Phys. Rev. B 8, 440 (1973). Transition metal perovskite crystals have been treated by T. WOLFRAM, E. A. KRAUT, F. J. MORIN: Phys. Rev. B 7, 1677 (1973).

2. Theory of Chemisorption

S. K. Lyo and R. Gomer

With 7 Figures

Chemisorption is defined, somewhat loosely, as the adsorption of atoms or molecules on surfaces with a binding energy in excess of 1 eV. In many cases energies as high as 3 or 4 eV are observed, so that the process clearly corresponds to electron sharing, i.e., chemical bonding. This bonding can be understood at least qualitatively in terms of the usual concepts of chemical bonding. It is not surprising that the principal lines of attack have been extensions of the LCAO-MO (Linear Combination of Atomic Orbitals—Molecular Orbitals) and valence bond methods, respectively. The former is considerably easier to handle at least in the Hartree-Fock approximation; the limits of validity of this approximation are also most easily understood in terms of LCAO-MO arguments. In its simplest form this approach consists of considering the formation of eigenstates of the system from a basis set consisting of eigenstates of the relevant band of the metal plus the relevant adsorbate orbital. Although this approach ignores the fact that this set is incomplete and usually the overlap between the adsorbate and metal states as well, it contains all the qualitative features of more refined theories and is very illuminating. We shall therefore devote considerable attention to it.

2.1. Qualitative Discussion

We start with a very pictorial preview of what we shall find. Consider an adsorbate atom with a filled level of energy E_a as it approaches a metal surface. As the atom comes near the metal the originally sharp level at E_a may broaden by interaction with the metal, i.e. by the fact that tunneling into or from the metal gives it a finite lifetime and hence a half-width

$$\Delta \cong \hbar/2\tau . \tag{2.1}$$

This situation is usually encountered if E_a is relatively close to the Fermi energy E_F, i.e. if the tunneling barrier is not too high. It may also happen, however, that interaction of the adsorbate state leads to splitting off of a relatively sharp localized bonding state near the bottom of the band, and formation of an antibonding state above the Fermi level. In the simplest model which considers the metal band states this level falls outside the band and is also sharp. More realistic theories show that it can in fact be very broad. In either case bonding occurs because there is a net lowering in the energy of all the filled states, i.e. those below E_F.

As in all bonding an essential feature is that two electrons can simultaneously be on the adsorbate; the intra-atomic Coulomb repulsion U of these electrons presents one of the principal difficulties of any calculation. In the Hartree Fock scheme the interaction of an electron of given spin with the average population of electrons of opposite spin is computed. It turns out that this approximation is valid [2.1] when $\pi\Delta/U > 1$ or if the separation of bonding and antibonding orbitals exceeds U. If these inequalities are reversed correlation becomes very important; for instance, a generalized valence bond approach must be used. Fortunately the value of U which would apply to the free atom is considerably reduced near a metal surface by screening effects which can be understood in terms of an image interaction. Disregarding any other effects, the effective ionization energy $-E_a$ is decreased at a distance x from the surface by an amount V_{im}, classically given by $e^2/4x$, since the resulting ion interacts attractively with the metal through its image charge; thus the level E_a is pushed up by V_{im}. On the other hand, the electron affinity is increased by V_{im} since an ion A^- would also interact attractively with the metal. Consequently the effective Coulomb repulsion becomes

$$U_{\text{eff}} = U - 2 V_{im}. \tag{2.2}$$

We should note however that recent quantum treatments [2.2, 3] of image interaction give results which deviate significantly from the classical value at small metal-ion separations.

For a quantitative discussion of the problem we introduce a total density of states,

$$\varrho(E) = \sum_m \delta(E - E_m) \tag{2.3}$$

and a local density of states at the adsorbate

$$\varrho_a(E) = \sum_m |\langle a|m\rangle|^2 \delta(E - E_m), \tag{2.4}$$

where $|\varphi_a\rangle \equiv |a\rangle$ and E_m is the eigenvalue corresponding to the eigenvector of the system with the Hamiltonian $\mathscr{H}$ (i.e. $\mathscr{H}|m\rangle = E_m|m\rangle$).

$\varrho_a(E)$ is the weighted sum of the density of states and measures a local property at the adsorbate.

ϱ, ϱ_a can be expressed very conveniently in terms of the Green's function of the metal-adsorbate system defined by

$$G(E - \mathcal{H} - i\alpha) = 1 \tag{2.5}$$

or equivalently by

$$G = \frac{1}{E - \mathcal{H} - i\alpha} \tag{2.6}$$

where α is an infinitesimally small positive quantity. Then

$$G_{mm} = \frac{1}{E - E_m - i\alpha} = \frac{(E - E_m) + i\alpha}{(E - E_m)^2 + \alpha^2} = P \frac{1}{E - E_m}$$
$$+ i\pi\delta(E - E_m), \tag{2.7}$$

where P indicates the Cauchy's principal part. It then follows from (2.3), (2.4), and (2.7) that

$$\varrho(E) = \frac{1}{\pi} \operatorname{Im} \{\operatorname{Tr} G\} \tag{2.8}$$

and

$$\varrho_a(E) = \sum_m \langle a|m\rangle \langle m|a\rangle \frac{1}{\pi} \operatorname{Im} \{G_{mm}\} = \frac{1}{\pi} \operatorname{Im} \{G_{aa}\}, \tag{2.9}$$

where Tr and Im mean trace and imaginary part of what follows. This is a central result, since the experimental information obtainable from field emission [2.4] and photoemission [2.5] is known to yield $\operatorname{Im}\{G_{aa}\}$ rather directly.

2.2. Newns-Anderson Model

We treat now the simplest LCAO-MO model in the Hartree-Fock approximation. As already pointed out, the basis consists of the relevant band states of the metal $|k\rangle$ and a single adsorbate wave function φ_a. The approach we shall follow here is largely that of NEWNS [2.6], in turn based on a formalism used by ANDERSON [2.7] to treat impurities *in* rather than *on* a metal. Except for the inclusion of self-consistency the results are entirely equivalent to those obtained earlier by more "chemical", i.e. wave function rather than Green's function methods, for instance by Grimley [2.8].

In the following we consider the chemisorption of hydrogen-like atoms on transition metals surfaces. The relevant wave functions are d-band wave functions of the metal and singly occupied, non-degenerate orbitals of the adsorbate.

The Hamiltonian for electrons of spin σ of the metal-adsorbate system is given in the Hartree-Fock approximation by

$$\mathcal{H}^\sigma = \mathcal{H}_m + V^\sigma, \qquad (2.10)$$

where $\mathcal{H}_m$ is the unperturbed metal Hamiltonian, and V^σ is the perturbation introduced by the adatom. Following Newns and Anderson we assume for simplicity that $\{|a\rangle, |k\rangle\}$ $(\mathcal{H}_m|k\rangle = E_k|k\rangle)$ form an orthornormal complete set (i.e. $\langle k|a\rangle = 0$, $\langle k|k'\rangle = \delta_{k,k'}$). The Hamiltonian (2.10) can then be expressed in terms of this basis set: We assume that

$$\mathcal{H}_{kk'} = E_k\delta_{kk'}, \ \mathcal{H}_{ak} \equiv V_{ak}, \ \mathcal{H}_{ka} \equiv V_{ka},$$
$$\mathcal{H}_{aa}^\sigma = E_a + V_{im} + \langle n_{-\sigma}\rangle U_{eff} \equiv E_{a\sigma}, \qquad (2.11)$$

where $\langle n_\sigma\rangle$ is the ground state expectation value of the population of σ-spin electrons on the adatom. V_{ak} describes the hopping of the electron between the adsorbate and the metal and is taken to be independent of spin. The first term of $\mathcal{H}_{aa}^\sigma$ represents the unperturbed adatom level, the second term the upward shift of the level due to the image charge, and the third term the "Coulomb raising" of the σ-spin level due to the presence of-σ-spin electrons.

We can now obtain the Green's functions by taking matrix elements of both sides of (2.5) with respect to the basis set $\{h\} \equiv \{k, a\}$:

$$\sum_{h''} G_{hh''}^\sigma (E - \mathcal{H}^\sigma - i\alpha)_{h''h'} = \delta_{hh'} \qquad (2.12)$$

which can be solved straightforwardly [2.7] and yields

$$G_{kk'}^\sigma = g_k\delta_{kk'} + g_k V_{ka} G_{aa}^\sigma V_{ak'} g_{k'} \qquad (2.13)$$

and

$$G_{aa}^\sigma = \cfrac{1}{E - E_{a\sigma} - \sum_k \cfrac{|V_{ak}|^2}{E - E_k - i\alpha}}. \qquad (2.14)$$

In (2.13) g_k is the Green's function of the metal $\left(\text{i.e. } g_k = \dfrac{1}{E - E_k - i\alpha}\right)$.
The last term in the denominator of G_{aa}^σ can be separated into real and imaginary parts using an identity similar to the third equality in (2.7), yielding

$$G_{aa}^\sigma = \frac{1}{E - E_{a\sigma} - \Lambda - i\Delta}, \tag{2.15}$$

where

$$\Delta(E) = \pi \sum_k |V_{ak}|^2 \delta(E - E_k) \tag{2.16}$$

and

$$\Lambda(E) = P \sum_k \frac{|V_{ak}|^2}{E - E_k} = \frac{1}{\pi} P \int_{-\infty}^{\infty} \frac{\Delta(E')dE'}{E - E'}. \tag{2.17}$$

Equation (2.14) enables us to find ϱ_a^σ as

$$\varrho_a^\sigma = \frac{1}{\pi} \operatorname{Im}\{G_{aa}^\sigma\} = \frac{1}{\pi} \frac{\Delta}{(E - E_{a\sigma} - \Lambda)^2 + \Delta^2}. \tag{2.18}$$

If the energy dependence of $\Lambda(E)$ and $\Delta(E)$ could be neglected, (2.18) would have a simple Lorentzian line shape centered on $E = E_{a\sigma} + \Lambda$, the shifted adsorbate level, with a halfwidth at half maximum of Δ, which would be related to the average, i.e., energy independent tunneling time by the "golden rule" expression

$$1/\tau = \frac{2\Delta}{\hbar} = \frac{2\pi}{\hbar} \sum_k |V_{ak}|^2 \delta(E - E_k) \tag{2.19}$$

as indicated in (2.1).

In general, the energy dependence of Δ and Λ cannot be neglected, and ϱ_a will be more complicated, even in the simple model depicted here. This fact is responsible for the variety of behavior which may occur. Equations (2.6) and (2.9) show that the energy eigenvalues E_m correspond to poles of G_{aa} on the real axis (i.e. without the imaginary part $i\alpha$ in (2.14)). Poles can also occur for (2.15) (i.e. keeping the imaginary part $i\alpha$) outside the band where $\Delta = 0$ if, from (2.15)

$$E - E_{a\sigma} - \Lambda(E) = 0 \tag{2.20}$$

has real roots. This situation corresponds to the intersection of the line $E - E_{a\sigma}$ vs. E with $\Lambda(E)$. Possible situations are shown in Fig. 2.1.

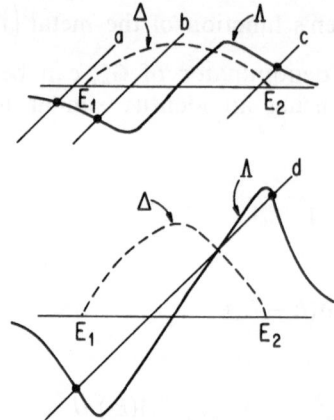

Fig. 2.1. $\Delta(E)$ and its Hilbert transform $\Lambda(E)$ are shown schematically. The roots of $E - E_{a\sigma} = \Lambda(E)$ are indicated by dots. A weak metal-adatom interaction gives rise to a sharp localized state either below (a) or above (c) the band or a virtual level in the band (b). A strong metal-adatom interaction causes a bonding anti-bonding splitting (d). E_1 indicates the band bottom and E_2 band top

It is seen that one (Fig. 2.1a and c), two (Fig. 2.1d), or zero (Fig. 2.1b) states may detach themselves from the band. It is not difficult to show that only one state below (above) the band generally corresponds to a localized state at the adatom formed through a very weak interaction, when the original unperturbed adatom level lies below (above) the band, while a state below (filled) and above (empty) the band corresponds to very strong chemisorption, i.e., the formation of a surface molecule between the adsorbate and its neighboring substrate atoms (Fig. 2.1d).

It is worthwhile to show this more explicitly. If the localized states are far above and below the band we may approximate $\Lambda(E)$ in their vicinity by

$$\Lambda(E) \cong (E - E_c)^{-1} \pi^{-1} \int \Delta(E') \, dE' , \tag{2.21}$$

where E_c is the energy of the band center, and the integral runs only over the band. Then we have from (2.20)

$$E - E_{a\sigma} - (E - E_c)^{-1} \pi^{-1} \int \Delta(E') \, dE' = 0 \tag{2.22}$$

or, from the definition of Δ, (2.16)

$$(E - E_{a\sigma})(E - E_c) = \sum_k |V_{ak}|^2 . \tag{2.23}$$

We next represent the states $|k\rangle$ in terms of a set formed by taking the most relevant atomic orbitals, say $5d$ orbitals in the case of tungsten, from each substrate atom in the metal. We shall pretend that these form a complete orthornormal set, within the manifold of the relevant substrate band (e.g. $5d$-band), which is of course only an approximation. It is possible to construct orthonormal sets along such lines, for instance Wannier orbitals (which are not localized enough for our purposes) but we shall ignore such refinements. With this approximation we have

$$V_{ak} \cong \sum_j \beta'_j \langle j|k\rangle,\tag{2.24}$$

where

$$\beta'_j = V_{aj}.\tag{2.25}$$

Then (2.20) becomes, since the $|k\rangle$ form a complete set (e.g. within the $5d$-band)

$$(E - E_{a\sigma})(E - E_c) = \sum_j |\beta'_j|^2\tag{2.26}$$

or

$$E_l = \frac{1}{2}\left[E_c + E_{a\sigma} \pm \sqrt{(E_{a\sigma} - E_c)^2 + 4\sum_j |\beta'_j|^2}\right].\tag{2.27}$$

These are just the bonding and antibonding levels of a surface molecule formed between the adsorbate and those substrate atoms for which $\beta'_j \neq 0$. Equation (2.20) may also have a root within the band where $\varDelta \neq 0$, as shown by b of Fig. (2.1). In this case the level has a finite width and corresponds to a virtual state.

It is interesting to consider the location of $E_{a\sigma}$ in more detail. We have already seen that it is given by

$$E_{a\sigma} = E_a + V_{im} + U_{\text{eff}}\langle n_{-\sigma}\rangle$$
$$= E_a + V_{im} + (U - 2V_{im})\langle n_{-\sigma}\rangle.\tag{2.28}$$

Thus, if the total electronic charge at the adsorbate is $1e$ and nonmagnetic (i.e. if $\langle n_\sigma\rangle = \langle n_{-\sigma}\rangle = 1/2$), then

$$E_{a\sigma} = E_a + U/2\tag{2.29}$$

which is independent of the image energy V_{im}.

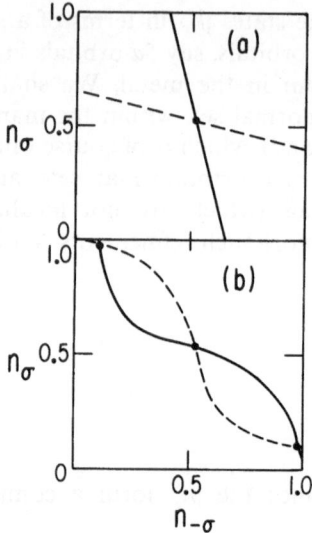

Fig. 2.2a and b. Illustration of occurrence of non-magnetic (a) and magnetic (b) solutions. The solid lines represent (2.30a) and the dashed lines (2.30b). The dots correspond to the self-consistent solutions

We must next evaluate $\langle n_\sigma \rangle$ and $\langle n_{-\sigma} \rangle$. In principle this can be done by noting that

$$\langle n_\sigma \rangle = \int_{-\infty}^{E_F} \varrho_a^\sigma \, dE, \tag{2.30a}$$

$$\langle n_{-\sigma} \rangle = \int_{-\infty}^{E_F} \varrho_a^{-\sigma} \, dE \tag{2.30b}$$

Since ϱ_a^σ contains $\langle n_{-\sigma} \rangle$ and vice versa the two equations (2.30) could be solved self-consistently for $\langle n_\sigma \rangle$ and $\langle n_{-\sigma} \rangle$. There will always be a root for which $\langle n_\sigma \rangle = \langle n_{-\sigma} \rangle$, the so called nonmagnetic solution. It may also happen that there are two symmetric magnetic roots $\langle n_\sigma \rangle = a$, $\langle n_{-\sigma} \rangle = b$, and $\langle n_\sigma \rangle = b$, $\langle n_{-\sigma} \rangle = a$, $a \neq b$. Figure 2.2 shows a typical nonmagnetic solution (a), and magnetic solutions (b). The solid lines represent (2.30a), and the dashed lines (2.30b). As is well known, the Hartree-Fock solution has an extremum property. Therefore if there is only one nonmagnetic solution, this should give an energy minimum. If there are two magnetic solutions and one nonmagnetic solution, then the former correspond to minima, and the latter to a maximum, because the end points (cf. Fig. 2.2b) cannot give minima. It is generally known that the Hartree-Fock approximation breaks down entirely when the solutions have magnetic roots, since this case corresponds to so much repulsion that correlation effects are too important to be treated adequately in the Hartree-Fock approximation. We will confine ourselves therefore, at least implicitly, to cases where only nonmagnetic solutions occur, but will continue to apply the validity criterion already discussed.

The density of states for the system is given by (2.8), (2.13), and (2.14)

$$\varrho^\sigma = \frac{1}{\pi} \, \mathrm{Im} \left\{ \sum_k G_{kk} + G_{aa} \right\}$$
$$= \varrho_m^0(E) + \frac{1}{\pi} \, \frac{\partial}{\partial E} \, \mathrm{Im} \left\{ \ln \left[E - E_{a\sigma} - \sum_k \frac{|V_{ak}|^2}{E - E_k - i\alpha} \right] \right\} \tag{2.31}$$

with $\varrho_m^0(E) = \sum_k \delta(E - E_k)$, a density of states for the metal. The change of the density of states due to the adsorption of the adatom is thus given by

$$\Delta\varrho^\sigma \equiv \varrho^\sigma - \varrho_m^0(E) = \frac{1}{\pi} \, \frac{\partial}{\partial E} \, \mathrm{Im} \{ \ln [E - E_{a\sigma} - \Lambda(E) - i\Delta(E)] \}. \tag{2.32}$$

Therefore $\Delta\varrho^\sigma$ is proportional to the derivative of the phase of the argument of the natural logarithm in (2.32). There are two different kinds of roots of $E - E_{a\sigma} - \Lambda(E) = 0$ in Fig. 2.1. One corresponds to negative slopes of $\Lambda(E)$, the other to positive slopes. The phase of the argument of the logarithm increases rapidly at the former and decreases at the latter roots contributing large positive and negative values to $\Delta\varrho^\sigma$, respectively (see, e.g., Fig. 2.7). Therefore the former give rise to resonances and the latter to antiresonances. The local density of states ϱ_a^σ has peaks at the resonances and dips at the anti-resonances as will be shown later (see Fig. 2.6).

We are now ready to evaluate the chemisorption energy as the difference between the system energy when the adsorbate is not interacting and when the interaction is turned on. This is given by

$$\delta E = \left\{ \sum_\sigma \int_{-\infty}^{E_F} E \varrho^\sigma \, dE - V_{im} \right\} - \left\{ \sum_{\sigma} \int_{-\infty}^{E_F^0} E \varrho_m^0(E) \, dE + E_a \right\}$$
$$- U \langle n_\sigma \rangle \langle n_{-\sigma} \rangle. \tag{2.33}$$

The first curly bracket represents the total system energy after chemisorption and the second before chemisorption. V_{im} must be subtracted as the interaction of the ion core with the metal; it takes care of the fact that we have added it to E_a. The U term appears with a negative sign since it has been counted twice in the sum over system states; the factors Σ_σ stands for the summation over spin. It is necessary to distinguish between the Fermi energy of the interacting and non-interacting systems although the difference is infinitesimally small. Equation (2.33) can be rewritten as

$$\delta E = \sum_\sigma \int_{-\infty}^{E_F^0} E(\varrho^\sigma - \varrho_m^0(E))\, dE$$
$$+ \sum_\sigma (E_F - E_F^0)\, \varrho^\sigma(E_F)\, E_F - E_a - U\langle n_\sigma\rangle \langle n_{-\sigma}\rangle - V_{im}\,. \tag{2.34}$$

By making use of charge conservation,

$$\sum_\sigma \int_{-\infty}^{E_F} \varrho^\sigma\, dE = \sum_\sigma \int_{-\infty}^{E_F^0} \varrho_m^0(E)\, dE + 1 \tag{2.35}$$

and multiplying expression (2.35) by E_F and substituting the resulting expression for $(E_F - E_F^0)\varrho(E_F)$ in (2.34) we see that

$$\delta E = \sum_\sigma \int_{-\infty}^{E_F} (E - E_F)\,\Delta\varrho^\sigma\, dE + E_F - E_a - U\langle n_\sigma\rangle\langle n_{-\sigma}\rangle - V_{im}. \tag{2.36}$$

If energy is counted from the reference zero of E_F and if (2.32) is used in (2.36), expression (2.33) is equivalent to an expression derived by Newns directly from G_{aa} by means of Levinson's theorem on the poles and zeros of a function, and the fact that the poles of G_{aa}^{-1} correspond to the E_k and its zeros to the E_m.

The quantity $\Delta(E) = \pi \Sigma_k |V_{ak}|^2\, \delta(E - E_k)$ plays an important role in the Newns-Anderson model: It enters into Λ, $\Delta\varrho$, ΔE etc. Unfortunately $\Delta(E)$ cannot be evaluated exactly at present. Nevertheless we can proceed further by introducing the concept of a group orbital (φ_g) [2.9, 10] which simplifies the problem. We assume that the adatom interacts with the metal through a group orbital φ_g (i.e. $V_{ak} \cong V_{ag}\langle g|k\rangle$), which is taken as a linear combination of some substrate atomic orbitals having large overlap with the adatom orbital. We therefore approximate Δ by

$$\Delta(E) \simeq \pi |V_{ag}|^2\, \varrho_g(E), \tag{2.37}$$

where ϱ_g, the surface density of states at φ_g is given by

$$\varrho_g(E) = \sum_k |\langle g|k\rangle|^2\, \delta(E - E_k). \tag{2.38}$$

We have thus couched the problem in terms of overlap integrals between φ_a and φ_g, and the projection of the total density of states of the metal on φ_g. The choice of φ_g depends on the bonding geometry of the adatom. The surface density of states ϱ_g is a very important quantity in the theory of chemisorption. Recently there have been a number of calculations [2.11, 12] of this quantity for transition metals and simple cubic metals.

2.3. Reformulation of the Theory Using a Complete Basis Set

The model we have just sketched has the advantage of great simplicity and shows the qualitative features to be expected from any LCAO-MO theory in a very transparent way. However, it suffers from the fact that it neglects overlap between the metal and adsorbate states, and that it uses a very incomplete set of states. Outside the metal, where the adsorbate is located, the bound states of the metal decay rapidly, but the continuum states (i.e. unbound states) have large amplitude, and should therefore be much more important than in the case of an inpurity *in* the metal. It therefore seems a logical extension of the model to include continuum states of the metal to treat chemisorption. However, we now run into a conceptual difficulty. If the set of metal states is complete, it must contain the adsorbate orbital, i.e. the set of metal states plus the adsorbate orbital is overcomplete.

Since the totality of eigenfunctions of the metal $\{k\}$ form a complete set, we can always expand $\mathscr{H}$ and G in it even if the resultant matrices are not diagonal. The density of states, for instance, would still be given by Tr G since the trace is invariant. The problem then is to bring $|a\rangle$ into the picture somehow. There are various methods of doing so, which turn out to be if not equivalent at least closely related. Perhaps the simplest to understand, if not to justify is an approximation due to PENN [2.13].

The approximation consists of assuming for V^σ in (2.10) that

$$V_{kk'}^\sigma = \sum_b \langle k|V^\sigma|b\rangle \langle b|k'\rangle \simeq \langle k|V^\sigma|a\rangle \langle a|k'\rangle, \tag{2.39}$$

where $\{|b\rangle\}$ are eigenstates of the free adsorbate of which $|a\rangle$ is the most relevant member. Essentially we are treating $|a\rangle$ as a complete set. This can be justified in terms of a single resonance orbital approximation [2.14]. The equation of motion of the Green's function is still given by (2.12) except that the set $\{h\}$ now comprises $\{k\}$, the totality of the bound and unbound eigenstates of the metal. Inserting (2.39) in (2.12), one obtains [2.13]

$$G_{kk'}^\sigma = g_k\delta_{kk'} + \frac{g_k V_{ka}^\sigma \langle a|k'\rangle g_{k'}}{1 - \sum_k \dfrac{\langle a|k\rangle V_{ka}^\sigma}{E-E_k-i\alpha}}. \tag{2.40}$$

The local density of states and the change of the total density of states are found from (2.40) using (2.8) and (2.9), respectively

$$\Delta\varrho^\sigma(E) = \frac{1}{\pi}\frac{\partial}{\partial E}\operatorname{Im}\left\{\ln\left[1 - \sum_k \frac{\langle a|k\rangle\, V_{ka}^\sigma}{E - E_k - i\alpha}\right]\right\} \tag{2.41}$$

$$\varrho_a^\sigma(E) = \frac{1}{\pi}\operatorname{Im}\sum_{kk'}\langle a|k\rangle\, G_{kk'}\langle k'|a\rangle = \frac{1}{\pi}\operatorname{Im}\left\{\frac{\sum_{k'}|\langle a|k'\rangle|^2\, g_{k'}}{1 - \sum_k \dfrac{\langle a|k\rangle\, V_{ka}^\sigma}{E - E_k - i\alpha}}\right\}. \tag{2.42}$$

We can get a more symmetric result by using a different approximation for $V_{kk'}^\sigma$. For this purpose we multiply (2.39) by $\langle a|k\rangle$ and sum on k, obtaining $V_{ak'}^\sigma \simeq \langle a|V^\sigma|a\rangle\,\langle a|k'\rangle$. Using this in (2.39), we have

$$V_{kk'}^\sigma \simeq \frac{\langle k|V^\sigma|a\rangle\,\langle a|V^\sigma|k'\rangle}{\langle a|V^\sigma|a\rangle}. \tag{2.43}$$

It is to be noted that (2.39) and (2.43) are not equivalent, because they are approximate results. Using (2.43) in the equation of motion of the Green's function, one obtains [2.13]

$$G_{kk'}^\sigma = g_k\delta_{kk'} + \frac{g_k V_{ka}^\sigma V_{ak'}^\sigma g_{k'}}{V_{aa}^\sigma - \sum_k |V_{ak}^\sigma|^2\, g_k}. \tag{2.44}$$

It also follows from (2.8) and (2.44) that

$$\Delta\varrho^\sigma(E) = \frac{1}{\pi}\frac{\partial}{\partial E}\operatorname{Im}\left\{\ln\left[V_{aa}^\sigma - \sum_k |V_{ak}^\sigma|^2\, g_k\right]\right\}. \tag{2.45}$$

Although the denominator of the second term in (2.44) appears to be entirely different from that of the Newns-Anderson result [cf. Eqs. (2.13) and (2.14)] one can put the former, after some algebra [2.13], into

$$G_{kk'}^\sigma = g_k\delta_{kk'} + \frac{g_k V_{ka}^\sigma V_{ak'}^\sigma g_{k'}}{E - E_{a\sigma} - \sum_k |V_{ak}'^\sigma|^2\, g_k}, \tag{2.46}$$

where

$$V_{ak}'^\sigma = \langle a|\mathscr{H}^\sigma - E|k\rangle = V_{ak}^\sigma + (E_k - E)\,\langle a|k\rangle. \tag{2.47}$$

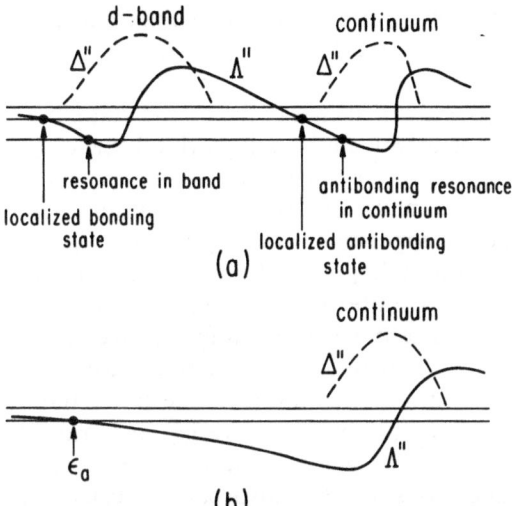

Fig. 2.3a and b. $\Delta''(E)$ and its Hilbert transform $\Lambda''(E)$ are shown when the adatom is near the surface (a) and at infinity (b). The roots of $(\bar{V}^\sigma)^{-1} = \Lambda''(E)$ are illustrated by circles

Equations (2.40) and (2.46) are equivalent to the results of KANAMORI et al. [2.15] and ANDERSON and McMILLAN [2.16] respectively.

The forms (2.40) and (2.44) are in fact rather similar to the Newns-Anderson result, except for the quantititatively significant difference that they include important contributions from continuum states with large amplitudes at the absorbate. In particular, (2.40) and (2.44) can give rise to resonances, antiresonances and localized states and so on. This can be seen qualitatively by approximating V_{ka}^σ and V_{aa}^σ by $\langle k|a \rangle \bar{V}^\sigma$ and $\bar{V}^\sigma$, respectively. With these approximations (2.40) and (2.44) become identical

$$G_{kk'}^\sigma = g_k \delta_{kk'} + \frac{g_k \langle k|a \rangle \langle a|k' \rangle g_{k'}}{\dfrac{1}{\bar{V}^\sigma} - \Lambda''(E) - i\Delta''(E)},\tag{2.48}$$

where

$$\Delta''(E) = \pi \sum_k |\langle a|k \rangle|^2 \delta(E - E_k)\tag{2.49}$$

and

$$\Lambda''(E) = \frac{1}{\pi} P \int_{-\infty}^{\infty} \frac{\Delta''(E')}{E - E'} dE'.\tag{2.50}$$

The general form of Δ'' and Λ'' will be similar to that of Δ and Λ, so that the existence of a localized state below the band of interest, for instance, corresponds to the intersection of Λ'' with the line $1/\bar{V}^\sigma$. Thus the stronger $\bar{V}^\sigma$ the lower in energy the bound state. It is interesting that the analogue of the empty antibonding orbital (the empty upper state in the Newns model) cannot be produced at all now, without invoking the existence of the continuum states, which continue Λ (or Λ'') in such a way as to make an upper intersection possible (Fig. 2.3). Thus overlap has pushed the antibonding orbital up in energy as expected.

If the potential is finite but the overlap with the d-band states is allowed to approach zero, i.e. if the atom is moved far from the surface, the intersection of $1/\bar{V}$ with Λ'' comes about entirely through the contribution of the continuum states. (For the latter the density of states increases beyond bound, but the overlap integral $\langle k|a\rangle \to 0$ as the de Broglie wavelength of the continuum states becomes smaller than the atomic dimensions of the adsorbate, because the positive and negative contributions from φ_k will cancel. This Δ and Δ'' for the continuum are bounded.) This illustrates again the point that the latter were necessary to form $|a\rangle$ in the first place.

For computational purposes it is convenient to use (2.40)–(2.42), and (2.36). The scattering potential V^σ is assumed to be

$$V^\sigma|a\rangle = (V_0 + V_{im} + U_{eff}\langle n_{-\sigma}\rangle)|a\rangle, \tag{2.51}$$

where V_0 is the ionic potential. The second term V_{im} describes the interaction of the valence electron with the metal electrons (i.e. exchange and correlation energy), and the third term is the Hartree-Fock intra-atomic repulsion at the adatom.

The wave functions corresponding to the continuum states of the unperturbed metal can be approximated by assuming a step-like model potential $W(z)$ given by

$$\begin{aligned} W(z) &= 0 & z &> 0 \\ &= -V_c & z &< 0, \end{aligned} \tag{2.52}$$

where $z > 0$ corresponds to the vacuum and $z < 0$ to the region inside the metal. $-V_c$ is the bottom of the conduction band measured from the vacuum level. With these approximations it is possible to carry out fairly realistic calculations. The main parameters entering the result are 1) the metal-adatom separation z_0, 2) the surface density of states ϱ_g which is introduced as before through a group orbital (φ_g) approximation, 3) the adatom-group orbital overlap integral $s = \langle a|g\rangle$, 4) image energy V_{im}, and finally a "Coulomb" parameter defined as

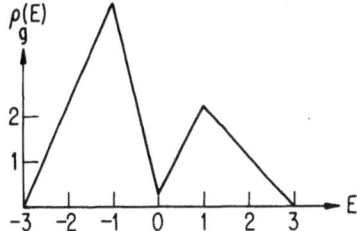

Fig. 2.4. Model surface density of states used for the numerical calculation. The total *d*-band width of tungsten is taken as 6 units, and the origin is at the band center. The Fermi level is approximately at 0.5

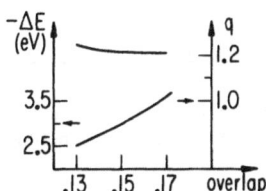

Fig. 2.5. The binding energy (lower curve) and the total electronic charge (in units of *e*) on the adatom (upper curve) as a function of overlap

$\eta = s^{-1} \int \varphi_g(r) [a_0/|r - r_a|] \varphi_a(r - r_a) d^3r$ ($a_0 \doteqdot$ first Bohr radius, $r_a \doteqdot$ adatom location).

A numerical computation has been [2.14] carried out for hydrogen adsorption on the (100) surface of tungsten, which has a body centered cubic structure. For this system the experimental binding energy [2.17] is 3.0 eV and the total electronic charge in the adatoms is $q = 1.0$ (a.u.). Field [2.18] and photoemission [2.19] measurements show resonance peaks at $0.9 \sim 1.1$ eV and 5.4 eV below the Fermi level E_F with approximate full widths at half maximum of 0.6 eV and 0.7 eV, respectively. The Fermi level [2.18] is at -4.64 eV below the vacuum on (100) surface.

In order to explain these data a model surface density of states, as shown in Fig. 2.4, was introduced. The full band width of tungsten ($\simeq 10$ eV) is taken as 6 units. The origin is at the band center and E_F is at 0.5. This corresponds approximately to the bonding of a hydrogen adatom above the middle of two nearest neighbor tungsten atoms on the 100 surface [2.11].

Figure 2.5 shows the self-consistently determined total charge at the adatom and the binding energy for three different values of the overlap integral $s = 0.13$, $s = 0.15$, and $s = 0.17$ for $V_c \simeq 12.2$ eV, $z = 1.5$ a.u., V_{im}

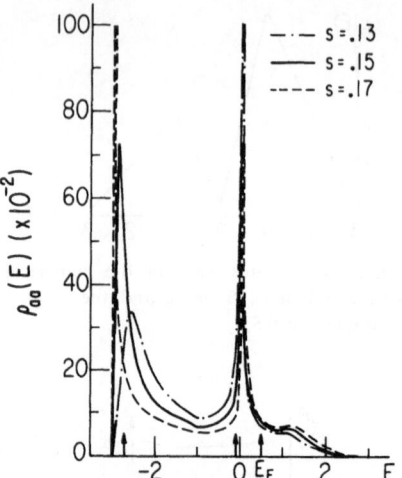

Fig. 2.6. The local density of states for three different overlaps. The energy scale is the same as in Fig. 2.4. The Fermi level, and the experimental values of resonances are indicated by arrows

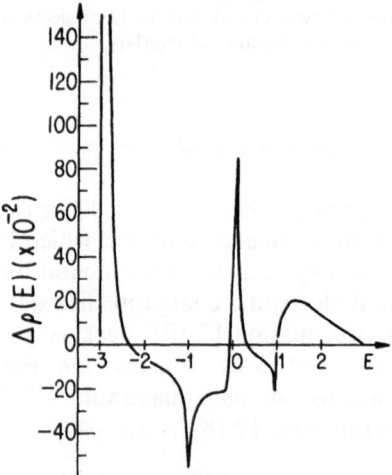

Fig. 2.7. The change of the total density
of states for an overlap $s = 0.15$

$= 5.5 \, \text{eV}$ and $\eta = 1.0$ [2.14]. It should be mentioned that s, z_0, and V_{im} cannot be uniquely determined at present. The solutions are non-magnetic. Figure 2.6 shows the local density of states for the same values of the parameters. It is seen that the overlap tends to split two resonance levels (i.e. bonding, antibonding levels) apart. The occurrence of the resonance near $E = 0$ is due to a minimum in the surface density of states there. Figure 2.7 shows the change of the total density of states $\Delta \varrho$ for $s = 0.15$ with other parameters unchanged. One observes the anti-resonances as well as the resonances. The results corresponding to $s = 0.15$ with the other parameters taken as above give reasonable agreement with the experimental data. As can be seen from (2.2), (2.51), and (2.36), the main effect of the image energy is to increase the binding energy approximately by $1/2\ V_{im}$ for a non-magnetic neutral binding (i.e. $\langle n_\sigma \rangle = \langle n_{-\sigma} \rangle = 1/2$), as in the present case. The calculation just presented suggests that a generalized LCAO-MO treatment in the Hartree-Fock approximation may be adequate for handling many cases of chemisorption. It should be pointed out, however, that the occurrence of a subsidiary resonance near E_F could only be explained by postulating a dip in the surface density of states. At the moment the existence of this dip is not firmly established, either experimentally or theoretically, and resolution of this question must await further work. It has recently been pointed out by HERTZ [2.20] that structure in ϱ_a including subsidiary peaks of the kind just discussed can also arise by treating spin fluctuation on the adsorbate and the metal in higher order than in Hartree-Fock. It is also conceivable that the subsidiary resonance near E_F arises from interaction of the metal with the hydrogen $2p$ state.

2.4. Adsorbate-Adsorbate Interaction

The interaction of adsorbates with each other through the mediation of the substrate is of considerable interest in connection with the coverage dependence of binding energies, and also the formation of periodic adsorbate arrays. We will not discuss here direct electrostatic inter-actions, i.e. dipole-dipole effects but restrict ourselves to the more subtle level splittings mediated by the substrate. This subject has been investigated by GRIMLEY [2.21] and more recently by EINSTEIN and SCHRIEFFER [2.10]. We give only the briefest sketch of the effect, following an unpublished treatment by LYO. We have already seen that the crucial quantity in calculating chemisorption energies is the quantity L defined by

$$L = V_{aa} - \sum_k |V_{ak}|^2\ g_k = V_{aa} - \langle a | V g\ V | a \rangle \qquad (2.53)$$

which enters (2.45).

If, say, two adsorbate atoms are being considered, so that we must define state vectors $|a_1\rangle$ and $|a_2\rangle$ corresponding to electrons on A_1

and A_2, respectively, it can be shown that L goes over into the determinantal form

$$L = \begin{vmatrix} V_{aa} - \langle a_1 | Vg\,V | a_1 \rangle & \langle a_1 | Vg\,V | a_2 \rangle \\ \langle a_2 | Vg\,V | a_1 \rangle & V_{aa} - \langle a_2 | Vg\,V | a_2 \rangle \end{vmatrix}, \qquad (2.54)$$

where $V = V_1 + V_2$. If the adsorbate atoms are far apart the coupling between them will be negligible and the off-diagonal elements will vanish. In that case $\ln L_{12}$ for the combined system becomes simply $2 \ln L_1$ as it should and the energy is linear in the number of adsorbate particles. Adsorbate-adsorbate interaction is thus contained in the off-diagonal matrix elements of L and can lead to increases or decreases in E, i.e. to attractive or repulsive interactions.

2.5. Valence-Bond (Schrieffer-Paulson-Gomer) Approach
[2.22, 23]

We have seen from the foregoing that the validity of the LCAO-MO approach in the Hartree-Fock approximation is limited to broad resonances, small U, or localized states straddling the substrate band. The basic problem is to take proper account of the kind of correlation which results from electrons hopping off the adsorbate into the metal. In principle this can be handled in LCAO-MO by including configuration interaction or by more sophisticated definitions of $|a\rangle$. A different approach is to exaggerate correlation from the beginning by setting up the analogue of a valence-bond wave function. It is well known that the LCAO-MO method exaggerates ionic contributions, while the Heitler-London function omits ionic terms altogether. Thus the latter approach has (excessive) correlation built into it. In the valence bond approximation it is customary to consider only electron pair bonds, the spin singlet leading to bonding because of the symmetric space part of the wave function. This is not an iron-clad rule but arises in most cases from quantitative considerations, i.e., the actual magnitudes of the Coulomb and exchange integrals. While HeH is not stable, either in the MO or valence bond schemes, He_2^+ is bonding in both. If the requirement for the electron pair bond is waived, the VB approximation becomes largely equivalent to the LCAO-MO scheme with postulated infinite U, which can be treated more or less along the lines outlined in the last section. If we insist, however, on the importance of spin pairing, the substrate metal as it stands is not a suitable partner since, at ordinary temperatures there are effectively no unpaired spins available. Consequently it is necessary to create electron-hole pairs by promoting electrons above the Fermi energy in order to create free spins which can then pair with the adsorbate spin to form a valence bond. One may think of the adsorbate spin as inducing spin in the substrate. This process would cost energy of course, were it not that the attendant bond formation leads to a net lowering.

It is not difficult to proceed slightly beyond this statement. The energy required to create a spin S on a metal surface atom is

$$\Delta E_{spin} = \frac{1}{2}\chi H^2 = \frac{(\mu_B S)^2}{2\chi}, \tag{2.55}$$

where μ_B is the Bohr magneton and χ a local spin susceptibility, defined by $\mu_B S = \chi H$, H being the (fictitious) magnetic field which induces S. If a full spin $S = 1/2$ is induced on a metal surface atom, a bond with the adsorbate can be formed, which lowers the energy by an amount W_m. If no spin were present, on the other hand, the adsorbate would interact repulsively with the surface, the energy being increased by W_r. If we interpolate between these limits we can write for the net energy change, regarded as a function of spin S

$$\Delta E(S) = (\mu_B S)^2/2\chi - 2(W_m + W_r)S + W_r \tag{2.56}$$

which reduces to W_r for $S = 0$ and $(\mu_B S)^2/2\chi - W_m$ for $S = 1/2$. By minimizing with respect to S we find the maximum decrease in energy as

$$\Delta E = -(2\chi/\mu_B^2)(W_m + W_r) + W_r. \tag{2.57}$$

It can be shown[1] that

$$2\chi/\mu_B^2 = (2W_b)^{-1}, \tag{2.58}$$

where W_b is the width of the relevant metal band (for approximately flat bands) so that finally

$$\Delta E = -\frac{(W_m + W_r)^2}{2W_b} + W_r. \tag{2.59}$$

---

1 The total magnetic moment $\chi_{tot} H$ induced by a field H in a free-electron-like metal by Pauli paramagnetism is $2\mu_B S\varrho(E_F)\Delta E$, where $\Delta E = (1/2)\mu_B H$ is the change in energy and $\varrho(E_F)$ is the density of states at the Fermi level not counting spin. Thus the total susceptibility is $\chi_{tot} = \mu_B^2 \varrho(E_F)/2$. If the density of states is roughly equated to $N/2W_b$ where N is the number of atoms in the metal the susceptibility per atom is $\chi = \mu_B^2/4W_b$.

This indicates that binding increases as the band gets narrower, in agreement with the fact that binding is strongest on transition metals.

A more quantitative formulation of this theory [2.22] is a rather difficult many-body problem, and goes considerably beyond the scope of this article. Since the basic approach abandons the concept of one-electron energy levels, the theory is also difficult to couch in the language of local densities of state. Very qualitatively, a level spectroscopy should indicate some disturbance in the substrate densities near E_F and should show an electron on the adsorbate near E_a. In the tight binding case, where the VB method also predicts a local state below the band, this quasi one-electron level could be shifted below E_a of the free atom.

2.6. Linear Response (Kohn-Smith-Ying) Method

An entirely different approach to chemisorption applied to date only to hydrogen has been taken by Kohn and his coworkers [2.24], discussed also in Chapter 1. In this method a bare proton is allowed to imbed itself in the electron gas at a metal surface, and the linear response of the resultant electron charge density is found self-consistently by going beyond the Fermi-Thomas approximation. The total energy is expressed in terms of the charge density and its gradient, and then minimized variationally with respect to the charge density. The metal is treated as a jellium. The only free parameter is the Wigner-Seitz radius. Thus the model ignores the band structure of the metal. For transition metals the validity of this approximation is doubtful. Although the calculated binding energy (relative to H^+ and M^-) is not very good, the method deals rather effectively with the charge density at and near the adsorbate and thus explains observed dipole moments rather well. In principle, the self-consistent potential which results can be used to calculate quasi-one-electron energy levels, and thus a local density of states can be extracted. In principle, the method is extendable to other cases, although actual calculations will undoubtedly be very difficult.

2.7. Concluding Remarks

This chapter has attempted to depict the current status of the theory of chemisorption. It seems appropriate to conclude with a short discussion of unsolved problems and possible future developments.

We have indicated that reasonably quantitative calculations of binding energy and local density of states at the equilibrium separation seem within our grasp. In the writers' opinion it is likely that the majority of cases can be handled by LCAO-MO schemes, probably in the Hartree-Fock approximation, because screening of the Coulomb repulsion U at the equilibrium distance reduces it so substantially. This is not to say that many important problems do not remain. Foremost among these are the calculation of self-consistent surface densities of state and dielectric response, i.e. image energy calculations for real transition metals. The first of these problems probably offers no major obstacles, and it is probable that the latter will also be solved before too long. Thus the problem of chemisorption at the equilibrium configuration seems, if not solved, at least soluble.

At large surface-adsorbate separations the problem appears equally soluble, if only because interaction energies are small and perturbative treatments are thus almost guaranteed to work. The situation seems quite different at intermediate separations. Here interaction is reasonably strong, but for cases of large U screening is almost certainly insufficient to justify the Hartree-Fock approximation. In these situations it seems very probable that the valence bond or possibly the linear-response methods developed by SCHRIEFFER and by KOHN, respectively, will turn out to be far more appropriate than the LCAO-method. It seems probable that much future work will concern itself with this region.

References

2.1. J. R. SCHRIEFFER, D. C. MATTIS: Phys. Rev. **140**, 1412 (1965).
2.2. D. M. NEWNS: Phys. Rev. B **1**, 3304 (1970);
D. E. BECK, V. CELLI: Phys. Rev. B **2**, 2955 (1971).
2.3. J. A. APPELBAUM, D. R. HAMANN: Phys. Rev. B **6**, 1122 (1972).
2.4. D. R. PENN, R. GOMER, M. H. COHEN: Phys. Rev. B **5**, 768 (1972).
2.5. D. R. PENN: Phys. Rev. Letters **28**, 1041 (1972).
2.6. D. M. NEWNS: Phys. Rev. **178**, 1123 (1969).
2.7. P. W. ANDERSON: Phys. Rev. **124**, 41 (1961).
2.8. T. B. GRIMLEY: Proc. Phys. Soc. (London) **90**, 751 (1967) and previous papers.
2.9. T. B. GRIMLEY: J. Vac. Sci. Technol. **8**, 31 (1971).
2.10. T. L. EINSTEIN, J. R. SCHRIEFFER: Phys. Rev. B **7**, 3629 (1973).
2.11. D. R. PENN: Surface Sci. **39**, 333 (1973).
2.12. D. KALKSTEIN, P. SOVEN: Surface Sci. **26**, 85 (1971);
F. CYROT-LACKMANN, M. C. DESJONQUERES: Surface Sci. **40**, 423 (1973);
R. HAYDOCK, M. J. KELLY: Surface Sci. **38**, 139 (1973).
2.13. D. R. PENN: Phys. Rev. B **9**, 839 (1974).
2.14. S. K. LYO, R. GOMER: Phys. Rev. B **10**, 4161 (1974).
2.15. J. KANAMORI, K. TERAKURA, K. YAMADA: Progr. Theor. Phys. **41**, 1426 (1969).
2.16. P. W. ANDERSON, W. L. MCMILLAN: In *Scuola internationale di fisica, Varenna, Italy*, ed. by W. MARSHALL (New York, Academic Press, 1967).
2.17. P. W. TAMM, L. D. SCHMIDT: J. Chem. Phys. **51**, 5352 (1969); **55**, 4253 (1971).

2.18. E. W. Plummer, A. E. Bell: J. Vac. Sci. Technol. **9**, 583 (1972).
2.19. B. Feuerbacher, B. Fitton: Phys. Rev. B **8**, 4890 (1973).
2.20. J. A. Hertz: To be published.
2.21. T. B. Grimley, S. M. Walker: Surface Sci. **14**, 395 (1969).
2.22. J. R. Schrieffer, R. Gomer: Surface Sci. **25**, 315 (1971).
2.23. R. H. Paulson, J. R. Schrieffer: Surface Sci. **48**, 329 (1975).
2.24. J. R. Smith, S. C. Ying, W. Kohn: Phys. Rev. Letters **30**, 610 (1973).

3. Chemisorption: Aspects of the Experimental Situation

L. D. Schmidt

With 13 Figures

Because of its obvious relevance to catalysis and corrosion, the characterization of chemisorption has been the major objective of most studies of solid surfaces. It is the intent of this chapter to try to present a brief overall picture of the current situation regarding the interpretations of experiments on chemisorption. This is a rapidly developing and changing subject. Many of the concepts which appear in textbooks as recently as ten years ago have now been shown to be incomplete or wrong, and even monographs tend to become outdated within a few years. In a single chapter it will be impossible to cover chemisorption completely or even to list more than a small fraction of the significant recent developments.

This chapter will therefore be restricted to a consideration of some important recent experimental results regarding adsorbate structures in chemisorption. At the beginning, we note those topics not covered. First, we shall restrict our discussion to the interpretations directly required by experiments rather than those involving theoretical analysis of such data because chemisorption theory has been covered in Chapter 2. Second, we shall consider mainly atomic structures rather than kinetics or electronic structures because these are also covered in other chapters. Third, we shall be concerned only with chemisorption rather than physical adsorption or metal adsorption even though there have been important recent discoveries regarding these adsorbates also. Fourth, we shall consider only chemisorption on metals because, while our knowledge of adsorption on insulators and semiconductors has advanced considerably, the techniques used and the apparent nature of the structures are quite different than on metals. Finally we shall present experimental results with little consideration of the experiments themselves except where experimental uncertainties make interpretations ambiguous. Most of the new experimental methods are covered in other chapters, and earlier ones (methods of work function measurement, field emission microscopy, etc.) have been reviewed extensively in previous monographs.

It is hoped that the discussion will be comprehensible to the reader who is not working in surface physics as well as informative to those who are familiar with the subject. The pertinent quantities used in describing experimental results are adsorbate density n, fractional saturation coverage θ, work function change $\Delta\phi$, activation energy of desorption E_d, sticking coefficient or probability of condensation s, and symmetry of the surface structure, designated $(j \times k)$, where j and k are integers denoting the sepeating distance on the surface in multiples of the solid lattice constant (see Section 6.5 for further discussion of nomenclature). These quantities are readily interpreted without a detailed understanding of the apparatus or experiments from which the quantities are obtained. Interpretation of such data is at times less than obvious, however, and a section is devoted to this.

By adsorbate structures we mean simply the locations of the adsorbate atoms on a surface. This may appear at first to be a very easy task requiring only diffraction measurements. This may be nearly so if LEED theory advances to the stage where adsorbate structure can be obtained from LEED experiments (which are well developed). However even when the structures of perfectly ordered configurations are known, other measurements will be required to interpret the complex structures which exist in many chemisorption systems and the incompletely ordered structures at less than saturation densities and at low or high temperatures.

This chapter is divided into sections covering some of the subjects on which recent advances have been most significant: crystallographic anisotropies, adsorption binding states, adsorbate-adsorbate interactions, and variations between similar substrates. The discussion could also be divided into a few prototype chemisorption systems (H_2, N_2, and CO on body centered cubic metals and CO on face centered cubic metals) because there are only a few of such systems for which data are extensive enough to permit generalizations.

3.1. Structures of Clean Surfaces

Any attempts at locating adsorbate atoms on a surface presuppose a knowledge of the positions of the substrate atoms. This in fact has been, and in some cases still is, a major hindrance to determination of adsorbate structures. Until the 1950's surfaces were generally considered to be structureless (completely disordered or else possessing only the minimum free energy plane) with concepts such as Langmuir's "isolated active areas" invoked to explain any obvious deviations from this assumption. In the 1950's the field emission and field ion microscopes, LEED, and widespread use of macroscopic single crystals and ultrahigh vacuum techniques revealed clearly that polyfacetted surfaces can in no sense be considered uniform. Chemisorption structures and kinetics are now recognized to vary strongly with surface orientation, and as a consequence the proportion of chemisorption papers concerned with polycrystalline and polyfacetted surfaces is rapidly diminishing to situations where primarily engineering information is desired or where single crystal preparation techniques are difficult.

Therefore we must first enquire as to the structure of clean solid surfaces. This involves both the symmetry of the surface atoms and thermal disorder. Symmetry and long range order are studied by LEED, while thermal disorder and the positions of individual atoms are investigated mainly by field ion microscopy. Long range order varies between different metals and between metals, semiconductors and insulators. The trend regarding long range order, at least of metals, appears well established however: field ion microscopy [3.1] clearly indicates that the high symmetry, low free energy planes of metals are *well ordered* with *very low defect densities*, while other planes of a given metal exhibit considerable disorder. Incidentally, most published field ion micrographs are irrelevant to this question because they are obtained after field evaporation, i.e. after removing the top few layers of disordered atoms. Observing a perfect plane in such a micrograph implies nothing about the macroscopic plane. Only after first cleaning a surface by heating in the absence of a field can one use field ion microscopy to infer atom positions on a macroscopic thermally annealed surface.

3.1.1. Body Centered Cubic Transition Metals

These metals — especially W, Mo, Ta, and Nb — are easily studied because they can be cleaned by heating alone or by heating in O_2 at low pressures. Their clean surfaces give only (1 × 1) LEED patterns, implying that the symmetry is identical to that of the ideal crystal plane obtained by cleaving a perfect crystal (Fig. 3.1). This does not give the possible vertical displacements of the surface atoms or the extent of thermal disorder. From field ion microscopy [3.1] it appaers that (110), (100), and (211) planes are perfect, the stepped surfaces consisting of (110) planes, and other planes such as (111) are rather disordered. However, chemisorption variations between these apparently disordered planes imply that specific adsorption sites exist, and that even these surfaces are not without order. For adsorbate structure determination on these metals it is therefore reasonable to assume that the structures are nearly those of the ideal perfect crystal planes, some of which are shown in Fig. 3.1.

Those of us studying chemisorption have been rather defensive of our apparent assumption that tungsten is the only metal in the Periodic

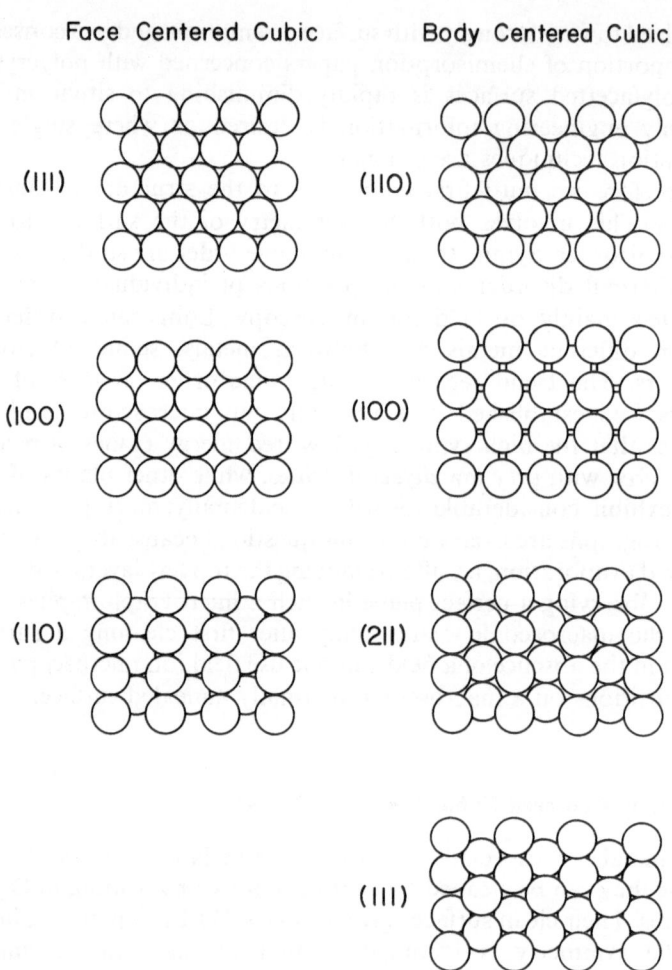

Fig. 3.1. Ideal surface atom positions for high symmetry planes of face centered cubic (fcc) and body centered cubic (bcc) metals. For most transition metals LEED and field ion microscopy indicate that clean surfaces exposing the planes have these structures and are nearly perfect with little disorder

Table. The major reason has admittedly been the ease of preparing clean surfaces of W, but the extensive measurements on this solid have in fact been quite useful because this is almost the only metal on which enough duplication of data exists to give confidence in the validity of results (but, as we shall see, a number of controversies still exist). Further, adsorption on W may prove to be a standard with which to calibrate instruments and to compare adsorption properties (relative densities and sticking coefficients, for example) on other surfaces. Also fortuitous for those studying tungsten is that it is the only metal which was reliably cleaned before 1970, except for vacuum deposited thin films. Auger electron spectroscopy (AES) has shown that many (perhaps most) earlier studies on metals such as Ni and Pt may, in fact, have been on surfaces which contained significant contamination with carbon, sulfur, or oxygen.

3.1.2. Face Centered Cubic Metals

These metals include many of the best catalysts (Pt, Ni, Pd, and Rh) and the noble metals (Cu, Ag, and Au). LEED indicates that (111) planes have (1 × 1) symmetry for all fcc metals. On Ni, Pd, and Cu the (100) planes also have (1 × 1) symmetry [3.2–4]. However, from there the situation becomes complex. The clean (100) planes of Pt, Au, and Ir exhibit a (5 × 1) symmetry [3.5]. It now appears from LEED intensity analysis that this arises from a nearly hexagonal overlayer of metal atoms on the ideal (100) plane. This layer is slightly compressed from the ideal (111) plane, and all atoms are not coplanar. Registry with the underlying (100) plane occurs at five lattice spacings in one direction and one in the other to yield a (5 × 1) LEED pattern. The (110) plane is the plane of next close packing, and, while on some metals it may have (1 × 1) symmetry, (110) Pt exhibits a (2 × 1) LEED pattern [3.6]. Probably alternate atoms along the rows are displaced or missing, but no clear picture of atom positions has emerged. The stepped planes of Pt, which consist of (111) planes and (100) steps, exhibit (1 × 1) LEED patterns with regularly spaced steps one atom in height [3.7].

3.1.3. Semiconductors

Silicon and germanium surfaces have been fairly well characterized because of their applications in solid state electronics. While vacuum cleaving may produce (111) planes with (1 × 1) periodicity, heating this or any other plane usually produces larger unit cell structures. The close packed (111) planes of Si and Ge exhibit (7 × 7) and (8 × 8) periodicities while the (100) planes both have (2 × 1) periodicities [3.8, 9]. Reasonable structures have been proposed to explain these observations [3.1]. The cause of the rearrangement of these surfaces from their bulk structures is the covalent bonding between atoms; the dangling bonds at the surface cannot be satisfied without large changes in the positions of the atoms in at least the first layer. Any detailed attempts at chemisorption structure determination on these surfaces requires a knowledge of the positions of the surface atoms and possible changes in these positions after adsorption. Additional experimental difficulties arise from the strong influences of trace impurities which alter electronic properties of these surfaces and from the fact that most gases have very low sticking coefficients on semiconductors.

3.1.4. Insulators

The structures of very few crystal planes of insulators are known with the exceptions of those prepared by low temperature vacuum cleaving, which are (1×1). A fundamental problem with these surfaces is that upon heating the atomic composition of the surface layer is not known. For oxides there is evidence that the metal oxygen ratio is not always that of the bulk and that it may be varied by heating or exposure to O_2 [3.10]. Oxide catalysts invariably have H_2O or OH^- groups on their surfaces.

3.2. Methods of Adsorbate Characterization

It is appropriate to comment briefly on some of the experimental techniques used to characterize adsorption because, while some are mentioned in other chapters, others are not. Also most of our present knowledge of adsorbate structures comes from techniques which have been widely used for ten years or more and now are rather "classical". The other chapters deal more with newer ones and those aspects of older ones which have not been widely exploited. Here our interest is in the utility and limitations of various techniques and the ways in which they complement each other for adsorbate structure determination.

The basic data one needs to characterize an adsorbate are 1) the number of binding states, 2) their saturation densities, both relative to each other and relative to the surface atom density, and 3) the periodicity of the adsorbate. For a simple system one might be able to construct a model for the adsorbate from such data. However almost without exception such a structure would be either nonunique or there would be gaps or contradictions in data. Therefore one would need additional information such a whether states are atomic or molecular, binding energies, sticking coefficients, and signs and magnitudes of dipole moments. These data, all readily obtained in most modern surface chemistry laboratories, are still insufficient in many cases, and one needs finally either a complete diffraction analysis or spectroscopic information on bonding and electronic structures.

3.2.1. Diffraction

LEED in principle is the most direct means of determining adsorbate structure. As discussed in Chapter 6, the positions of diffraction spots immediately give the size and orientation of the unit cell, and analysis of spot intensity versus electron energy gives in principle the positions of atoms in the unit cell and the distances between adsorbate and substrate atoms. Unfortunately LEED has only been used extensively to determine perodicities of adsorbates and atom positions on clean surfaces, but LEED theory now appears to be capable of yielding information on positions of adsorbate atoms as well.

In addition to perfectly ordered adsorbates, one must also describe the imperfectly ordered configurations obtained at various temperatures and at coverages other than saturation. The applicability of LEED theory for this purpose is uncertain, but measurements of diffraction spot size and intensity can still yield considerable information about the degree of order. The variation in the intensity and width of fractional order adsorbate beams with coverage in a particular state yields information concerning adsorbate-adsorbate interactions [3.11–13]. If intensity $I \sim \theta$, the adsorbate probably exists as islands or domains, while if I increases at greater than the first power of coverage, random occupation is inferred. The width of adsorbate beams $\Delta w_{\frac{1}{2}}$ gives the approximate degree of order, $\Delta w_{\frac{1}{2}} \simeq 2/l + w_0$, where w_0 is the instrumental resolution and l is the distance over which order persists. Typical instrument resolutions are 0.01 $\mathring{A}^{-1}$, so that one can determine the degree of order for $l \leqq 100$ $\mathring{A}$.

The variations of adsorbate beam intensities and widths with substrate temperature also gives information on the order of an adsorbate. At sufficiently low deposition temperature the adsorbate should be immobile and thus incapable of ordering, and at high temperature thermal disorder should produce a decrease in spot intensities [3.13]. We shall see that all of these phenomena have been observed.

All of this analysis presupposes that there are nonintegral diffraction beams. In their absence any analysis by LEED appears questionable. It should also be emphasized that perfectly ordered structures (states saturated and annealed) are probably the exception rather than the rule. One must use considerable caution in interpreting LEED data for other than the simplest systems because it is at present impossible to determine from intensity measurements the fraction of an adsorbate which is in a given configuration. Sharp and intense diffraction beams may be observed when only a small fraction of the adsorbate occupies sites with the corresponding configurations with the rest disordered or in binding states with (1×1) periodicity. Care should also be exercised in specifying adsorbate temperatures because many adsorbates which exhibit multiple lattice spacing periodicities may disorder at higher temperatures. Since most LEED experiments have been carried out at 300 K or above, many ordered structures may have been missed entirely. Examples of low temperature ordered configurations are alkali metals [3.14, 15] and CO on fcc metals [3.2–4]; many structures are observed below room temperature but few or none above.

3.2.2. Kinetics

Kinetics are probably the most used and most productive methods for characterizing adsorbate structures. Thermal desorption and especially flash desorption directly yield binding states, relative saturation densities, and desorption rate parameters [3.16, 17]. Desorption kinetics is covered thoroughly in Chapter 4. Analysis of desorption data is generally straightforward, but problems such as induced heterogeneity, bulk solution, and desorption of nonvolatile species must be considered [3.17].

Condensation kinetics provides an indirect means of characterizing adsorption because sticking coefficients depend on the state being populated, and the coverage dependence in a state is controlled by the mechanism of condensation — direct condensation or condensation through a precursor intermediate [3.17]. The magnitude of the sticking coefficient is an important experimental parameter in all measurements because, if too small, it may be difficult to populate a state by gas condensation at low pressures.

Surface diffusion yields information on the potential experienced by an adsorbate as it moves laterally along a surface. Surface diffusion processes govern annealing and order-disorder transformations and may be rate limiting steps in desorption of a dissociated species, in bimolecular reactions, and in bifunctional catalysis. Measurement of surface diffusion rates on single crystal planes are exceedingly difficult, and most data has been obtained using field emission microscopy. Early results have been summarized [3.19–21] and there have been few recent studies. Most information is obtained only as average diffusion coefficients over a field emitter at nominal coverages with qualitative comments on temperatures required for equilibration on particular planes. Almost completely lacking are precise measurements of surface diffusion coefficients on single crystal planes as a function of densities in individual binding states. The major exception is the elegant work of EHRLICH and coworkers on surface self-diffusion using field ion microscopy [3.20, 21]. Field desorption limitations have thus far prevented comparable studies for other than metallic adsorbates. Another study which illustrates qualitatively the crystallographic variations in diffusion coefficients is the work of ENGEL and GOMER who measured temperatures for diffusion of different binding states of O and of CO on most planes of a W field emitter [3.23].

Activation energies vary from 10–20% of binding energies, for many atomic adsorbates; the low values are encountered on atomically, smooth the high values on atomically rough planes, as far as is known. For CO on W the ratio can approach 100% on loosely packed planes.

3.2.3. Auger Electron Spectroscopy (AES)

In the few years since the utility of AES for surface study was demonstrated [3.25], it has rapidly become an essential instrument in surface physics laboratories. With AES and a mass spectrometer to monitor partial pressures in a vacuum system the experimenter can make a complete chemical analysis of an adsorption system — both gases impinging on a surface and the chemical composition of the surface itself. The characteristics of AES which make it so useful are that all elements except hydrogen can be detected with roughly comparable sensitivities, and the height of an AES peak for a surface species is apparently proportional to its density for monolayer or submonolayer densities. The latter assumes that peak shape changes due to chemical shifts can be accounted for and that the adsorbate is confined to the surface layer. Both of these characteristics are predicted theoretically and verified experimentally in all systems so far examined.

Since all of the adsorption systems to be discussed in this chapter have been shown by AES to contain less than a few percent of a monolayer of contaminants (except for possible chance overlap with substrate peaks), it is virtually certain that all of these studies were on clean surfaces. Such a statement could not be made unequivocally for any surface as recently as five years ago.

3.2.4. Miscellaneous Techniques

While LEED and desorption measurements provide the most used and most versatile means for characterizing adsorbate structures, other techniques are useful and sometimes necessary. For adsorbates which do not give stable gas molecules upon desorption, work function changes, condensation rates, AES, and electron stimulated desorption (ESD), [see Chapter 4], are the only methods for determining binding states and densities. Notable use of these techniques has been in O_2 on W which desorbs largely as O atoms and oxides; work function changes and ESD have yielded a fairly detailed picture of this adsorption system [3.24].

Spectroscopic measurements are now recognized as among the most powerful for characterizing adsorbates because they are sensitive to electronic structures, charge distributions, and types and symmetries of bonds. Infrared spectroscopy has been used for many years to examine high area surfaces and more recently for adsorption of CO on well characterized single crystal planes [3.26, 27]. New techniques such as ion neutralization spectroscopy (INS), X-ray photoelectron spectroscopy (XPS), ultraviolet photoelectron spectroscopy (UPS), and field emission energy distributions promise to yield considerable information on electronic structures and symmetries of adsorbates. These are still being developed as surface analytical techniques and are discussed in Chapter 5. It should be noted that spectroscopies are most useful when detailed information on states, densities, and periodicities are available. We shall concentrate on the latter because other chapters deal with the applications of spectroscopy to particular surface and adsorption systems.

3.3. Crystallographic Anisotropies

This section is concerned with the variations in adsorption properties — binding states, condensation and desorption kinetics, and saturation densities — with the crystallographic orientation of the substrate. In the last decade it had become abundantly clear to experimentalists that these variations are in many cases even more significant between crystal planes of a given substrate than between the same planes of different metals. However, models of chemisorption are still proposed in which the substrate is structureless, and elaborate attempts at interpretation are still proposed for data obtained on polycrystalline surfaces.

The adsorption system we shall choose to illustrate crystallographic anisotropies is N_2 on W. The first systematic attempt at observing variations between planes was made on this system by DELCHAR and EHRLICH [3.28], and there have since been a number of studies which are all in fairly good agreement. Recent references to work on the various planes are summarized in Table 3.1; additional references can be found from these.

On all planes listed in Table 3.1 there appears to be one major tightly bound state which obeys second order desorption kinetics and is atomic. On most planes additional tightly bound states are observed which have much lower sticking coefficients than those indicated, and there are a number of weakly bound or γ states which are molecular. We shall

Table 3.1. Nitrogen on tungsten

Plane	E_{d_0} [kcal/mole]	S_0	N_0 [a] [atoms/cm²]	Ref.
(110)	79	0.004	1.8×10^{14}	[3.29]
(100)	79	0.4	5×10^{14}	[3.28,30–32]
(211)		<0.01	—	[3.28]
(111)	75	≤0.04	—	[3.28]
(210)	~75	0.25	2.2×10^{14}	[3.33]
(310)	~75	0.28	3.2×10^{14}	[3.33]

[a] Saturation density of tightly bound atomic states.

consider mainly the most tightly bound states as they best illustrate crystallographic anisotropies.

First we note that the activation energies of desorption are experimentally almost indistinguishable on all planes, 80 ± 5 kcal/mole. The binding energies of the atoms, E_A, related to the desorption activation energies of the molecule by the relation

$$E_A = \tfrac{1}{2}(E_{dA_2} + D),$$

are higher than this, ~ 150 kcal/mole, are even closer to each other because of the dominance of D, the dissociation energy of N_2, in the above expression.

Table 3.1 and Fig. 3.2 show that the saturation densities (in the tightly bound states) differ by a factor of at least 3 for the different planes and that initial sticking coefficients differ by a factor of ~ 100. There are distinct differences in the shapes of the $s(\theta)$ curves. On (110), $s \sim (1-\theta)^2$ as predicted by a site occupation mechanism requiring two vacant adjacent sites for condensation, while on (100), (210), and (310) s is initially constant as predicted by condensation through a precursor intermediate

$$N_2(\text{gas}) \rightleftarrows N_2^*(s) \rightarrow 2N(s), \tag{3.1}$$

where $N_2^*(s)$ is a weakly adsorbed species on the surface. Various precursor models similar to that implied by (3.1) have been developed [3.31], and, for N_2 on (100)W, $s(\theta)$ and its dependence on surface temperature can be fitted quantitatively by such expressions.

Nitrogen on the (100) plane has been studied by LEED [3.30, 31] and shown to produce a $c(2 \times 2)$ pattern as expected if N atoms occupy every other site on the surface, Fig. 3.3. The adsorbate beam intensities

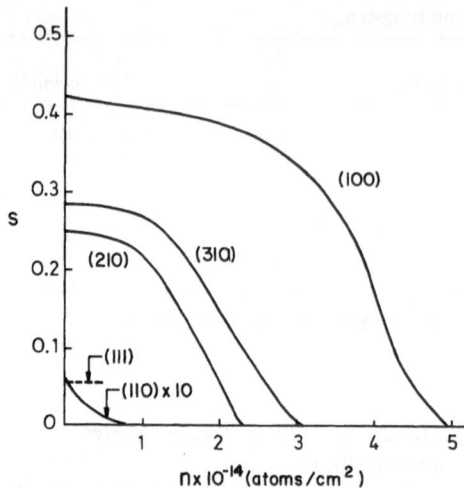

Fig. 3.2. Sticking coefficient versus adsorbate density for N_2 on several crystal planes of W. These results show that there are large variations in both saturation densities and sticking coefficients between different crystal planes

increase at least linearly with coverage, indicating *attractive* adsorbate-adsorbate interactions and island formation [3.32, 13]. The adsorbate-caused beams increase in intensity and sharpen when the surface is heated above 300 K after deposition, suggesting that ordering is a thermally activated process. This will be discussed in more detail later under adsorbate-adsorbate interactions.

The integral diffraction beams appear to remain sharp at coverages and temperatures where the adsorbate beams are broadened by disorder [3.30, 32]. This indicates that adsorption sites probably have four-fold symmetry, as shown in Fig. 3.3 rather than the two-fold symmetry of bridge bonded sites. Nitrogen atoms could, of course, be either on top of the surface W atoms rather than in the sites indicated, but the higher coordination in the latter suggests this location.

3.3.1. Stepped Surfaces

The (210) and (310) planes expose flat planes with four-fold sites identical to those on the (100) plane with steps consisting of substrate atom configurations identical to those on the (110) plane. ADAMS and GERMER [3.33] showed that the LEED pattern from the (210) plane after cleaning was as indicated in Fig. 3.3 with regularly spaced steps.

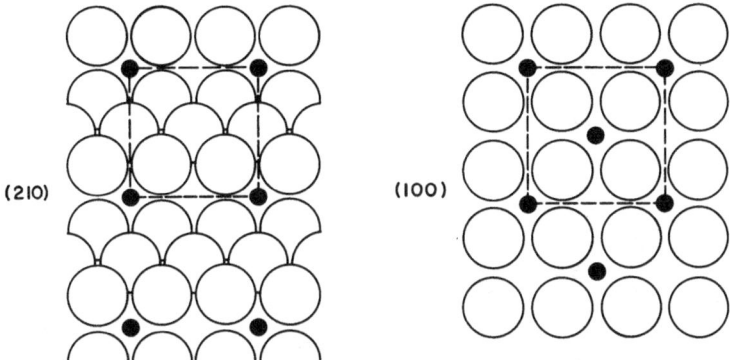

Fig. 3.3. Proposed adsorbate structures for N_2 on the (100) and (210) planes of W. Nitrogen atoms occupy alternate sites of four-fold coordination on both planes, and on the (210) plane atom positions are correlated between rows

ADAMS and GERMER used LEED, work function change, and flash desorption to study nitrogen on the stepped (210) and (310) planes to compare with the (100) plane. Their results are summarized in Figs. 3.2 and 3.4 and Table 3.1. Deposition of N_2 on (310) at 300 K produced weak and broad $p(2 \times 1)$ adsorbate beams. Heating to 900 K increased the order in the [100] direction along the steps (resulting in streaks), and heating to 1200 K produced sharp and intense adsorbate beams with the $p(2 \times 1)$ structure. The structure implied by these results are shown in Fig. 3.3. When completely ordered, N atoms occupy alternate four-fold sites just as they do on the (100) plane. However, there is also a correlation of sites *between rows*. Above 900 K the adsorbate orders both along and between rows, but below 900 K ordering occurs only along rows. Similar behavior is noted on the (310) plane although the unit cell is not rectangular. A most remarkable feature of these results is the long range correlation between rows, a distance of over 10 Å on the (210) plane.

Stepped surfaces of clean platinum also exhibit regularly spaced monatomic steps [3.7]. Ordering of adsorbates is observed on these planes although the adsorbates examined are more complex, and the interpretation of these processes is less clear than for nitrogen on W.

In Fig. 3.4 are shown work function changes versus fractional coverage in the tightly bound states. It is evident that the sign of the dipole moment on the (100) plane is opposite that on the (210) and (301) planes. A simple explanation [3.28, 33] for these variations is that the varying positions of the image planes can produce a reversal in $\Delta\phi$ if the N atom on the stepped planes is below the image plane. This points up the difficulty of interpreting work function changes in terms of adsorbate charges. The magnitude and even the sign of the change depends on the positions

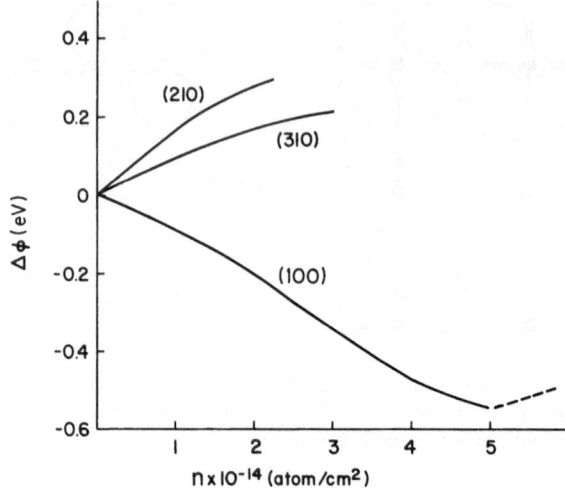

Fig. 3.4. Work function change versus adsorbate density for N_2 on W

of the atomic and the electronic configurations, neither of which is as yet established for any adsorption system.

Little mention has been made of weaker binding states for N_2 on W [3.28, 31]. These have lower sticking coefficients than the most tightly bound states, and experimental data are consequently much more subject to artifacts from contamination and electron impact effects. One much studied binding state is the β_1 state on (100) W. This is almost as strongly bound as the β_2 state but desorbs with first-order kinetics and has a dipole moment opposite that of the β_2 [3.34]. It apparently may be populated readily either by electron impact, by adsorption of NH_3, or in the presence of contaminants, but has a very low sticking coefficient otherwise [3.31]. This state has been the subject of much speculation, but determination of its structure awaits additional, perhaps spectro-scopic, examination. For all but the tightly bound states there are possible discrepancies in saturation densities because of the experimental problem of saturating a low s binding state. ADAMS and GERMER [3.33], while noting the problem, assumed that only (100) type sites were populated on (210) and (310) planes. Higher densities would, of course, be obtained if (110) sites and weak binding γ sites are populated.

Crystallographic anistropies in both s and saturation density are probably greater for N_2 on W than for any other well characterized chemisorption system. This is perhaps because of the large dissociation energy of N_2 and the availability of three valence electrons in the nitrogen atom. However, as we shall see in the following sections, even H_2 exhibits anisotropies which can in no sense be neglected.

3.4. Binding States

While there have been many careful and comprehensive studies of particular gases on single crystal planes of metals using a variety of techniques, only for a few systems have corresponding data been obtained on several crystal planes of a given metal. Thus comparison of binding states, saturation densities and binding energies is possible in only a few systems. As just discused, nitrogen on W is one such system, but here there is only one binding state and binding energies are approximately identical on all planes examined. The two examples we shall consider here are H_2 on W and CO on W. For these gases there are between two and five states on each plane so far examined, and, although each system has been the subject of over twenty papers, both exhibit sufficient complexity to require many more for understanding. Other particularly well documented systems are CO on Ni, Pd, and Cu. These will be considered in later sections.

3.4.1. Hydrogen on Tungsten

In Table 3.2 and Figs. 3.5 through 3.7 we indicate a summary of data on the (110), (100), (211), and (111) planes [3.35–46]. Flash desorption reveals two major states on the fairly close packed (110), (100), and (211) planes, and at least four states on the (111) plane. All states appear to be "simple" in that they exhibit nearly constant desorption pre-exponential factors and activation energies. Values of pre-exponential factors are approximately those predicted theoretically [3.31, 35], $\sim 10^{-2}$ cm^2 molecule^{-1} sec^{-1} for second-order desorption and 10^{13} sec^{-1} for first-order desorption. Relative densities in the different states are also known fairly accurately from the areas under the flash desorption peaks [3.35], and these are indicated in Table 3.2. The desorption spectra, binding energies, orders of desorption, and relative densities have for all planes been reproduced in two or more laboratories [at least five for (100)W] and there is little question of their validity.

Work function changes with coverage have also been measured by many investigators [3.40], and these are summarized in Fig. 3.6. Again there is quantitative reproducibility except for some early work in which CO contamination was possible or where perhaps some states were not

Table 3.2. H_2 on W

Plane	State	E_{d_0} [kcal/mole]	Relative[a] saturation density	LEED structure	Ref.
(100)	β_2	32	1.0	$c(2 \times 2)$	[3.35, 40]
	β_1	26	$2.0 \pm .2$	(1×1)	
(110)	β_2	33	1.0	$p(2 \times 1)$	[3.35, 40–42]
	β_1	27	$1.0 \pm .1$	$p(2 \times 1)?$	
(211)	β_2	38	1.0	(1×1)	[3.35, 40, 43]
	β_1	20	$1.0 \pm .3$	(1×1)	
(111)	β_4	37	1.0		[3.35, 40]
	β_3	30	$1.5 \pm .2$		
	β_2	22	$1.5 \pm .2$		
	β_1	14	$1.0 \pm .2$		

[a] Density relative to that of state designated as 1.

saturated. There are fairly distinct breaks in these curves which in most cases coincide with coverages where flash desorption shows that one state saturates and another is being populated.

Relative sticking coefficients versus relative coverages [3.44, 40] also are fairly well established, as shown in Fig. 3.7. Each plane has a characteristic $s(\theta)$ curves breaks sometimes observed at coverages where the state saturates and another is being populated.

LEED patterns have also been observed for many of these planes, although some studies have employed substrate temperatures above 300 K and may therefore have missed ordered structures which could exist at lower temperatures. On (100) W Estrup and Anderson [3.36] showed several years ago that H_2 exposure produces a $c(2 \times 2)$ structure which attains maximum intensity at $\sim \frac{1}{4}$ monolayer. At this coverage the adsorbate beams split and their intensity diminishes until at saturation a (1×1) periodicity is again observed. On (110) W early LEED investigations concluded that only (1×1) periodicity occured, but recently Maticek using high energy electron diffraction (RHEED) [3.41] found a (2×1) structure with maximum intensity at $\sim \frac{1}{2}$ monolayer. Only (1×1) periodicities are reported on (211) W, and no recent observations appear to exist for (111) W.

One might conclude from this apparent agreement that the experimental situation for H_2 on W is settled of and that structure determination only awaits model building and testing. However, we shall see that this is by no means the case. A major problem is stoichiometry. While

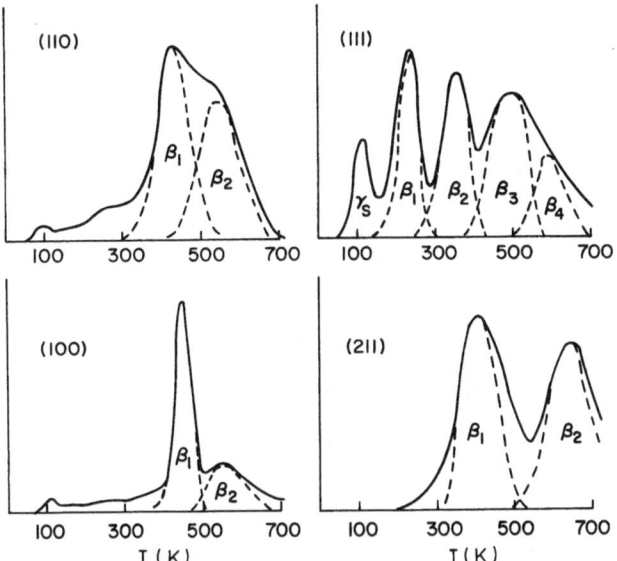

Fig. 3.5. Flash desorption spectra for H_2 on several high symmetry planes of W

relative densities on a given plane are readily obtained (Fig. 3.5), comparisons of densities on different planes are more difficult, and absolute densities are in most experimental systems almost impossible to measure. This is because absolute density determination requires measurement of system pressures, volume, and pumping speeds. While it is easy to estimate these quantities, all three are susceptible to systematic errors.

On (100)W ESTRUP and ANDERSON [3.36] estimated a saturation density of 2.0×10^{14} atoms/cm² by noting that a $c(2 \times 2)$ pattern, requiring at least one H atom for two surface W atoms (1.0×10^{15} atoms/cm²), occurred at approximately one-fourth of the saturation coverage. TAMM and SCHMIDT [3.35] estimated a density by assuming that the β_2 state produced the $c(2 \times 2)$ LEED structure (5×10^{14} atoms/cm²); the β_1 state has twice this density to give a total of 1.5×10^{15} atoms/cm². Note that both of these estimates are *indirect*, one from an assumed density for a given LEED pattern and the other for a density to fit the stoichiometry of flash desorption spectra. The only direct measurement of density is that of MADEY [3.45] who, using a calibrated flux molecular beam method, reported a saturation density of $2.0 \pm 0.3 \times 10^{15}$ atoms/cm². This apparently fits the ESTRUP and ANDERSON density although a recalibration of this flux gives a slightly lower value [3.45]. Also a recent measurement of the H/O ratio from H_2O in the

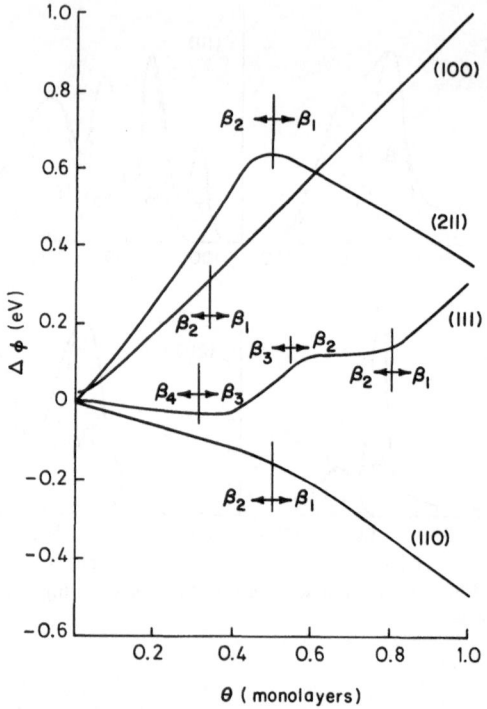

Fig. 3.6. Work function change versus fraction of saturation coverage θ of H_2 on several high symmetry planes of W. Approximate coverages where one state saturates and population of another begins are indicated on the figure. These in most cases correlate fairly well with breaks in the work functions

author's laboratory [3.45] yielded a value of $1.4 \pm 0.3 \times 10^{15}$ atoms/cm^2 for the saturation density of H_2 on (100)W. The unfortunate conclusion of these measurements appears to be that the saturation density of H_2 on (100)W is between 1.5×10^{15} and 2.0×10^{15} atoms/cm^2. This makes it impossible to distinguish between the two proposed structures from present data.

For (110)W the model [3.35] which appears to explain the two atomic states of equal density, the existence of a (2×1) periodicity for one state and a (1×1) peridicity for both, and a saturation density of approximately one H atom per surface W atom is shown in Fig. 3.8. The β_2 state occupies every other site and the β_1 state occupies the remaining sites.

It is not yet established whether there are two distinct states on the (110) plane at saturation. There is a distinct increase in the ESD [3.42] yield of H$^+$ and a decrease in $\Delta\phi$ [3.40] at the coverage where the β_1

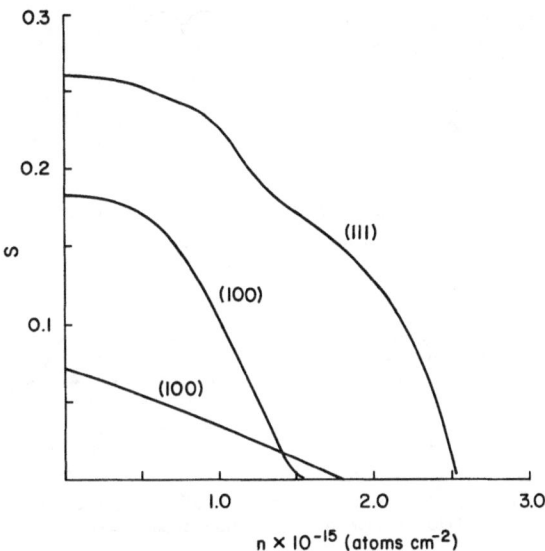

Fig. 3.7. Sticking coefficient s versus density of H_2 on several crystal planes of W

state begins to populate. However, the sticking coefficient exhibits at most a small break at this coverage [3.44]. All of these results are probably compatible with either one distinct state or two if the properties of one single state vary with coverage. The strongest evidence appears to be the RHEED observation [3.41] of a $p(2 \times 1)$ structure at saturation. This can only be explained with distinct β_1 and β_2 states, although it has been suggested [3.42] that the β_1 state may not have been saturated in these experiments.

On (100)W two distinct structure models have been proposed. That of ESTRUP and ANDERSON [3.36] assumes H atoms occupy bridge bond positions, as shown in Fig. 3.8. This obviously leads to a $c(2 \times 2)$ periodicity at one-fourth of saturation and (1×1) periodicity at saturation. However, it does not explain the existence of the two binding states with a 2:1 ratio observed in flash desorption. To do this, TAMM and SCHMIDT [3.35] proposed the model shown in Fig. 3.8. Here at one-third of saturation (assumed sites of four-fold symmetry) a $c(2 \times 2)$ periodicity is predicted. Then if the β_1 state consists of H_2 in the remaining sites with four-fold symmetry, one obtains the 2:1 ratio as observed. This model was proposed merely as the simplest one which gives a 2:1 ratio, occupies only one type of adsorption site, and leaves none of these sites vacant at saturation. One problem is that it does not predict a

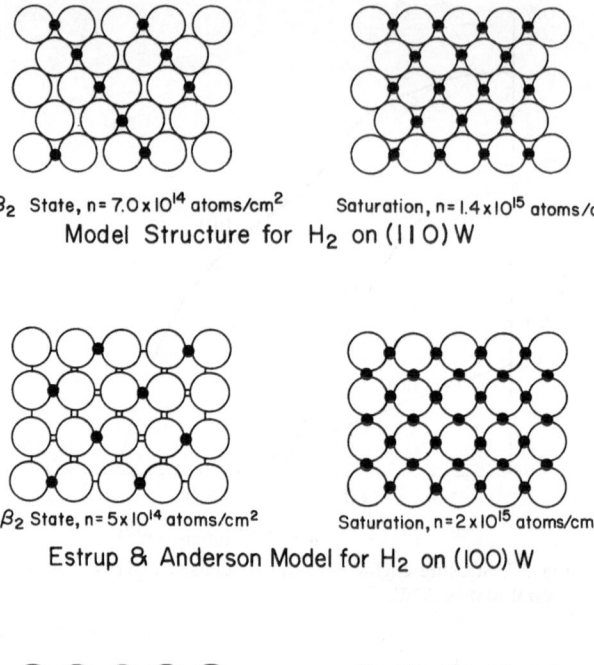

β_2 State, n= 7.0 x 10^{14} atoms/cm^2 Saturation, n= 1.4 x 10^{15} atoms/cm^2

Model Structure for H$_2$ on (110) W

β_2 State, n= 5 x 10^{14} atoms/cm^2 Saturation, n= 2 x 10^{15} atoms/cm^2

Estrup & Anderson Model for H$_2$ on (100) W

β_2 State, n= 5 x 10^{14} atoms/cm^2 Saturation, n= 1.5 x 10^{15} atoms/cm^2

Tamm & Schmidt Model for H$_2$ on (100) W

Fig. 3.8. Proposed adsorbate structures for H$_2$ on the (110) and (100) planes of W. On (110) there are two states of equal density which fill sequentially. On (100) two models have been proposed but neither simultaneously satisfies LEED and flash desorption observations

(1 × 1) periodicity at saturation unless the adsorbate in the two sites were equivalent or disordered.

LEED should in principle be capable of distinguishing between models. If sites had four-fold coordination the substrate beams should not broaden at low coverage where the adsorbate beams are broad, while with two-fold sites some of the substrate beams should broaden also. No broadening of substrate beams has been noted [3.36, 37, 43], but experimental accuracy may not be sufficient to conclusively eliminate two-fold sites. Also the ESTRUP and ANDERSON model predicts a maximum

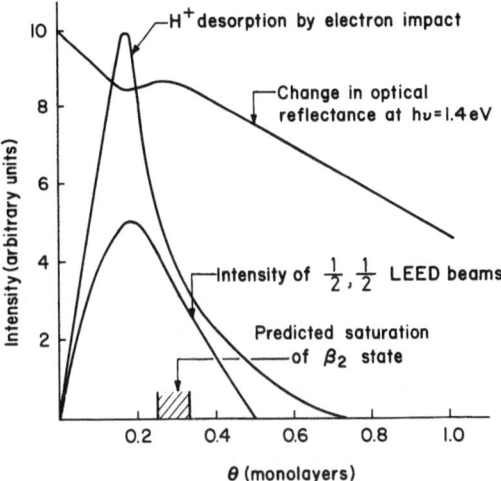

Fig. 3.9. Coverage variation of several properties of H_2 on (100)W. A distinct change at $\theta = 0.16$ is noted in the electron impact desorption cross section, in the intensity of adsorbate beams in LEED, and in the optical reflectivity. This coverage is smaller than those predicted by either structure model for this system, Fig. 3.8

ordering of the $c(2 \times 2)$ structure at 0.25 monolayers while that of TAMM and SCHMIDT determines a maximum at 0.33 monolayers. Results [3.37] indicate that the maximum is attained at 0.16 monolayers (Fig. 3.9) and that the spots begin to split *before* either model predicts a saturated state.

Confirmation that a change is occurring at low coverage comes from electron stimulated desorption [3.45] of H^+. As shown in Fig. 3.9, the H^+ current reaches a maximum at $\theta \simeq 0.16$, a coverage at which the β_2 state is only about one-half populated. Additional evidence for distinct but complex changes in the adsorbate in this coverage interval comes from field emission energy spectra [3.49], photoemission spectra, [3.50, 51] and optical reflectivity [3.52] (Fig. 3.9). These are discussed in Chapter 5.

There appears to be no model which simultaneously gives the 2:1 ratio of state densities and the $c(2 \times 2) \rightarrow (1 \times 1)$ symmetry. This is so even if one uses any combination of sites and any density desired. Therefore it appears that one of the above assumptions must be wrong or the experiments have been improperly interpreted. First it is possible that the 2:1 ratio of states is accidental and not related to specific sites on the surface. ADAMS [3.47] has examined the influence of various neighbor interactions and, while he could predict two peaks in flash desorption on this basis, these peaks do not, with reasonable assumptions, correspond to either a 2:1 ratio nor first-order desorption kinetics from the low temperature peak. First-order desorption from the β_1 state would be observed if the adsorbate were molecular or if desorption occurred only from adjacent pairs of atoms. The latter is certainly more likely, and the H_2 sketched in Fig. 3.8 is merely meant to be a representation of this.

The second possibility is that the LEED data have been misinterpreted. It is entirely possible that at saturation the hydrogen is sufficiently disordered to give a (1×1) periodicity (meaning no long-range order) without requiring one H atom per W atom. The surface diffusion coefficient of hydrogen is fairly high even at zero coverage (mobile at ~ 200 K) [3.21], and at saturation hydrogen could be mobile enough to disorder even at 78 K, the lowest temperature so far examined. A related possibility is that the periodicities observed in LEED are not associated simply with adsorbate atom periodicity but rather with electronic periodicities not correlated with atomic positions. The interpretation of diffraction from a low mass species is especially difficult because it originates almost entirely from multiple scattering.

A final possibility is bulk solution of hydrogen [3.37]. Hydrogen is only slightly soluble in W, but, since it dissolves endothermically, the solubility increases as the temperature increases. This could conceivably cause the apparent contradiction in LEED and desorption data, but no detailed explanations along these lines have as yet been proposed.

The differences in work function changes between the (110), (100), (111), and (211) planes, Fig. 3.6, are striking. They imply significant differences in atomic charges and/or dipole moments of the adsorbates. The decrease in ϕ with hydrogen coverage on the close packed (110) plane, where the H atom is most likely to be outside of the image plane, has the obvious interpretation that the H atom is *positively charged*. The other planes are atomically more open, and the adsorbate could be below the image plane to produced the increase observed on most planes. The Helmholtz equation is of course wholly inadequate to discuss this problem. The point we wish to emphasize is that neither the magnitude nor the sign of the charge on the H atom are experimentally established. Experiments which measure atomic charges (XPS) or adsorbate local densities of state (UPS) and calculations on charge distributions at surfaces appear to be the only way to conclusively answer these questions for this or any other adsorption systems.

3.4.2. Carbon Monoxide on Tungsten

A review of chemisorption experiments would be incomplete if it did not include CO on W [3.53, 54]. This system may exceed H_2 on W as the most often examined adsorption system, but the most significant discovery from all of these studies is probably an appreciation of the complexities which may exist in chemisorption. CO on W differs from H_2 and N_2 on b.c.c. metals and CO on fcc metals in that it is not reversible: properties are not a function of temperature and coverage alone but depend also on the temperature-time history of the adsorbate. This is not unique to CO on W, and a similar situation is observed for O_2 on all metals, for systems which exhibit bulk solution, and probably most of the complex coadsorption systems found in catalysis. The feature that separates CO on W from these is that experimentors have been determined to understand this supposedly straightforward system which has been thought to exhibit neither compound formation, bulk solutions, or surface reactions between species.

We shall not attempt a historical review [3.53, 54] but merely note that studies by REDHEAD [3.55] showed that there were more than 3 high-temperature or β states and several low-temperature or α states on polycrystalline W. Some time previous to this, field emission experiments had indicated that no carbon or oxygen residues remained after repeated adsorption and desorption cycles of CO [3.56], and this was interpreted (perhaps wrongly) as proof that CO on W does not dissociate. GOMER and his coworkers used field emission microscopy [3.57, 59] and more recently macroscopic surfaces [3.60] in an extensive investigation of this system. They have clearly demonstrated that if a surface saturated with CO at low temperature is heated to ~ 400 K, $\sim 50\%$ of the CO desorbs. If then CO is readsorbed, approximately the same coverage is obtained but the work function changes sign. They termed the original configuration the virgin state which, at least, partially distinct from the α and β states obtained upon readsorption. From considerable evidence from field desorption, dipole moment, and electron impact desorption GOMER and coworkers have confirmed earlier speculations that these states of CO must exist in several forms: linear with C bonded to the surface, bridge bonded either through the C atom or through both C and O, and perhaps dissociated. This latter was not a popular position until a few years ago, but now considerable evidence points to at least partial dissociation. Typical flash desorption spectra from (110) and (100) surfaces [3.60–62] saturated at room temperature or below are shown in Fig. 3.10. We shall confine our discussion to the (110) and (100) planes as these have been studied more extensively. It will also be convenient to consider first the β states and then the α and virgin states because, while the former are nearly reversible, the latter certainly are not.

β *States.* These are the states which desorb above 900 K. All are probably electronegative, and have very low electron impact desorption cross sections. As shown in Fig. 3.10 and Table 3.3, about two thirds of

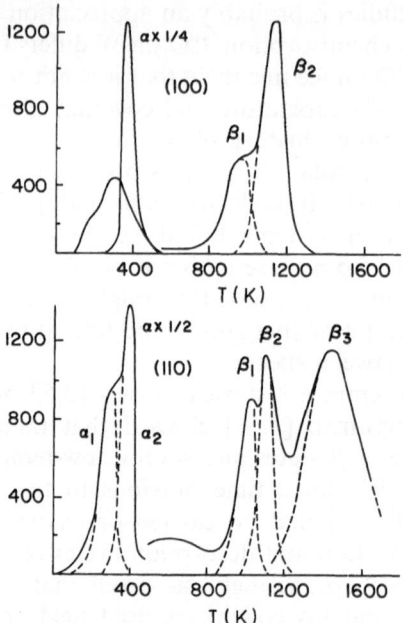

Fig. 3.10. Flash desorption spectra of CO
on the (110) and (100) planes of W

Table 3.3. CO on W

Plane	States	E_{d_0} [kcal/mole]	Fraction of total density	Saturation density [molecules/cm²]	Surface atom density [atoms/cm²]	Ref.
(100)	β_3	75	0.43 ⎤			
	β_2	64	0.10 ⎥	$6 \pm 1 \times 10^{14}$	1.0×10^{15}	[3.61]
	β_1	57	0.09 ⎥			
	α's	25–30	0.37 ⎦	$3.8 \pm 1 \times 10^{14}$		
(110)	β_2	66	0.3 ⎤	$\sim 7 \times 10^{14}$	1.42×10^{15}	[3.60, 62]
	β_1	50	0.1 ⎦			
	α's	15–20	0.6	$\sim 8 \times 10^{14}$		

this adsorbate appears as a single state in flash desorption with the rest
appearing as one or two smaller lower temperature states. The desorption
temperature is significantly higher on (100).

Absorbate densities have been measured using effusion methods for
(110)W and a calibration against O_2 saturation using AES for (100)W.
As shown in Table 3.3, these indicate saturation densities in the β states

of approximately one CO molecule for every two surface W atoms on both planes.

There are no (or perhaps very weak) LEED patterns observed upon adsorption at room temperature [3.63–65], but heating to ~ 900 K produces a $c(2 \times 2)$ structure on (100)W and a complex sequence of structures on (110)W. ESTRUP and ANDERSON [3.65] postulated that the $c(2 \times 2)$ structure was obtained from a state they termed β_H after the other, termed β_L, was desorbed. More recent flash desorption studies show that these states are not as simple as first assumed [3.61, 67]. In both cases the results suggest that reconstruction (motion of the W atoms and possibly incorporation of CO into the top layer of the surface) occurs upon heating to near the desorption temperature.

Next we consider the question of whether CO is dissociated in the β states. KING and coworkers [3.66] showed that the breaking of the CO bond and its reforming in desorption is thermodynamically feasible in spite of the high bond energies of C–O, C–W, and O–W. Also CLAVENNA and SCHMIDT [3.61] showed that the β_3 state on (100)W desorbed with second-order kinetics. This was explainable as dissociative adsorption or as desorption of pairs of molecules, an earlier interpretation used to explain the isotope exchange of isotopically labelled atoms in the β states [3.67]. Finally UPS experiments by BAKER and EASTMAN [3.68] and by PLUMMER [3.69] (see Chapter 5) showed that the spectrum from the β states was almost identical to that of C and O adsorbed separately while that of α and virgin was similar to that of gaseous CO. It appears from this work that some dissociation occurs upon adsorption even at room temperature. Recent work on the recombination of CO from C and O coadsorbed on (100)W show that this is an efficient process [3.70] and that no C or O residues are found on W because neither can dissolve appreciably in the bulk below the desorption temperature. Residues are observed on Ta because bulk solution of C occurs below the desorption temperature.

The simplest interpretation consistent with the data on the β states of CO on the planes (several states in non-integral ratios to substrate densities, complex LEED structures upon heating, and low desorption cross sections) is that considerable dissociation occurs to yield C–W and O–W bonds; these species must then recombine to desorb as CO. Since C and O should have low surface diffusion coefficients (CO is known to be almost immobile up to the desorption temperature) [3.21], the states observed depend on the probabilities of having neighboring pairs of atoms or requiring surface diffusion. The ESCA spectra of the β states on (100)W reveal distinct core electron energies of C and O which seem to correlate with the β_1, β_2, and β_3 states observed in flash desorption [3.71].

α *States*. On both the (100) and (110) planes the saturation density of CO which desorbs below 800 K is approximately one CO molecule for two surface W atoms (Table 3.3). This density is approximately the same for adsorption at 100 or 200, and at 300 K these states can be almost saturated by exposures to high CO pressures [3.60, 62, 72, 73]. Slightly different desorption spectra and perhaps slightly higher amounts are obtained upon readsorption after heating a saturated surface, but this has not been quantified, and we shall term all states in this temperature regime α states. Extensive and fairly reproducible work function data for CO may be found in the references cited. However, work function changes can not be simply considered because values depend on the surface temperatures before, during, and after deposition. Dipole moments in some cases appear to actually change sign upon heating with no accompanying desorption.

The most thorough investigations of the α states of CO are those of KOHRT and GOMER [3.60] for (110)W and YATES and KING [3.72, 73] for (100)W. These are in qualitative agreement with earlier conclusions from measurements on polycrystaline W by EHRLICH [3.74], GOMER and coworkers [3.57–59], and MENZEL [3.75].

On (100)W a state shown in Fig. 3.10 has been observed to consist of two states with almost identical desorption temperatures and roughly equal saturation densities [3.72, 73]. The major experimental distinction between the states is that electron stimulated desorption of the higher binding energy α_2 state produces desorption of primarily O^+ ions while CO^+ ions are produced upon electron bombardment of the α_1 state.

The flash desorption spectra of the α states from both (100) and (110) planes exhibit high temperature "tails" which can be fitted by simple desorption kinetics from one or several binding states and which are not caused by slow pumpout of the desorbed gas. These apparently result from partial conversion into the β states while the surface is being heated. KOHRT and GOMER [3.60] showed that isothermal desorption rates from the β_1 and β_2 states on (110) yielded the same desorption parameters as did flash desorption experiments, while the α and virgin states yielded very low pre-exponential factors and activation energies. Activation energies in Table 3.3 are obtained assuming a pre-exponential factor of 10^{13} sec^{-1} for those states obeying first-order kinetics.

There is as yet no clear consensus as to the structures of these adsorbates. The CO and O^+ yields from the α_1 and α_2 states on (100) could indicate that the α_1 state is bonded to one W atom while the α_2 state is bridge bonded. KOHRT and GOMER [3.60] have discussed possible bonding configurations and electronic structures. It has recently been shown [3.71] that adsorption at 100 K produces ESCA spectra which are distinct from those of either the α or β states. This supports the earlier hypothesis from work function measurements that there exists at low temperature "virgin" states which convert into α or β states upon heating [3.57–59].

3.5. Adsorbate-Adsorbate Interactions

In this section we consider the repulsive and attractive interactions between adsorbate molecules (see also Section 2.4), the exclusion of sites, and phase changes in adsorbed layers. Before 1960 the prevalent picture was that repulsive interactions generally accounted for the observed reduction in heats of adsorption with coverage and the deviations from linearity of work function changes with coverage. Use of single crystals to avoid crystallographic averaging and LEED to directly measure adsorbate periodicities and the degree of order has revealed considerable complexity which was quite unanticipated.

3.5.1. Repulsive Interactions

Experiments on single crystal planes have shown that repulsive adsorbate-adsorbate interactions on metal surfaces are generally very weak and that the variation of adsorption energy with coverage observed on polyfacetted surfaces is primarily due to the existence of multiple binding states on different planes and multiple states on a given plane. To the author's knowledge there are no examples where the binding energy in a single state on a metal surface varies by more than 5 kcal/mole between zero coverage and saturation, and in most cases the variation is immeasurably small. As examples the atomic β_1 and β_2 states of H_2 on (110)W exhibit variations of 2.0 and 3.3 kcal/mole, respectively, between zero and saturation densities, and the first order β_1 state of H_2 on (100)W exhibits a variation of no more than 0.3 kcal/mole, or less than 1% of the desorption activation energy [3.35]. Data for H_2 on W are in fact upper bounds obtained from analysis of shapes of flash desorption spectra; the apparent variations in E_d could be entirely due to variations in pre-exponential factors with coverage or the presence of other binding states on the edges of the crystals.

Heats of adsorption of CO on various planes of Ni and Pd have been measured accurately using adsorption isotherm methods [3.2, 3, 76–78]. The most tightly bound states on (100)Ni, (111)Ni, and (111)Pd (E_{d_0} between 30 and 40 kcal/mole) decrease by no more than 1 kcal/mole at saturation. On (100)Pd, however, E_d, decreases by 2 kcal/mole at saturation of the most tightly bound state. Variations at higher coverages are

Table 3.4. Examples of adsorbate-adsorbate interactions

Effect	State	E_{do} [kcal/mole]	ΔE_{sat} [kcal/mole]	LEED pattern	Ref.
Repulsion	β_1 H on (100)W	25	<0.3	(1×1)	[3.44]
	β_2 N on (100)W	78	<5	$c(2 \times 2)$	[3.30, 3.1]
	β_1 and β_2 H on (110)W	33, 26	<3.0	(2×1)	[3.41, 44]
	CO on (100)Pd	38	3		[3.3]
	CO on (100)Ni	30	<0.4		[3.2]
	CO on (100)Cu	14	<0.4		[3.4]
Island formation	H₂, N₂ and Th on (100)W			$c(2 \times 2)$	[3.13, 30, 36]
Order-disorder transition	H₂, N₂ and Th on (100)W			$c(2 \times 2) \leftrightarrow (1 \times 1)$	
	CO on (100)Wi and (100)Cu			$c(2 \times 2) \leftrightarrow (1 \times 1)$	[3.2, 4]
One-dimensional order	N₂ on (210) and (310)W			streaks	
Short-range order	CO on (100)Pd			rings	[3.3]
Out-of-registry transition	CO on (110)Ni			$c(2 \times 2) \leftrightarrow$ hexagonal	
	CO on (100)Cu and (100)Pd			$c(2 \times 2) \leftrightarrow$ complex	[3.3, 4]
Structure change	H₂ on (100)Mo			$c(2 \times 2) \leftrightarrow (4 \times 2)$	[3.82]

larger for these systems due to phase transitions to other configurations as will be discussed later.

The general trend appears to be that for adsorbate-adsorbate distances of two and, in some cases, one lattice spacing, repulsive interactions are at most a few kcal/mole. This is so even when work functions vary by typically 0.2–0.5 eV upon saturation of the state. It appears that electron screening efficiently reduces electrostatic interactions to a small fraction of those predicted assuming no screening. Also the major consequence of adsorbate-adsorbate repulsion may not be a variation in heat of adsorption with coverage but rather the exclusion of bonding sites and induced heterogeneities observed as multiple binding states.

3.5.2. Attractive Interactions and Ordered Structures

Many systems exhibit LEED periodicities greater than those of the substrates. For the tightly bound state of CO on Ni, Pd, and Cu it appears that there is essentially random occupation of sites until the adsorbate density is large enough for core repulsions to become important. For example, as shown in Fig. 3.11, CO on (100)Ni and (100)Cu orders into a $c(2 \times 2)$ structure because the close packed density of CO molecules does not allow every site to be occupied to produce a (1×1) structure. The maximum temperature at which this ordered configuration exists, indicated in Fig. 3.12, increases from 80 K at low coverage to 400 K as the CO density approaches saturation. The existence of a $c(2 \times 2)$ structure at very low coverages on (100)Ni suggests that there may be weak attractive interactions between CO molecules. On (100)Cu ordering occurs only near saturation of the $c(2 \times 2)$ structure and this is preceeded by a ring pattern showing a liquid-like adsorbate configuration.

There are several examples where attractive interactions lead to island formation. Both H_2 and N_2 on (100)W produce $c(2 \times 2)$ structures and the intensities of the half-order adsorbate beams increase approximately linearly with coverage [3.30, 36, 37]. Random occupation of sites predicts [3.12, 13] that these intensities should increase as $\sim \theta^2$; with island formation the intensity should be proportional to θ. The adsorbate beams for both gases also appear to be fairly sharp even at low coverages, indicating that islands are quite large, $> 20 \, \text{Å}$ for H_2 on (100)W at 0.1 of saturation of the β_2 state. It appears that the adsorbate at less than saturation density exists as rather large patches in which every other site is occupied (Fig. 3.8) with essentially bare regions between them. There are two equivalent positions for H atoms in the $c(2 \times 2)$ structure, (by either model) and this results in antiphase domains which presumably prevents the islands from becoming even larger. As

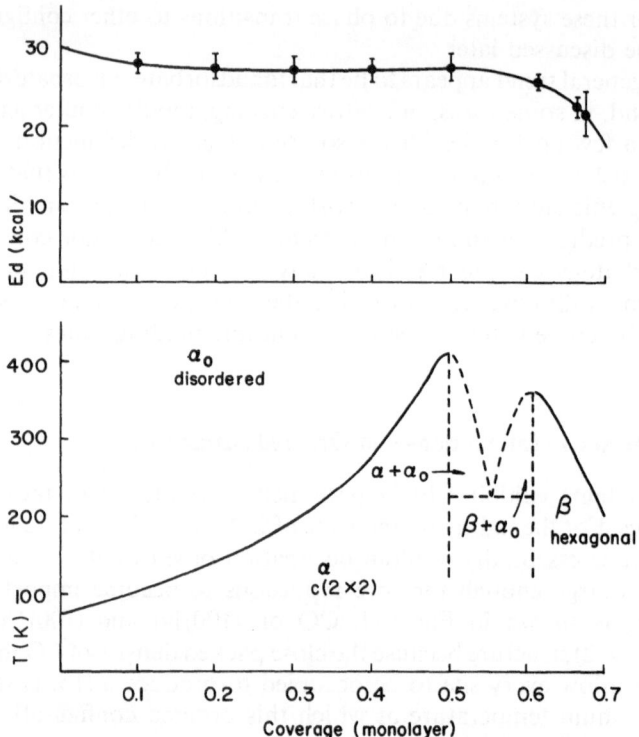

Fig. 3.11. Heat of adsorption versus coverage and proposed phase diagram for CO on (100)Ni. A sequence of ordered structures is observed by LEED at different coverage and temperature

the coverage of H_2 approaches saturation of the β_2 state, these beams split continuously and diminish in intensity. This is interpreted as ordered additional atoms in the $c(2 \times 2)$ structure [3.36].

For CO on (100)Ni and on (100)Cu the $c(2 \times 2)$ structure apparently results from the fact that the adsorbate is too large for closer packing. For H_2 and N_2 on (100)W the atomic diameters are probably small enough that this is not the cause of the $c(2 \times 2)$ being the maximum density, (However, it should be noted that effective adsorbate sizes depend on the charges on the atoms, and these are not well established.) The $c(2 \times 2)$ structures for H and N have been interpreted as resulting from bonding with W atoms in the plane of the surface. Saturation of these bonds limits the adsorbate density, and $s - d$ hybridization offers a possible mechanism for island formation [3.31, 35]. Analogous theories of indirect interactions also predict island formation.

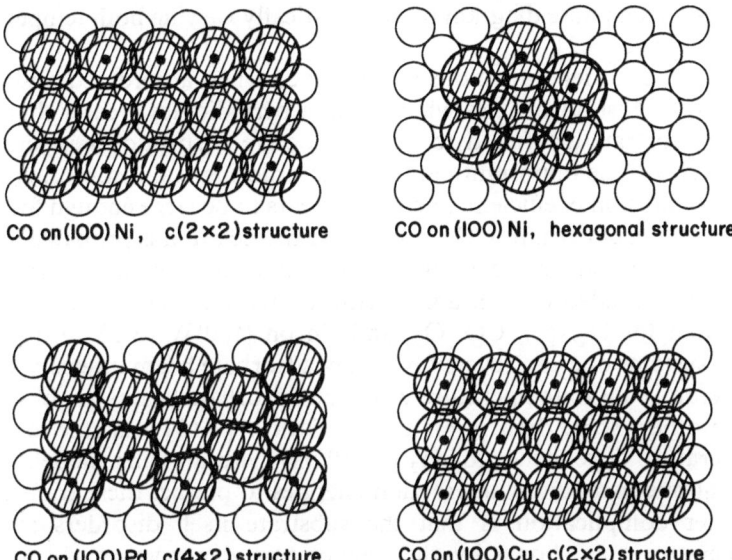

CO on (100) Ni,   c(2×2) structure

CO on (100) Ni, hexagonal structure

CO on (100) Pd, c(4×2) structure

CO on (100) Cu, c(2×2) structure

Fig. 3.12. Proposed structures for CO on (100) planes of Ni, Pd, and Cu. On Ni and Cu adsorbates give $c(2 \times 2)$ structures and presumably occupy sites of four-fold coordination while on Pd a $c(4 \times 2)$ structure exists which probably occupies sites of two-fold symmetry. Distinct sequences of higher coverage structures exist on each metal as discussed in the text

3.5.3. Two-Dimensional Phase Transitions

There are now a number of fairly well documented examples of structure transitions of adsorbates. These frequently appear to occur over fairly small ranges in temperature or density, and in some cases heats of transition have been measured. However none appears to be clearly a first-order phase transition. Experiments rely mainly on LEED with desorption kinetics or isotherms to determine coverages and heats of adsorption. Interpretation of LEED is made difficult by experimental problems with instrument resolution (seldom greater than 0.01 Å^{-1}), temperature variations over a crystal (temperatures are only with difficulty measured to better than $\pm 20°$, and $\pm 50°$ is probably typical), and surface imperfections due to disorder, trace impurities, and orientation inaccuracies.

Transitions are mostly of the order vs. disorder or registry vs. out-of-registry type as observed by following LEED patterns as either the temperature is varied at constant coverage or as the coverage is varied at constant temperature. In only one case has an investigator attempted to construct a phase diagram by systematically varying both temperature and coverage [3.2].

Order-disorder transitions are observed when an ordered adsorbate is heated. Most adsorbates which exhibit periodicities greater than that of the substrate are found to disorder before desorption temperatures are reached. The important parameter in ordering and disordering is the surface diffusion coefficient, and this appears to be large enough for most systems (except perhaps CO on W where dissociation is probable for the tightly bound states), to allow rapid equilibrium on the surface. The variation of adsorbate beam intensities with temperature has been measured for H_2, N_2, CO, O_2, and Th on (100)W [3.12, 13, 37]. For many systems one can reversibly follow adsorbate beam intensity versus temperature up to temperature where no adsorbate beams are observed, but in others some desorption occurs. ESTRUP [3.13] was able to fit the temperature dependence for H_2 on (100)W with an Ising model, but in general one must know interaction energies to predict these transitions. Another complication is that the substrate itself disorders at high temperatures, as shown by the temperature dependence of substrate beam intensities.

As discussed previously, N_2 on the stepped (210) and (310) planes of W exhibits two dimensional ordering at low temperatures, one dimensional ordering along rows at higher temperatures, and probably no ordering at very high temperatures [3.33]. Similar behavior has been noted for alkali metals on (211)W which has a rowed structure (Fig. 3.1).

Carbon monoxide on fcc metals exhibits a number of complex transitions which have been examined in detail by TRACY [3.2, 4], by ERTL and coworkers [3.76, 77], and by TAYLOR and ESTRUP [3.78]. CO exhibits registry with substrate lattice sites at low coverages, but it can be forced out of registry at higher coverages into roughly hexagonal adsorbate layers. For CO on (100)Ni TRACY showed that a $c(2 \times 2)$ structure exists at low coverage. This saturates at one CO molecule for every two surface Ni atoms, and at higher coverages the half-order beams split as the adsorbate is forced into a hexagonal structure. This can be further compressed until the heat of adsorption falls to such a low value that no more adsorbate can be added. These data are summarized in Fig. 3.11, while Fig. 3.12 shows the adsorbate structures proposed by TRACY. The phase diagram in Fig. 3.11 has only qualitative significance in many regions, but shows the general behavior to be expected from such systems. There is a definite reduction in the heat of adsorption as the hexagonal layer is compressed but at most a small one for the $c(2 \times 2)$-hexagonal transition. TRACY also speculates that there is a eutectic point where $c(2 \times 2)$, hexagonal, and disordered phases coexist. This elegant experiment clearly shows the desirability of making measurements below room temperature because the LEED pattern observed for $T \geq 300$ K would probably be (1×1) throughout, and it would not be possible to even form the hexagonal structure because its heat of adsorption is too small for typical pressures used in ultrahigh vacuum systems. A apparently simple and uninteresting system becomes remarkably complex when all states and all possible ordered structures are examined.

It should be noted that these are not actually infinite two-dimensional systems because they always contain high concentrations of imperfections which make the applicability of theories of infinite two-dimensional phase transitions questionable. For finite systems the transition temperature may be strongly affected by these defects. In three-dimensional systems one routinely has impurity concentrations much less than 1%; on surfaces this is probably seldom attained with present standards of cleanliness. It is found that a few percent of CO on (100)W completely destroys the $c(2 \times 2)$ LEED structure of H_2; evidently each CO molecule nucleates a domain of H_2 around it and suppresses long range ordering [3.37].

3.6. Adsorption on Similar Metals

One of the ways of deciding the types of bonds formed in chemisorption is to examine their sensitivity to small changes in the electronic or lattice properties of the solid. Adjacent transition metals in the periodic table frequently possess the same crystal structures so that differences in adsorption should be attributable to differences in lattice constant, number of valence electrons, spatial variations of electron densities, etc. Ideal candidates for this purpose are the bcc metals — W, Mo, Ta, and Nb — and fcc metals — Ni, Cu, Pd. [Most other metals either cannot be cleaned readily or do not have (1×1) surface periodicities.] In this section we shall summarize some comparative studies of these metals for chemisorption of H_2, N_2, and CO. Extensive comparisons of adsorption and catalysis on various metals have been carried out for many years but here we consider only adsorption on single crystal planes. Variation between planes may in fact be as great as that between metals, and therefore comparisons on polycrystalline wires or thin films appear to have only qualitative significance.

3.6.1. H_2, N_2, and CO on W, Mo, and Ta

In Fig. 3.13 are shown saturation flash desorption spectra for these gases on the (100) and (110) planes of W, Mo, and Ta. The original references [3.35, 79–82] give quantitative behavior, but the figure immediately

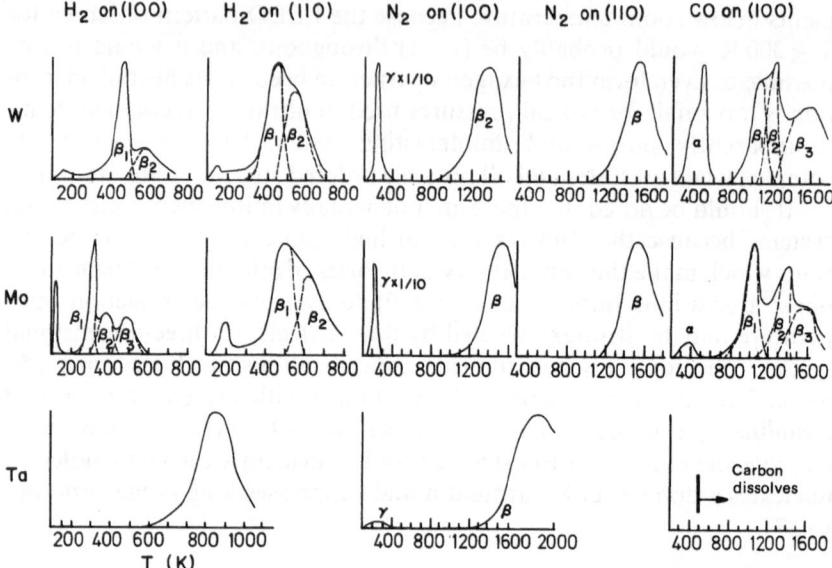

Fig. 3.13. Flash desorption spectra of H_2, N_2, and CO on (100) and (110) planes of W, Mo, and Ta. Close similarities are observed between W and Mo in most cases while on Ta heats of adsorption are higher and bulk solution occurs

indicates the numbers of states (number of peaks), their densities on a plane (relative areas), and desorption activation energies (roughly proportional to desorption temperature). Those peaks which are approximately symmetric about the maximum obey second-order desorption kinetics and those with larger slopes on the high temperature side obey first-order kinetics.

Bulk Solution. Tantalum shows poor correspondence with W and Mo because H, N, and C all dissolve – H even below 80 K and N and C at ~ 500 K. Desorption of H_2 and N_2 exhibits peaks if the heating rates are properly chosen, and these indicate much higher heats of adsorption on Ta than on W or Mo. Carbon from CO dissolves below the desorption temperature on (100)Ta, and therefore no desorption of CO occurs [3.82].

Number of Binding States. Here there is correspondence between planes of W and Mo for H_2 on (110) and N_2 on (110) and (100), but there is an additional state [3.80] of H_2 on (100)Mo which is not found [3.35] on (100)W. The β_1 states of H_2 on (100) planes both obey first-order desorption kinetics and have a saturation density ratio of 2:1 for $\beta_1 : \beta_2$. To return briefly to the question regarding the structure of H_2 on (100)W, we note that there appears to be 25 % more H_2 on Mo than on W. This suggests adsorption in the sites of two-fold coordination (Fig. 3.8) with 25 % of these unoccupied on W (the TAMM and SCHMIDT density with the ESTRUP and ANDERSON configuration). Bulk solution of this H_2 in W could be the cause of this difference. Also the LEED patterns are different with a $c(4 \times 2)$ observed on (100) Mo rather than the $c(2 \times 2)$ on (100) W. HUANG and ESTRUP have recently observed complex changes in the LEED patterns at low temperatures which have thus far not been explained.

There is little correspondence between the states of CO on (100) W [3.61] and (100) Mo [3.70]. There are three β states, and the β_3 states both obey second-order desorption kinetics. However, the density ratios bear no simple relationship to each other. The absence of stoichiometric ratios fits the suggestion made earlier that the desorption states observed on W may not be completely distinct states on the surface but rather occur because of surface diffusion limited recombination of C and O. Different surface diffusion coefficients (probably higher on Mo than on W) could produce different apparent amounts in the various peaks.

Relative Binding Energies. There are definite differences in binding energies on comparable planes of W, Mo, and Ta. Both H_2 and N_2 are more strongly bound on Ta than on W or Mo. Hydrogen is more strongly bound on the (100) plane of W than of Mo, but on the (110) plane bonding is stronger on Mo. Bonding of N_2 is slightly stronger on both planes of Mo than on W.

These differences have not been explained theoretically. Tantalum metal with approximately $4d$ electrons in its valence shell appears to adsorb all three gases more strongly than does W or Mo which have approximately $5d$ electrons. However, the larger lattice constant of Ta may allow the adsorbates to form stronger bonds (as well as dissolve). The differences between H_2 on W and Mo are more subtle as perhaps expected by virtue of their essentially identical lattice constants and similar electronic structures. Bonding is strongest on the close packed plane of Mo, and equal binding energies are observed on the (100) and (110) planes of W. This does not appear to be explainable with any simple pictures of surface energies or the spatial extensions of valence wave functions.

3.6.2. CO on Ni, Pd, and Cu

TRACY'S comparative study [3.2–4] of CO adsorption on the (100) planes of these metals has been considered in the previous section in relation to phase changes at high coverages. The heats of adsorption are much higher on the transition metals Ni and Pd than on the noble metal Cu, although chemical bonds are definitely formed on all three surfaces. On (100)

planes of Ni and Cu $c(2 \times 2)$ structures are observed while on (100) Pd a
$c(4 \times 2)$ structure occurs. TRACY interprets this as sites of four-fold
coordination on Ni and Cu but two-fold sites on Pd. The CO–CO
interaction is repulsive on Pd, attractive on Ni, perhaps slightly attractive
on Cu. Similar studies have been reported on other planes of fcc metals
[3.76–78] but not with sufficient detail to permit close comparison.

As with adsorption on bcc metals, no theories as yet appear close to
being able to interpret this behavior.

3.6.3. Comparison between fcc and bcc Substrates

This section would not be complete without mention of the fact that
adsorption in general appears to be stronger, as well as more complicated
on bcc substrates than on fcc metals. The reasons for this are probably
related to the fact that fcc structures are more closepacked. This means
first, that surface densities of state are broader than on bcc metals which
in turn will tend to reduce adsorption energies, and second that the
geometric possibilities, for different binding states are more restricted.

3.7. Summary

We have attempted to be rather conservative with respect to interpretations of chemisorption experiments, first, to be able to cover a number of
topics in chemisorption and, second, because adsorbate structures are
still somewhat speculative even for the simplest and most studied
systems. For example, in spite of the extensive measurements on CO
on the (100) planes of Ni, Pd, and Cu, the positions of adsorbate atoms
with respect to the substrate atoms and the types of bonds can only be
inferred indirectly. These appear to be among the simplest chemisorption
systems in that they involve weak bonds, one state, and no dissociation,
bulk solution, or motion of substrate atoms.

From the many new experimental techniques and extensive experiments on chemisorption there has evolved a confidence that quantitative
and reproducible experiments are possible in spite of the unexpected
complexities which have been observed. It also appears that in order to
completely specify chemisorption structures it will be essential to employ
spectroscopic techniques and all of the capabilities of diffraction as well
as all of the now established tools for surface study. The assistance of
theoreticians is also sorely needed; experiments appear to be far ahead
of theory, with few solid interpretations existing for the numbers of
states, binding energies, types of bonds, surface phase transitions, etc.

The ultimate goals of understanding adsorption and reactions of
complex molecules such as hydrocarbons appear to the author to be

approximately as far away as was the understanding of H_2 on W thirty years ago. Published studies of H_2 adsorption existed at that time just as do those of hydrocarbon adsorption and reactions today; but interpretations of both seem to be so speculative as to be completely untestable. All of the modern experimental techniques and a number of new ones will undoubtedly be required before it will be possible even in principle to specify all adsorption structures.

References

3.1. E. W. MÜLLER: Advan. Electron. Phys. **13**, 83 (1960).

3.2. J. C. TRACY: J. Chem. Phys. **56**, 2736 (1972).

3.3. J. C. TRACY, P. W. PALMBERG: J. Chem. Phys. **51**, 4852 (1969).

3.4. J. C. TRACY: J. Chem. Phys. **56**, 2748 (1972).

3.5. Recent summaries may be found in T. A. CLARKE, R. MASON, M. TESARI: Surface Sci. **40**, 1 (1973); and J. T. GRANT: Surface Sci. **18**, 228 (1969).

3.6. H. P. BONZEL, R. KU: J. Chem. Phys. **58**, 4617 (1973).

3.7. G. A. SOMORJAI: Catalysis Rev. **7**, 87 (1972).

3.8. F. JONA: IBM J. Res. Dev. **9**, 375 (1965).

3.9. J. J. LANDER, J. MORRISON: J. Appl. Phys. **34**, 1403, 2298, 3317 (1963).

3.10. T. M. FRENCH, G. A. SOMORJAI: J. Phys. Chem. **74**, 2489 (1970).

3.11. J. E. HOUSTON, R. L. PARK: Surface Sci. **26**, 286 (1971).

3.12. P. J. ESTRUP, E. G. MCRAE: Surface Sci. **25**, 1 (1971).

3.13. P. J. ESTRUP: In *The Structure and Chemistry of Solid Surfaces* (Wiley, New York, 1969), p. 19-I.

3.14. A. G. NAUMOVETS, A. G. FEDORUS: Sov. Phys.-JETP Letters **10**, 6 (1969); Surface Sci. **21**, 426 (1970).

3.15. J. M. CHEN, C. A. PAPAGEORGOPOULOS: Surface Sci. **21**, 377 (1970).

3.16. P. A. REDHEAD: Vacuum **12**, 203 (1962).

3.17. L. D. SCHMIDT: Catalysis Rev. **9**, 115 (1974).

3.18. J. M. BLAKELY: Progr. Mat. Sci. **10**, 1 (1963).

3.19. H. P. BONZEL: In *Structure and Properties of Metal Surfaces* (Maruzen, Tokyo, 1973), p. 248

3.20. G. EHRLICH, F. G. HUDDA: J. Chem. Phys. **44**, 1039 (1966).

3.21. R. GOMER: Disc. Farad. Soc. **28**, 23 (1959)

3.22. G. AYRAULT, G. EHRLICH: J. Chem. Phys. **57**, 1788 (1972).

3.23. T. ENGEL, R. GOMER: J. Chem. Phys. **50**, 2428 (1969).

3.24. T. E. MADEY: Surface Sci. **33**, 355 (1972).

3.25. L. A. HARRIS: J. Appl. Phys. **39**, 1419 (1968).

3.26. J. PRITCHARD, M. L. SIMS: J. Catalysis **19**, 427 (1970).

3.27. J. T. YATES, JR., D. A. KING: Surface Sci. **30**, 60 (1972).

3.28. T. A. DELCHAR, G. EHRLICH: J. Chem. Phys. **42**, 2686 (1965).

3.29. P. W. TAMM, L. D. SCHMIDT: Surface Sci. **26**, 286 (1971).

3.30. P. J. ESTRUP, J. ANDERSON: J. Chem. Phys. **46**, 567 (1967).

3.31. L. R. CLAVENNA, L. D. SCHMIDT: Surface Sci. **22**, 365 (1970).

3.32. D. L. ADAMS, L. H. GERMER: Surface Sci. **26**, 109 (1971).

3.33. D. L. ADAMS, L. H. GERMER: Surface Sci. **27**, 21 (1971).

3.34. H. F. WINTERS, D. E. HORNE: Surface Sci. **24**, 587 (1971).

3.35. P. W. TAMM, L. D. SCHMIDT: J. Chem. Phys. **51**, 5352 (1969); and **54**, 4775 (1971).

3.36. P. J. Estrup, J. Anderson: J. Chem. Phys. **45**, 2254 (1966).
3.37. K. Yonehara, L. D. Schmidt: Surface Sci. **25**, 238 (1971).
3.38. J. T. Yates, Jr., T. E. Madey: J. Chem. Phys. **54**, 4969 (1971).
3.39. M. R. Leggett, R. A. Armstrong: Surface Sci. **24**, 404 (1971).
3.40. B. D. Barford, R. R. Rye: J. Chem. Phys. **60**, 1046 (1974), see references cited for earlier work function measurements.
3.41. K. J. Matysik: Surface Sci. **29**, 324 (1972).
3.42. D. A. King, D. Menzel: Surface Sci. **40**, 399 (1973).
3.43. D. L. Adams, L. H. Germer, J. W. May: Surface Sci. **22**, 45 (1970).
3.44. P. W. Tamm, L. D. Scmidt: J. Chem. Phys. **52**, 1150 (1970); and **55**, 4253 (1971).
3.45. T. E. Madey: Surface Sci. **33**, 355 (1972); **36**, 281 (1973), and private communication.
3.46. G. Sanders, L. D. Schmidt: Unpublished results.
3.47. D. L. Adams: Surface Sci. **42**, 12 (1974).
3.48. T. E. Madey: Surface Sci. **36**, 281 (1973).
3.49. E. W. Plummer, A. E. Bell: J. Vac. Sci. Technol. **2**, 583 (1972).
3.50. B. Feuerbacher, B. Fitton: Phys. Rev. Letters **29**, 786 (1972).
3.51. E. W. Plummer: To be published.
3.52. G. W. Rubloff, J. Anderson, M. A. Passler, P. J. Stiles: Phys. Rev. Letters **32**, 667 (1974).
3.53. R. R. Ford: Advan. Catalysis **21**, 51 (1970).
3.54. R. Gomer: Jap. J. Appl. Phys. Suppl. 2, Pt. 2, 213 (1974).
3.55. P. A. Redhead: Trans. Faraday Soc. **57**, 641 (1961);
      G. Ehrlich, T. W. Hickmott, F. G. Hudda: J. Chem. Phys. **28**, 506 (1958).
3.56. R. Klein: J. Chem. Phys. **31**, 1306 (1958).
3.57. L. W. Swanson, R. Gomer: J. Chem. Phys. **39**, 2813 (1963).
3.58. A. E. Bell, R. Gomer: J. Chem. Phys. **44**, 1065 (1966).
3.59. T. Engel, R. Gomer: J. Chem. Phys. **52**, 1832 (1970).
3.60. C. Kohrt, R. Gomer: Surface Sci. **24**, 77 (1971); and **40**, 71 (1973).
3.61. L. R. Clavenna, L. D. Schmidt: Surface Sci. **33**, 11 (1972).
3.62. C. S. Steinbruchel: Ph. D. Thesis, University of Minnesota (1974).
3.63. J. W. May, L. H. Germer: J. Chem. Phys. **44**, 2895 (1966).
3.64. E. Bauer: Colloq. Intern. CNRS (1969) **187**, 111 (1970).
3.65. J. Anderson, P. J. Estrup: J. Chem. Phys. **46**, 563 (1967).
3.66. C. G. Goymour, D. A. King: J. Chem. Soc. Faraday Trans. **169**, 736, 749 (1973); Surface Sci. **35**, 246 (1973).
3.67. T. E. Madey, J. T. Yates, Jr., R. C. Stern: J. Chem. Phys. **42**, 1372 (1965).
3.68. J. M. Baker, D. E. Eastman: J. Vac. Sci. Technol. **10**, 223 (1973).
3.69. E. W. Plummer: To be published.
3.70. Y. Viswanath, L. D. Schmidt: J. Chem. Phys. **59**, 4184 (1973).
3.71. J. T. Yates, Jr.: To be published.
3.72. J. T. Yates, D. A. King: Surface Sci. **36**, 739 (1973).
3.73. J. T. Yates, D. A. King: Surface Sci. **32**, 479 (1972); and **38**, 114 (1973).
3.74. G. Eherlich: J. Chem. Phys. **34**, 39 (1961).
3.75. D. Menzel: Ber. Bunsenges. Phys. Chem. **72**, 591 (1968).
3.76. G. Ertl, J. Koch: *Adsorption Desorption Phenomena* (Academic Presss, New York, 1972), p. 345; Z. Naturforsch. **25**, 1906 (1970).
3.77. H. Conrad, G. Ertl, J. Koch, E. E. Latta: Surface Sci. **43**, 462 (1974).
3.78. T. N. Taylor, P. J. Estrup: J. Vac. Sci. Technol. **10**, 26 (1973).
3.79. H. R. Han, L. D. Schmidt: J. Phys. Chem. **75**, 227 (1971).
3.80. M. Mahnig, L. D. Schmidt: Z. Phys. Chem. **80**, 71 (1972).
3.81. S. M. Ko, L. D. Scgmidt: Surface Sci. **42**, 508 (1972).
3.82. C. Huang, P. J. Estrup: To be published.

4. Desorption Phenomena

DIETRICH MENZEL

With 12 Figures

The term "desorption" signifies the rupture of the adsorption bond and the resulting removal of adsorbed particles from the surface. This can be accomplished in different ways. If the temperature of the system is high enough that a sizeable fraction of the adsorbate complexes has energies above the desorption energy in the Maxwellian distribution and can therefore leave the surface, we speak of thermal desorption (TD). Electron impact can lead to transitions to excited or ionized states of the adsorbate whose potential energy at the equilibrium distance of the ground state is higher than that of the respective free particle, so that ions or neutrals can be desorbed; this process is called electron impact desorption (EID) or electron-stimulated desorption (ESD). Similar processes can be brought about by excitation by light; we then speak of photodesorption (PD). The impact of ions or fast neutrals can cause the removal of adsorbates in different ways; these processes could be called ion impact desorption (IID). Finally, a very high electric field (of the order of one V/Å) can bend down the ionic curve of the adsorbate so far, that rapid tunneling occurs from the ground state to the ionic state, and the resulting adsorbate ion is carried away from the surface immediately; these processes are termed field desorption (FD).

Measurements of desorption processes, especially those of thermal and electron impact desorption, have widely been used to define adsorbate states and to measure their populations. As to the kinetic features, both experimental accuracy and theoretical understanding are more difficult to obtain, as is always the case for a time-dependent process. The efforts undertaken to overcome these difficulties are very worthwhile, however, as increased information can help in the understanding of properties of the adsorbate like the binding energy and the energetic coupling to the substrate (TD, EID, and IID), the existence of and transitions to and from excited adsorbate states (EID and PD), and the shape of the ground state potential curves (FD). Furthermore, the practical importance of desorption processes is large: TD is an important step in heterogeneous catalysis; TD as well as EID, PD, and IID are significant processes in vacuum production and measurement, and in accelerators, plasma machines and fusion reactors; EID is a frequent disturbing effect in surface investigations because of the preference for slow electrons as a surface probe; TD and IID can be used to clean samples.

In the following the different processes will be considered with the main emphasis on TD and—to a lesser degree—EID. Because of the limited space and the smaller amount of information available, the other processes will only be briefly discussed. Experimental methods, typical results, and the prevalent theoretical explanations will be covered by examples rather than comprehensively. Emphasis will be placed on modern investigations of well-defined surfaces and adsorbate layers, which have mostly been done on metal surfaces. An attempt will be made to discuss open questions and current trends of experimental as well as theoretical work, as far as space permits.

4.1. Thermal Desorption

Thermal desorption constitutes a normal chemical reaction, in which one of the reaction partners is a solid surface. In the simplest case of independent desorption of single particles, it is similar to unimolecular decomposition, but more complicated mechanisms (association, diffusion, cooperative effects) have to be expected. Analysis can therefore be similar to that of chemical kinetics in general [4.1].

In the following, a discussion of the usual approach used to analyse data is given first; then a description of the most important techniques used and subsequently a survey of typical rate data follow; and finally the current status of the theories of TD is sketched.

4.1.1. Desorption Mechanisms and Rate Parameters

As usual in chemical kinetics, the general dependence of the rates R_d (desorbing particles per unit time and surface area) on the concentration of adsorbed particles per unit surface area, N, and the temperature T,

$$R_d = -dN/dt = f(T, N) \tag{4.1}$$

is analysed in terms of an Arrhenius equation

$$R_d = k_m \cdot N^m = N^m \cdot k_m^0 \cdot \exp(-E_m/RT), \tag{4.2}$$

where k_m is the rate constant, m is the formal order, E_m the activation energy and k_m^0 the pre-exponential factor, frequency factor or simply prefactor. The coverage-dependence is usually assumed to be contained totally in the N^m-term, and the temperature dependence in the exponential term, at least as a first approximation. The rationale for the assumption of such a simple Arrhenius or Polanyi-Wigner equation is the basic concept that the coverage term is produced by the number of particles taking part in the critical step, the pre-exponential is equal to the frequency of attempts of the system to move in the direction of reaction (assumed independent of T to first approximation), and the exponential term represents the relative number of these attempts having the necessary minimum energy. The general case of (4.1) can then be recovered by making E and k^0 functions of N and T, if necessary. As these equations represent macroscopic descriptions of a complicated sequence of microscopic processes, there need not be any simple correspondence between mechanism and order, as known from chemical kinetics. An attempt is usually made, however, to interpret a measured order as indicative of the mechanism of the rate-determining elementary step. This will usually be only the case for very simple sequences, or if that step is clearly much slower than all the others. For instance, if desorption of independent single particles occurs throughout the coverage range,

$$A(ad) \rightarrow A(gas) \tag{4.3}$$

the reaction rate at constant temperature will be given by

$$dN/dt = k_1 \cdot N \tag{4.4}$$

and the reaction will be first order. In this very simple concept the prefactor would be given by the number of attempts to leave the surface, i.e. the frequency of vibration v of the adatom; the activation energy would equal the sum of adsorption energy and activation energy of adsorption, which for nonactivated adsorption is equal to the adsorption energy (see, however, Subsection 4.1.4). If two like atoms recombine to desorb as a diatomic molecule,

$$2A(ad) \rightarrow A_2(ad) \rightarrow A_2(gas) \tag{4.5}$$

the reaction can be first order [if the desorption of $A_2(ad)$ is the rate-determining step or if $A_2(gas)$ is formed directly from 2 neighboring immobile atoms], at least for a certain coverage range, or it can be second order (if the adatoms can diffuse over the surface, and their recombination is rate-determining). In the latter case, the values of E and k^0 can be expected to depend on the degree of mobility. If the adsorbate layer can be regarded as a two-dimensional gas, k^0 can be given by the collision frequency in this gas, and E_d by the desorption energy, so that

$$k_2 = d_A \cdot (\pi k T/M)^{1/2} \exp(-E_d/RT), \tag{4.6}$$

where d_A is the collision diameter, and M the mass of A. If the diffusion is much slower, but still occurs, then for small N

$$k_2 = (v/N_s) \exp\left[-(E_{diff} + E_d)/RT\right],\tag{4.7}$$

where N_s is the number of sites on the surface, and E_{diff} the activation energy of surface diffusion [4.2].

In the case of dissociative adsorption, both desorption of atoms and molecules can, in principle, be expected. A simple energy consideration can show which one will take place at not too small coverages [4.3]. If Q is the bond energy per adatom and D the dissociation energy per mole of the gaseous molecule, then the desorption energy of atoms will be Q and that of molecules $2Q - D$ (assuming nonactivated adsorption). Desorption of atoms will then dominate, if $2Q - D > Q$, or $Q > D$. This condition is fulfilled, e.g. for oxygen on W, but not for hydrogen on W. At high temperatures, when the coverage is very small, the argument is not valid, of course, as not enough collisions between atoms occur; H_2 can therefore be atomized on hot filaments [4.4].

Many more complicated situations can be imagined, leading to integral or fractional orders of desorption. The latter are difficult to analyse since the data can always be interpreted in terms of a dependence of E and/or k^0 on N. Even for integral experimental orders, the mechanism can be ambiguous. For instance, if a layer of adsorbed atoms A consists of two different adsorbate states, one of which is mobile and the other immobile, and desorption occurs by recombination of one of each species, so that each desorption step is followed by conversion of another mobile to an immobile species, desorption will be first order until all mobile species are used up, and will then stop. After a temperature increase to make the second species mobile, desorption will resume according to second order [4.2].

This shows that no simple unequivocal conclusions about mechanisms can be made from reaction orders. As always in chemical kinetics, the logic points in the opposite direction: A mechanism can only be shown to be compatible with, but not proven by the kinetics. Nevertheless, the reaction order is one of the principal kinetic parameters, and the determination of m, E, and k^0 are the main objective of kinetic measurements. Apart from their bearing on the kinetics itself, they are also of interest for the understanding of the stationary properties of the adsorbate. From the integral

$$\int_0^\infty R_d \, dt = N\tag{4.8}$$

the number of adsorbed particles present can be obtained; E yields the adsorption energy in simple cases, and N and k^0 are important for the questions of dissociation and mobility. The methods for the determination of these parameters will be considered in the next subsection.

4.1.2. Experimental Methods of Thermal Desorption

The classical method in chemical kinetics is the isothermal measurement of the change of concentrations of reactants and/or products with time, after a sudden disturbance which makes the starting condition unstable. The reaction order can be derived from the rate-law obtained; measurements at different temperatures yield the activation energy and (with smaller accuracy because of the large extrapolation involved) the prefactor. Such isothermal measurements have also been performed for desorption, although not very often, compared to temperature-programmed methods. This is probably due to the fact that step functions in temperature or coverage to cause the initial unstable situation are not easy to accomplish in adsorption systems, so that the starting-point is not well-defined (this is not really necessary, though, as procedures exist for the analysis of rate data without clear start [4.5]). On the other hand, the temperature-programmed method to be described below is quite easy to use and at the same time gives direct information on adsorbate states and coverages, at least in principle.

The usual procedure is to follow the change with time elapsed after the step change, of any quantity x which is connected to the coverage in a known way [$x = f(N)$, for $T =$ const], so that

$$dx/dt = (df/dN) \cdot (dN/dt) \tag{4.9}$$

can be used to evaluate directly (4.2). Using temperature steps, KOHRT and GOMER [4.6] measured the rate of desorption from a single crystal by placing it in front of a field emission tip and recording the coverage accumulated on this tip (from the change of the field emission current) under conditions where every molecule arriving at the tip would stick to it; their evaluation used the procedure of [4.5] (see Fig. 4.1). PETER-MANN [4.7], and JELEND and MENZEL [4.8] used the EID ion current which under certain conditions is proportional to the coverage (see Section 4.2 below), to follow the change of coverage of a certain adsorbate state after step heating. ENGELHARDT and MENZEL [4.9] used the work function change due to adsorption which is also proportional to coverage in certain cases. The amount left on the surface after a certain

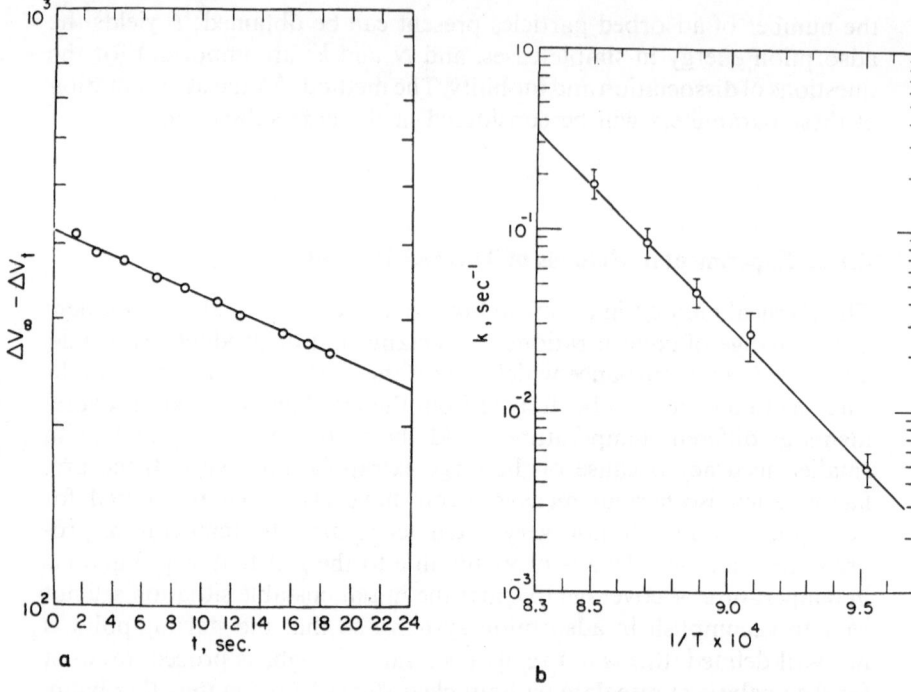

Fig. 4.1. (a) Isothermal desorption kinetics of β_2-CO from W (110), measured by recording the CO molecules condensed on a field emission tip in front of the crystal, by the change of voltage necessary for a certain emission current. (b) Arrhenius plot of data obtained in this way. (After [4.27])

isothermal heating time, measured by integral flash desorption, was used by TAMM and SCHMIDT [4.10] and others to construct isothermal kinetics.

An isothermal desorption method using pressure steps is the measurement of mean dwell times of the adsorbate on the surface at high temperatures, using modulated molecular or ionic beams. For a first-order desorption reaction, the inverse of the rate constant k_1 is equal to the mean dwell time τ, so that

$$\tau = \tau_0 \exp(E/RT). \tag{4.10}$$

With the assumption of constant τ_0 $(=1/k_1^0)$, this equation goes back to FRENKEL [4.11]. If a chopped or modulated molecular or ionic beam is directed onto a surface under conditions of reversible adsorption, so that desorption can occur, and the re-emitted particles are detected in a

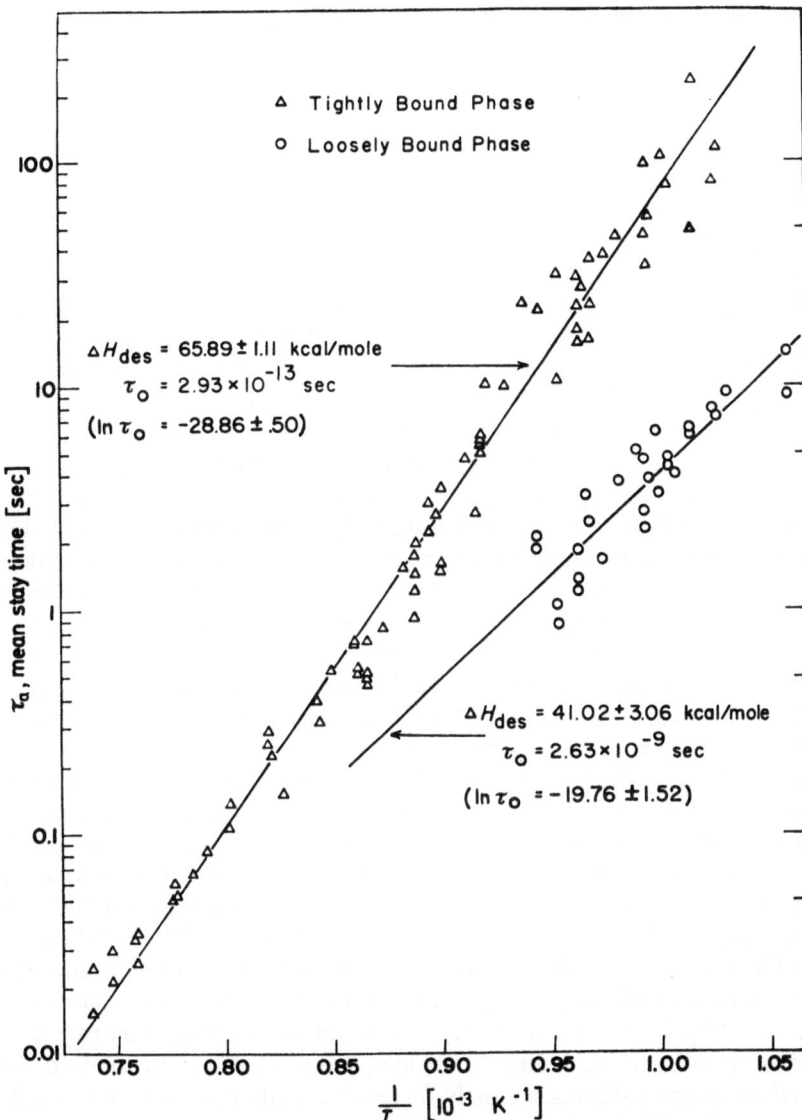

Fig. 4.2. Arrhenius plots for the mean dwell times of Ag on W (110), obtained with the molecular beam decay method. Two states in different coverage regimes are shown. (After [4.19])

mass spectrometer, the desorption kinetics can be either deduced from the decay time of the desorption signal after the beam has been shut off [4.12, 13], or from the phase shift of the output pulses against the input pulses [4.14]. By varying the temperature of the sample, the data for an evaluation according to (4.10) can be obtained. Because of the ease of detection in this case the method has so far been only used for ionic desorption [4.12, 14–17] and for desorption of species which are easily condensed on cooled walls (metals like Cd [4.13] and Ag [4.18, 19], oxides and halides [4.20]). In the first case, direct passage of the desorbing flux into a mass spectrometer is possible; in the second, detection by ionization in a single pass through an electron impact ionizer is possible with good s/n, since no scattering from the cooled walls occurs. Figure 4.2 shows an example. The method allows a less ambiguous determination of the k_1^0 and E values than the flash method (see below), provided only one or two adsorption states are present with first-order behavior throughout (for a discussion of the relative merits of the two methods, see [4.19]).

The method most frequently used to date is temperature-programmed desorption, mostly called flash filament desorption (as it was first performed by rapidly heating a filament or "flashing" it) or simply flash desorption (Ehrlich [4.21], Redhead [4.22]), although often rather slow heating rates are used nowadays. The basic idea of this method which goes back to the "glow-curve" method in thermo-luminescence [4.23], is to continuously heat the substrate and follow the rate of desorption as a function of temperature. The rate at any time will be determined by the temperature and the amount of gas left on the surface after the previous desorption; it will then show a maximum at the point, where the increase due to the temperature rise is equal to the decrease from the diminished coverage. Most investigations use the amount of gas evolved to measure the rate (see below), although other parameters coupled to the surface coverage and therefore to the desorption rate (work function change, Auger peak height, EID current) can be used. Essentially the same method has been utilized by Cvetanovic and Amenomiya [4.24] for the investigation of catalysts. The main reasons for the popularity of the the pressure-rise method seem to be that a) a picture of the different adsorption states on the surface can be obtained very rapidly, by correlating every maximum of the desorption rate with a distinct adsorption state; b) an integration of the rates to yield the total coverage, (although mostly in relative units only) can very easily be done [Eq. (4.8)]; c) approximate values of m, E, and (within limits) k^0 can be obtained in a few measurements. The main disadvantages are the often resulting ambiguity of the kinetic parameters, and the fact that distinct desorption peaks need not correspond to distinct states, and vice versa; this will be discussed below.

The basis of the method has been discussed by a number of authors [4.21–26]; the subsequent formulation follows that given by Redhead [4.22].

Inserting a linear temperature rise with time

$$T = T_0 + \beta \cdot t, \tag{4.11}$$

as used by most authors, into (4.2) leads to

$$-dN/dt = R_d(t) = N^m(t) k_m^0 \exp[-E_m/R(T_0 + \beta t)] \tag{4.12}$$

which is equivalent to

$$-dN/dT = (N^m \cdot k_m^0/\beta) \exp(-E_m/RT). \tag{4.13}$$

If E and k^0 are assumed to be independent of coverage N, then the temperature T_p at which the rate curve shows a peak, can be obtained from (4.13). It is found that

$$E_1/RT_p^2 = (k_1^0/\beta) \exp(-E_1/RT_p) \quad \text{for} \quad m = 1 \tag{4.14}$$

and

$$E_2/RT_p^2 = (2N_p \cdot k_2^0/\beta) \exp(-E_2/RT_p) \quad \text{for} \quad m = 2. \tag{4.15}$$

These equations are equivalent to

$$\ln(T_p^2/\beta) = E_1/RT_p + \ln(E_1/k_1^0 \cdot R) \tag{4.16}$$

and

$$\ln(T_p^2 \cdot N_p/\beta) = E_2/RT_p + \ln(E_2/2k_2^0 \cdot R), \tag{4.17}$$

where N_p is the coverage at the rate maximum and $2N_p \approx N_0$ for $m = 2$. Formulae for the peak shapes can be obtained from these equations with some approximations [4.21−26].

A similar, and in formal respects simpler, treatment is possible for the case of a reciprocal temperature rise

$$1/T = 1/T_0 - \alpha \cdot t. \tag{4.18}$$

The reader is referred to REDHEAD [4.22] or CARTER [4.25] for the formulation. Instead of a continuous temperature rise, stepwise heating can also been used. For the equivalence of this procedures to flash desorption, see [4.6].

Again, any variable connected to the coverage [Eq. (4.9)] can be used to measure the $R_d(t)$-curves. For instance, the change of work function has been used in this way [4.9]. Condensation of the desorbing particles on a cryogenically cooled field emission tip, as described above, has also been used [4.6, 27], as well as radioactive tracer methods [4.28]. The advantage of these techniques is the absence of disturbing wall effects.

The most frequently used variable connected to R_d, however, is the pressure rise in the surrounding vacuum system resulting from desorption. The pumping equation of a vacuum system is given by

$$dP/dt = (kT/V) \cdot (dN_g/dt) = Q/V - S \cdot P/V, \qquad (4.19)$$

where P is the pressure in the system, V its volume, N_g the number of gas particles contained in it, k the Boltzmann constant, Q the influx of gas (e.g. in Torr · l/sec), and S the effective pumping speed. Before the start of the desorption run, the base pressure is constant, so that

$$Q_0 = S P_0. \qquad (4.20)$$

If we assume that no change of pumping speed (e.g. by adsorption on the walls or readsorption on the sample) or of influx (e.g. by displacement from the walls or pump) occurs during desorption, simple balancing requires that the rate of desorption be equal to the rate of removal of gas from the system by the pumps plus the rate of increase of gas content of the system:

$$A \cdot R_d(t) = S \cdot \Delta P/kT + (V/kT)(dP/dt), \qquad (4.21)$$

where A is the surface area of the substrate, and $\Delta P = P - P_0$ the pressure increase in the system over the background pressure. This can also be written as

$$dP/dt + \Delta P/\tau_p = a R_d(t), \qquad (4.22)$$

where $\tau_p = V/S$ is the characteristic pumping time of the system and $a = AkT/V$. Two limiting cases are of interest:

1) For small pumping speeds, i.e. if τ_p is large compared to the heating time, the second term can be neglected and $R_d(t) \propto dP/dt$.

2) For large pumping speeds, i.e. if $\tau_p \to 0$, the second term overpowers the first, and $R_d(t) \propto \Delta P$.

Corrections for situations close to, but not exactly equal to one of these cases have been given by REDHEAD [4.22]. Situations between the extremes can best be analyzed using a network of operational amplifiers, simulating (4.22), so that $R_d(t)$ can be directly obtained. In modern uhv systems with large pumping speeds, Case 2 is usually realized for not too large heating rates; the pressure-time curves can then be used directly for the evaluation of R_d. The total amount of liberated gas in relative units is obtained directly from an integration of the $P(t)$-curves; for absolute determinations, an accurate knowledge of S would be required, which is very difficult to obtain, especially for the high S case. In earlier work total pressure measurements were used for the recording of flash filament spectra; in recent years most authors use small mass spectrometers to record only the masses wanted, which eliminates the possibility of disturbances by gases displaced from the walls and makes visible possible disturbing coadsorbates (e.g. CO in hydrogen). Even then, there remain a number of experimental problems, since interactions of the gas with the measuring device as well as with pumps and walls can still occur (when the pressure drops again, re-evolution from the walls is possible; scattering of gas from the walls can make the gas density anisotropic, so that it cannot be analysed in terms of a pressure change). Careful lay-out of the experiments and analysis of the results is necessary to exclude such disturbances (see the discussions given by MCCARROLL [4.29], HOBSON and EARNSHAW [4.30], PETERMANN [4.31]). For metallic adsorption which leads to desorption of ions, many of these difficulties can be circumvented by drawing these directly into a mass spectrometer for detection, without the build-up of a partial pressure [4.32]. For the application to fine-grained catalysts [4.24], the desorbed gas is carried away by a flow of inert gas. Temperature-inhomogeneity across the sample can also lead to difficulties, especially for direct Ohmic heating (see EHRLICH [4.21]).

While earlier work mostly used polycrystalline samples (filaments, ribbons), well-defined single crystal faces are being used increasingly in the recent years. Care has to be taken in that case to keep the percentage of surface area with other orientations than the main face as small as possible; in order to assure uniform heating, crystal discs are often suspended on thin wires and heated by light irradiation or electron impact [4.10, 33]. Figure 4.3 shows a famous example of a ·flash spectrum from a polycrystalline filament; in Fig. 4.4 some examples of flash spectra from single crystal faces are given.

The resulting $R_d(t)$-curves obtained by any of these methods can be analysed according to (4.14)–(4.17) or the peak shape formulae obtained from them. Equations (4.14) and (4.16) show that for first-order desorption with constant E and k_1^0, T_p is independent of the initial coverage for linear temperature-programming; the peak is asymmetric about T_p. The relation between E and T_p is then almost linear and for $10^{13} < k_1^0/\beta < 10^8 (K^{-1})$ is given to $\pm 1.5\%$ by

$$E_1/R T_p = \ln(k_1^0 \cdot T_p/\beta) - 3.64 . \tag{4.23}$$

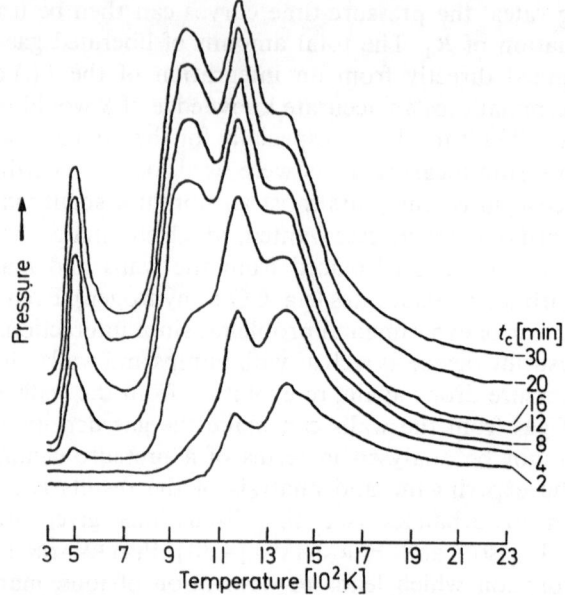

Fig. 4.3. Flash desorption spectrum of CO from a polycrystalline filament, recorded by the total pressure change. 4 states can be recognized. (After [4.34])

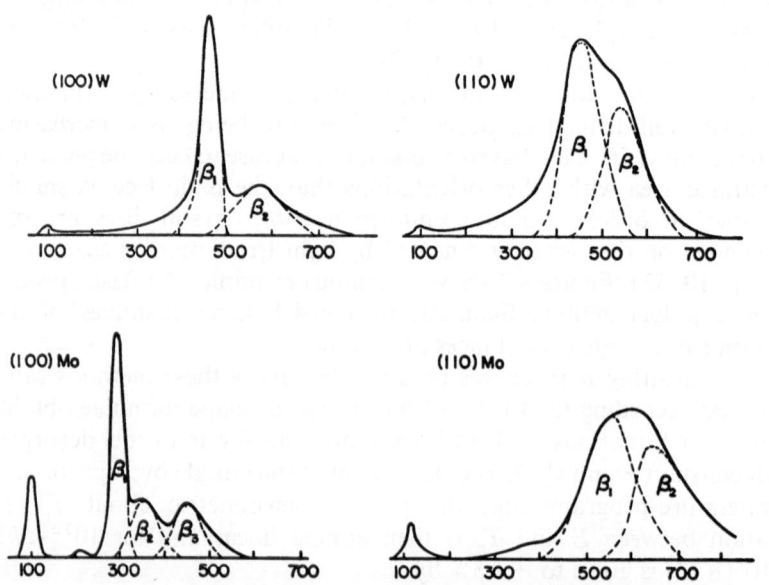

Fig. 4.4. Flash desorption spectra for hydrogen on some W and Mo faces, recorded mass spectrometrically. Resolution into partial peaks is indicated. (After [4.35])

For an evaluation of E from one T_p-value only, a value of k_1^0 must be assumed. This has been done quite often, assuming $k_1^0 \approx 10^{13} \sec^{-1}$. A real determination of E and k^0 is possible from measurements of desorption spectra at different heating rates and evaluation according to (4.16). In order to obtain an acceptable accuracy, β must be varied by two powers of ten or more, which is rarely possible (if β is very small, the accuracy is limited and wall or re-adsorption effects can be very strong; if β is too large, multiple peaks are not resolved any more [4.36]). Variation of β is also important in order to make sure that no redistribution between states occurs during heating, which could falsify the kinetics. If only one peak is present, then a reasonably accurate value of k_1^0 can be deduced from its half-width, which increases with decreasing k_1^0, see e.g. [4.37].

Irrespective of m, flash spectra can be evaluated by Arrhenius plots of the measured rates directly, for constant coverage; these are found by back-integrating the flash-spectra to the desired N-value [4.32].

For second-order desorption T_p is seen to decrease with increasing coverage [Eqs. (4.15) and (4.17)]. However, such behavior can also occur for $m = 1$, but E decreasing with increasing N. The peak should be symmetric about T_p for $m = 2$, but this can usually be checked only for single peaks.

A test for m is obvious from (4.16) and (4.17), if E is constant: a plot of $\log(N_0 \cdot T_p^2/\beta)$ versus $1/T_p$ must yield a line parallel to the ordinate for $m = 1$ and a straight line (with slope E_2/R) for $m = 2$ (see Fig. 4.5). Plots according to (4.14) and (4.15) can also be used for the determination of m and E. If none of these plots yields a straight line, either E or k^0 or both vary with N, and m cannot be unequivocally determined; wall effects can also be responsible. In such cases and especially for multiple, not totally resolved peak structures, computer analyses can be performed [4.35–39], which can attempt to take account of the experimental disturbances at the same time [4.29]. These are usually done under the assumption that the total spectrum is a sum of independent contributions of different binding states; nonadherence to simple assumptions (values of m, constancy of E_m and k_m^0 values) is then analyzed in terms of coverage-dependent E and/or k^0 values. Because of the large number of parameters involved, such resolutions into partial peaks (e.g. 6 peaks in the case of CO on polycrystalline W [4.38]) and coverage-dependent rate parameters [4.39] can become quite ambiguous. PISANI et al. [4.36] have recently given an illuminating example. For the desorption spectra of nitrogen on polycrystalline W, which consist of two overlapping peaks, they were able to fit a large number of data (120 desorption spectra) equally well assuming first- or second-order desorption for one of the peaks, which of course leads to quite different E and k^0 values also. In the same system,

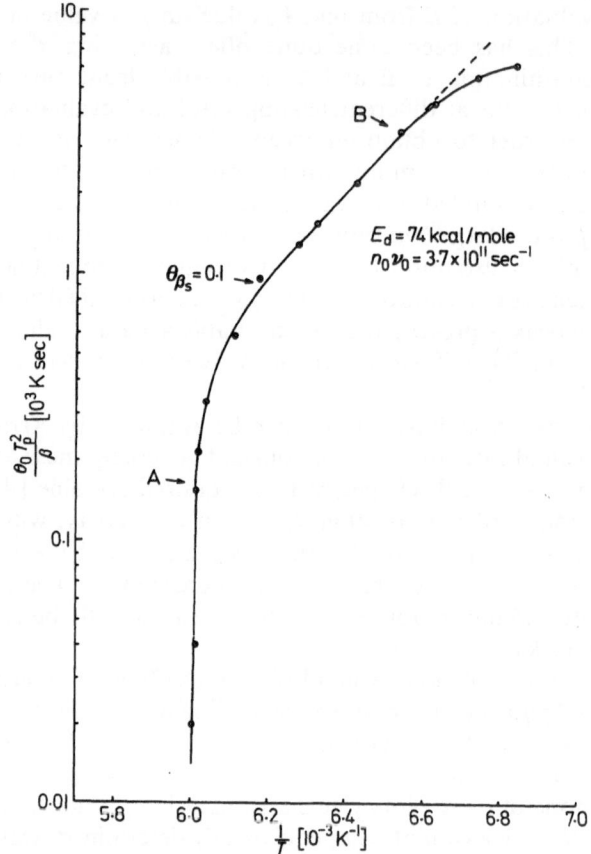

Fig. 4.5. "Second-order plot" according to (4.17) for flash desorption of β_3-CO from W (100). Ranges of first (A) and second order (B) behavior are obvious; deviations occur at higher coverages. (After [4.39])

Madey and Yates [4.40] had earlier encountered complex desorption kinetics, which they analyzed in terms of a variation of m with N (from 4 to 1). Clavenna and Schmidt [4.41] explained similar behavior they found on W (100) by a variation of k_2^0 with N and gave a model for it. Dawson and Peng [4.42] could not distinguish between 6 different models (with $m = 1$ or 2, E between 20 and 80 kcal/mol and a corresponding spread of k^0) for a certain state in this system.

The possibility of conversion between states during heating-up must also be considered carefully, see e.g. [4.43, 44]. Furthermore, the assumption of independent states has to be considered critically as well. For polycrystalline samples, it is sensible to expect surface sites with different binding energies, so that an interpretation of multipeak spectra assuming independent contributions from different binding states is in principle acceptable. However, more than one peak is usually found in desorption from low-index single crystal faces, too. (For instance, the desorption spectra of CO on W (100) [4.39] and other W faces are very similar to those on polycrystalline W [4.34, 45].) If these peaks are clearly separated, then the assumption of distinct binding states is usually warranted here, too; these can also be due to the presence of adsorbed species rather than to a-priori different sites (e.g. α- and β-CO on W). Then two or more binding states coexist simultaneously, as distinct entities in the composite layer. However, the composite layer can also consist of only one binding state only, if there is a variation of desorption energy with coverage strong enough to cause multiple peaks. For a long time the first explanation was assumed to be correct in all cases, and multipeak spectra have therefore been interpreted as re-presenting the equivalent number of distinct binding states. Detailed analyses in recent years (TOYA [4.46], GOYMOUR and KING [4.47], ADAMS [4.48]), have shown, however, that the variation of E with N need not be very strong to lead to distinct peaks, so that a repulsive inter-action between adsorbed particles can easily lead to such an effect. For instance, the two β-peaks of hydrogen on W (100) are now thought by some authors to be due to interaction in the adsorbed layer consisting of one distinct species only and not to two distinct binding modes side by side (but see Chapter 3 for alternative views). Such interactions can also lead to unexpected desorption orders (see Subsection 4.1.1).

This discussion had the purpose to show that the flash desorption method, while being a very valuable means for the investigation of desorption kinetics, is also fraught with experimental and interpretational difficulties. While in principle all information on m, E, and k^0 are contained in the desorption peak positions and shapes, provided enough spectra with varied β and N_0 values have been obtained, in practice it is often difficult if not impossible to unfold this information unambiguously. The safest information is usually that on E (however, see [4.36]), and on m for clear first order. The ambiguities are especially large for over-lapping peaks. This should be borne in mind when using kinetic data obtained by flash desorption.

4.1.3. Some Results of Thermal Desorption Measurements

In the following some results of desorption parameters, mostly for well-defined polycrystalline or single crystal metal surfaces, will be given and discussed.

Table 4.1. Some rate parameters for the desorption of metallic adsorbates from metals ($m = 1$ in all cases)

Substrate	Adsorbate	Desorbed species	Method	State	k_1^0 [sec^{-1}]	E_1 (kcal/mol)	Ref.
W (poly)	Cs	Cs$^+$	CB	clean	$1 \cdot 10^{12}$	47	[4.15]
W (poly)	Cs	Cs$^+$	CB	contam.	$1.7 \cdot 10^{10}$	36	[4.15]
Re (poly)	Cs	Cs$^+$	CB	clean	$5 \cdot 10^{12}$	44	[4.15]
Re (poly)	Ba	Ba^{2+}	CB	clean	$1.7 \cdot 10^{13}$	109	[4.15]
W (poly)	K	K$^+$	CB	clean	$4.2 \cdot 10^{12}$	54.4	[4.17]
W (poly)	K	K$^+$	CB	C-contam.	$1.7 \cdot 10^{13}$	54.4	[4.17]
W (poly)	Cd	Cd	CB	1	$1.1 \cdot 10^{10}$	41	[4.13]
				2	$7.1 \cdot 10^9$	21	[4.13]
W (110)	Ag	Ag	CB	1	$3.5 \cdot 10^{12}$	66	[4.19]
				2	$3.8 \cdot 10^8$	41	[4.19]
W (poly)	Ag	Ag	CB	clean	$3.6 \cdot 10^{12}$	67	[4.18]
W (poly)	Ag	Ag	CB	C-contam.	$8.2 \cdot 10^{12}$	73	[4.18]
W (poly)	Ag	Ag	CB	O-contam.	$1.4 \cdot 10^{12}$	47	[4.18]
W (poly)	Au	Au	CB	clean	$8.2 \cdot 10^{13}$	105	[4.18]
W (poly)	Au	Au	CB	C-contam.	$1.7 \cdot 10^{14}$	109	[4.18]
W (poly)	Au	Au	CB	O-contam.	$4.7 \cdot 10^{11}$	46	[4.18]
Ni (111)	Na	Na$^+$	TP	$\theta = 0 - 0.5$	10^{13}	$58 \rightarrow 32$	[4.32]
Ni (110)	Na	Na$^+$	TP	$\theta = 0 - 0.3$	$10^{13} - 10^{10}$	$50 - 32$	[4.32]

CB   Chopped beam.
TP   Temperature programmed.

For desorption of metallic adsorbates from metal surfaces, both as ions and neutrals, the kinetics seem to follow first order in all cases. Since the corresponding adsorption is always nonactivated, the activation energies of desorption correspond to the binding energies of the species concerned. "Normal" prefactors of $\approx 10^{13}$ sec^{-1} are often found, although lower and higher values have also been reported. Some selected values are given in Table 4.1. An interesting effect is the strong change observed by contamination of the surface, which in part only affects the pre-exponential factor. More complicated results were found for flash desorption of Cu and Au from the basal plane of graphite [4.49]. The behavior of peak temperatures indicated $m < 1$, and the peaks could be fitted well assuming constant E up to a certain coverage and $m = 0.5$; E increased at higher coverages. This behavior seems to be caused by the much stronger interaction between the metal atoms than between metal atoms and C surface, which leads to island formation and to the E increase with coverage. The rate-determining step for desorption is then probably the dissociation of a metal atom from the border of an island, which explains the observed fractional order. This mechanism is also supported by the results of condensation measurements. A somewhat

Table 4.2. Some rate parameters for the desorption of weakly bound adsorbates ($m = 1$ in all cases)

Substrate	Adsorbate	Method	k_1^0 [sec^{-1}]	E [kcal/mol]	Ref.
W (111)	Xe	TP	10^{15}	9.3	[4.51]
W (100)	CH$_4$	TP	$3 \cdot 10^{12}$	7	[4.37]
CO/W (poly)	H$_2$	Iso	10^3	7.5	[4.8]
He (Constantan)	He	RF	$2 \cdot 10^7$	$6 \cdot 10^{-2}$	[4.52]
SiO$_2$ (untreated)	Ar	TP	10^2	$0.9 - 1$	[4.53]
SiO$_2$ (freshly abraded)	Ar	TP	1	$0.9 - 1$	[4.53]
W (110)	v-CO	Iso	$10^4 - 5 \cdot 10^5$	10	[4.27]

TP Temperature-programmed desorption.
Iso Isothermal desorption.
RF Rapid flash.

similar case has been found for oxygen on Ag (110), where the measured zeroth-order desorption has been explained by desorption from the end of O-atom chains [4.9]. Zeroth order is also found for desorption from oxide layers [4.50].

Complicated desorption spectra for Na on different single crystal faces of Ni have been obtained by GERLACH and RHODIN [4.32]; only the low coverage peaks were therefore analysed. These seemed to obey first order kinetics; the observed shifts of T_p with coverage were attributed to changes of E and/or k^0 with coverage. For Na on Ni (111), Arrhenius plots gave constant k^0 values, while E decreased continuously by about 40 % over the coverage range examined. On the (110) face there was an abrupt change of E by about the same amount at a coverage, where a change of superstructure was observed in LEED, but here k^0 also changed in the same range. The data could also be analysed to yield constant k^0 values, however.

First-order desorption is usually also found for physisorbed or weakly chemisorbed gases. Some results are shown in Table 4.2. The variation of prefactors is seen to be considerable here, which might partly be due to spurious effects. The case of the weakly bound coadsorbed hydrogen on W, however, seems to be reinforced by the fact that the isosteric adsorption energy measured independently agrees very well with the activation energy E [4.8]. The possible reasons for the variation of k^0 values will be discussed in Subsection 4.1.4.

Complicated results are often found for electronegative gases on polycrystalline substrates. The complicated spectrum of CO on W [4.38] and the difficulties and ambiguities for nitrogen on W [4.36, 40–42] have been mentioned above. Interesting kinetic parameters can be derived from the work of SCHMIDT et al. for the simple gases H$_2$, N$_2$, and

Table 4.3. Some rate parameters for the desorption of electronegative adsorbates from metal surfaces

Substrate	Adsorbate	Method	State	Order m	k_1^0 [sec^{-1}]	k_2^0 [cm^2/sec]	E [kcal/mol]	Ref.
W (100)	H_2	TP	β_1	1	$10^{13\pm1}$		26	[4.10]
			β_2	2		$4\cdot10^{-2\,a}$	32	
	N_2	TP	β_2	2		$0.23^{\,a}$	74	[4.41]
			γ	1	(10^{13})		9–10	
	CO	TP	$\beta_3(\theta<0.2)$	1	(10^{13})		93 ± 5	[4.39]
			$\beta_3(\theta>0.2)$	2		$\approx10^{-3}$	74 ± 5	
			β_2	1	(10^{13})		62 ± 4	
			β_1	1	(10^{13})		57 ± 4	
			α	1	(10^{13})		21 ± 4	
W (110)	H_2	TP	β_1	2		10^{-2}	27	[4.54]
			β_2	2		$1.4\cdot10^{-2\,a}$	33	
W (111)	H_2	TP	β_1	2			14	[4.54]
			β_2	2			22	
			β_3	2		(0.01)	30	
			β_4	2			37	
			γ	1	(10^{13})		6–11	
W (poly)	H_2	TP		2		$2\cdot10^{-3\,a}$	35	
Ir (poly)	H_2	TP		2		$2\cdot10^{-2\,a}$	24	[4.55]
Rh (poly)	H_2			2		$1\cdot10^{-3\,a}$	18	
Re (poly)	H_2	TP	β_2	2		$(7\pm3)\cdot10^{-4}$	30	[4.56]
Mo (100)	H_2	TP	β_1	1	(10^{13})		16	[4.57]
			β_2	2		(0.05)	20	
			β_3	2		$5\cdot10^{-2}$	27	
Ni (111)	H_2	TP		2		$0.2–0.3^{\,a}$	23	[4.58]
W (110)	CO	Step	β_2	1	$5\cdot10^{11}$		66	[4.27]
		Iso			$1.2\cdot10^{12}$		69	
		Step	β_1	1	$3\cdot10^9$		40	
		Iso			$3\cdot10^{12}$		55	
		Step	v	1	10^4		9.5	
		Iso			$5\cdot10^5$		10	
W (poly)	CO	TP	1	1	$1\cdot10^{12}$		49	[4.38]
			2	1	$1\cdot10^{12}$		53	
			3	1	$2\cdot10^{12}$		60	
			4	1	$7\cdot10^{12}$		67	
			5	1	$1\cdot10^{13}$		75	
			6	2		0.01	78	
W (110)	O_2	Step	>1900 K	1	$2\cdot10^8$		92	[4.6]
		Iso	>1900 K		$3\cdot10^9$		92	
		Iso	<1900 K	1	$2\cdot10^7$		69	[4.6]
W (poly)	O_2		1900–2400 K	1	$3\cdot10^{13\pm1}$		129 ± 8	[4.59]
		Step	1600–1800 K	1	$3\cdot10^{16\pm1}$		120 ± 5	
			1300–1600 K	1	$3\cdot10^{16\pm1}$		109 ± 5	

Table 4.3 (continued)

Substrate	Ad-sorbate	Me-thod	State	Order m	k_1^0 [sec^{-1}]	k_2^0 [cm^2/sec]	E [kcal/mol]	Ref.
W (poly)	O_2	TP	Oxide layer	0	$k_0^0 = 10^{34}$ molec/cm^2 sec		100 ± 10	[4.50]
			1600–1800 K	1	$10^{14 \pm 2}$		110 ± 10	
Pt (poly)	O_2	TP		2		6	58	[4.60]

TP   Temperature-programmed desorption.
Step Step desorption.
Iso   Isothermal desorption.

k^0-values in brackets: assumed.
Most other k^0-values from curve-fitting.

[a] Values for low coverages, from plots according to (4.18).

CO on single crystal faces of W and Mo (see Table 4.3). Most strongly bound states of hydrogen on all faces of both metals showed second-order kinetics, as would be expected for associative desorption from an atomic layer. On the (100) faces a strongly bound state exhibiting first-order behavior was also found, which was originally assumed to be due to adsorbed molecules. Other work suggests, however, that this state is also atomic and in fact that the full layer consists of one species only, the two peaks arising from interactions between the adsorbed atoms, as discussed above. The spectra could be fitted very well by assuming constant E and k^0 values for all first-order states (with $k_1^0 = 10^{13}$ sec^{-1}) and for the second-order states at low coverage (with k_2^0 around 10^{-2} cm^2/sec). For the fitting of the higher coverages of the latter either a linear decrease of E or an increase of k^0 with increasing coverage had to be assumed. The same behavior has been found for hydrogen on Ir, Rh [4.55], Re [4.56], and Ni [4.58]. The change of k^0 with coverage has been explained [4.41] in terms of a random-walk surface diffusion process as the rate-determining step. This yields the expression

$$k_2^0 = (l^2 \cdot v_0)/[1 - (l_c/l) \cdot \theta^{1/2}]^2 , \qquad (4.24)$$

where l is the diffusion jump length, v_0 the vibrational frequency of the adsorbed atoms, and l_c the critical separation for recombination. Good fits could be obtained with this equation also, again showing the ambiguity of fitting. For low coverages $k_2^0 = l^2 \cdot v$, which is ≈ 0.01 cm^2/sec $\cdot$ atom, as found. This makes it understandable that about the same E-values are often obtained from the same R_d-values, assuming first- or second-order desorption: as N_0 is of the order 10^{15} atoms/cm^2, $N_0 \cdot k_2^0 \approx k_1^0$ for normal

k^0-values. The values from the work of KOHRT and GOMER [4.6, 27] for the desorption of different binding states of oxygen and CO on W (110) faces, which are listed in Table 4.3, have been obtained both by isothermal and step desorption with a method described in Subsection 4.1.2. Isothermal measurements gave the clearer picture. Their results for oxygen desorption disagree strongly with the more recent work of KING et al. [4.59], which is also listed in Table 4.3. These authors argue that the very low k^0-values of KOHRT and GOMER are due to a mixture of different states contributing to their Arrhenius plots.

The data of Tables 4.1–3 show a tendency towards a coupling of E and k^0 in the sense of a "compensation effect" [4.61]: large values of E are accompanied by large k^0-values, and vice versa. This relation has been shown to be linear ($\log k^0 \propto E$) in several cases, for instance, for multiple binding states [4.62] and continuously varied contamination [4.63]. The explanation of this effect is not clear at present (for one interpretation, see DEGRAS [4.64], who also lists more desorption parameters); in some cases it might even be an artifact.

Finally, results of another kind will be mentioned. Several authors [4.65–69] measured the angular distribution of hydrogen molecules desorbing from metal surfaces, both after diffusion through the metal and after deposition from a molecular beam. Strongly anisotropic distributions were observed, which were peaked in the surface normal up to the 9th power of the cosine of the angle to the surface normal. Recently, BRADLEY et al. [4.67] showed conclusively that this behavior only occurs with surfaces containing strongly bound impurity atoms; from clean surfaces of Ni, Fe, Pt, and Nb, they obtained the expected cosine distribution. The reason for the different behavior of contaminated surfaces is not understood. One explanation for peaked distributions could be that the two hydrogen atoms forming the desorbing molecule have to move exactly in phase when leaving the surface, in order to be successful; why such an effect should be enhanced by contamination, is not clear. Another possibility is an activation barrier for adsorption [4.67]. These investigations constitute the first step towards a differential understanding of desorption. Extension to other systems and to energy resolution besides angular resolution would be the desirable next step. First experiments of this kind have been reported [4.52, 66]; non-Maxwellian velocity distributions have been found.

4.1.4. Theories of Thermal Desorption

Some interpretational approaches to desorption kinetics have already been mentioned above. Virtually all treatments try to retain essentially the form of (4.2) as a starting point; in most cases m is set equal to 1 or 2, corresponding to the true molecularity, and additional coverage-dependences are expressed in terms of changes of E and/or k^0. In most cases, E can be assumed to correspond to the adsorption energy plus the activation energy of adsorption for activated adsorption [see, however, the case of (4.7)]. The value of E is therefore essentially a static property, which has to be obtained from a quantum-mechanical theory of the ground state (see Chapter 2). The basic problem of kinetics concerns then the value of k^0.

The simplest approach already mentioned in the beginning and in the discussion of (4.10) assumes k^0 to be independent of N and T and, for unimolecular desorption, to be equal to the frequency of vibration of the adsorbed atoms perpendicular to the surface. While this treatment seems to meet with success for a considerable number of cases for which k_1^0 is between 10^{12} and $10^{13}\ \mathrm{sec}^{-1}$, this is a coincidence, as will become obvious from the following.

Most detailed analyses use the transition state theory developed mainly by EYRING [4.70]. This theory is widely applied in chemical kinetics and other transport theories; for its application to desorption, see (Ref. [4.2, p. 141]) and (Ref. [4.70, p. 347]). In this approach, one assumes the existence of a state of the system, called the activated complex, which corresponds to the critical configuration for reaction. Apart from a certain probability of reflection (described by the transmission factor κ), the transition state corresponds to the point of no return in the movement of the system towards reaction in configuration space. For activated reactions it corresponds to the saddle point on the energy hypersurface, but it can be shown that the existence of such a point is not necessary to make the treatment valid. The desorption rate is then given by

$$R_\mathrm{d} = \kappa \cdot N^{\pm} \cdot v^{\pm}, \tag{4.25}$$

where $N^{\pm}$ is the surface concentration of activated complexes and $v^{\pm}$ the frequency of their disintegration, which at least for unimolecular desorption can again be set equal to the frequency of vibration normal to the surface. The crucial assumption is that the transition state is in thermodynamic equilibrium with the reactants (apart from the degree of freedom corresponding to the reaction coordinate), so that the dimensionless equilibrium constant for its formation can be written for desorption as

$$K^{\pm} = (N^{\pm}/N_\mathrm{s})/(N^m/N_\mathrm{s}^m) = N^{\pm} \cdot N_\mathrm{s}^{m-1}/N^m$$
$$= (f^{\pm}/f^m)\exp(-E_0/RT), \tag{4.26}$$

where N_s is the number of sites per unit area.

The f's are the complete partition functions of transition state and adsorbate, respectively, after extraction of the zero-point energy difference E_0. If one extracts the critical degree of freedom from f^+, which leaves f_+, and replaces it by either a translation along the reaction coordinate or a fully excited vibration [4.70], comparison with (4.2) leads to

$$k_m^0 = \kappa \cdot (kT/h)\,(f_+/f^m) \cdot N_s^{1-m}. \tag{4.27}$$

As

$$K^+ = \exp(-\Delta G_0^+/RT) = \exp[-(\Delta H_0^+ - T\Delta S_0^+)/RT], \tag{4.28}$$

where ΔH_0^+, ΔS_0^+, and ΔG_0^+ are the changes of standard enthalpy, entropy and free enthalpy, respectively, of formation of the transition state, (4.27) is equivalent to

$$k_m^0 = \kappa(kT/h)N_s^{1-m}\exp(\Delta S_0^+/R). \tag{4.29}$$

The meaning of "normal" k^0-values becomes obvious from this equation. Since $kT/h \simeq 2 \cdot 10^{12}$ to $2 \cdot 10^{13}$ sec^{-1} for the range 100–1000 K, $k_1^0 \approx kT/h \simeq 10^{13}$ sec^{-1} and $k_2^0 \simeq kT/h \cdot N_s \simeq 10^{-2}$ cm^2 sec^{-1} will be found, if $\kappa \approx 1$ and $f_+/f^m \approx 1$ (or $\Delta S^+ \approx 0$). The latter implies identical localization or configurational complexity for reactants and transition state. If the transition state is much more mobile than the reactants, k^0 will be larger than normal; a factor of up to 10^4 can be understood in this way. This explanation has been invoked for the large k_1^0-value of Xe on W (111) (Table 4.2) [4.51]. The large k_1^0-values for desorption of O atoms and of tungsten oxides from W (Table 4.3) have been explained by the assumption of complicated reactant complexes, but simple activated complexes; this will also lead to $\Delta S^+ \gg 0$ [4.59]. k^0-values below the normal ones must have the opposite meaning, i.e. a transition state which is less mobile or more complex than the reactants. Examples of calculations for different models are found, for instance, in [4.3, 59, 62, 71]; values up to 130 cal/K · mol have been obtained. Another possible explanation for very small k^0-values is a very small transmission factor [4.7, 8]. ARMAND et al. [4.72] have discussed the entropy changes in terms of the phonon properties of the surface. The coverage dependence of k^0 for detailed mechanisms can be evaluated statistically using the proper partition functions [4.3]; an interesting application has recently been given by GOYMOUR and KING [4.47].

For independent particle (first-order) desorption several other treatments have been developed. GOODMAN [4.73] has given a one-dimensional classical theory starting from Slater's theory of unimolecular decomposition [4.74]. He obtained k_1^0-values of the order of 10^{13} sec^{-1}

for several systems, in agreement with experimental results. In essence, this theory assumes an equilibrium distribution also.

This assumption is avoided in the "diffusion model" of chemical kinetics of KRAMERS [4.75] who used a classical Fokker-Planck equation for the movement of the representative point in phase-space for an analysis more general than that of EYRING. An important feature of this model is the coupling between the reactants and the heat bath (which in the case of desorption is the solid), which is represented by a friction coefficient η, equal to the inverse time constant for energy flow between heat bath and reactants. For intermediate η-values, the Eyring result is retained; for very high as well as very low η-values, compared to the characteristic frequencies, the pre-exponentials become much smaller than in the Eyring case, however. The physical meaning is that for low η, the coupling becomes so small that the rate-determining step becomes the energy transfer. In the usual treatment this could be accounted for by treating the energy transfer step as a separate step in the reaction sequence; the Boltzmann-distribution will then be essentially cut off at E_0. If η is large, even near the point of no return, the motion along the reaction coordinate is impeded, and low prefactors result, too.

SUHL and coworkers [4.76] have applied this treatment to surface reactions. They show that the coupling can contain contributions not only from phonons, but also from electrons; the latter will be strongest for metals. An estimate [4.77] suggests that under normal conditions both terms can be of comparable magnitude, but special conditions could lead to a predominance of one term, and also to an overall small coupling. Possibly some of the very low k_1^0 values in Table 4.2, which apply to weakly bound species, are due to very small η-values. SUHL [4.78] has also suggested that a strong change of k^0 could be expected under conditions of increased fluctuation amplitudes in the solid; this could, for instance, be the case near second-order phase transitions; he has cited some experimental evidence. Such effects could also be accounted for within transition state theory, but with a very small κ-value [4.7].

With a different diffusion approach, PAGNI and KECK [4.79] analyzed the time evolution of a non-equilibrium ensemble of adsorbed particles to obtain desorption kinetics. In that treatment, the two time constants of equilibration (energy transfer) and of relaxation of the surface population appear separately. They essentially find an Arrhenius equation with a slightly temperature-dependent prefactor, which leads to a compensation effect [4.61]. Their equations can fit a number of experimental results, but give an incorrect prediction for the case of Xe/W [4.51], for example.

The theoretical approaches discussed so far have all been basically classical treatments. Quantum mechanical approaches considering desorption due to phonon transfer from the solid to the bound adparticle have been given by LENNARD-JONES and coworkers [4.80] and recently by BENDOW and YING [4.81]. The Lennard-Jones theory was limited to one dimension and to one-phonon processes, which leads to unrealistic results (GOODMAN [4.82]). BENDOW and YING give a three-dimensional theory including multi-phonon processes; their treatment can be used to extract energy and angular distributions of desorbing particles. It is only applicable to situations, however, where the desorbing particle leaves the surface in one jump, i.e. where there are no bound vibrational levels of the adsorbate. For Ne desorbing from a layer of Xe on graphite at low temperatures, they deduce a temperature-dependence of the rate according to an Arrhenius law; at higher temperatures deviations occur because of the Bose distribution of phonons. Strongly peaked energy distributions (around E_0) and angular distributions (around the surface normal) are found. The prefactor is calculated to be $\approx 10^5$ sec^{-1}, which is of the order measured for some low temperature states (Table 4.2). Recently published results [4.52] for the desorption of He seem to agree with these predictions.

4.2. Electron Impact Desorption

Early reports about effects in ion gauges and mass spectrometer ion sources, which could only be explained by assuming that ions are emitted from surfaces under electron bombardment, were the first indication of this effect (for references see the reviews [4.83–86]). Closer investigation started in the early 60's: MOORE [4.87] showed mass spectrometrically that O^+ ions are emitted from oxygen layers on Mo surfaces under electron impact and found a direct proportionality between electron and ion current over 7 orders of magnitudes; REDHEAD [4.88] investigated the energy distribution of the O^+ emission from oxygen and CO on Mo by total current measurements and derived total cross sections; MENZEL and GOMER [4.89, 90] measured desorption cross sections for H, O, CO on W field emitters and used the effect for the differentiation between adsorption states. On the basis of these results, REDHEAD [4.88] and MENZEL and GOMER [4.89] independently proposed a mechanism which was able to explain qualitatively and, in part, quantitatively all observations, and which seems to be generally accepted today. Since then, a large number of investigations has appeared on this effect [4.85, 91]. In the following we shall describe first the main experimental methods and procedures, then summarize the general experimental results and give some examples, and then discuss their theoretical explanation. It will become obvious that EID has a threefold importance: as a basic physical process, as a probe for the investigation of adsorbates and as a disturbing feature in other surface work and in vacuum technology.

4.2.1. Experimental Methods of EID and Evaluation of Data

Basically, two different approaches exist to the investigation of EID. The particles liberated from a surface by electron impact are either detected directly (without or with mass separation and energy and angular resolution), or some parameter connected with the surface coverage is monitored and the change induced by electron impact is deduced from its time dependence. Thus the total ion current liberated has been measured with a simple collector assembly and its energy distribution determined by retarding grids, e.g. [4.88, 92, 93]; mass analysis of the ion current with small magnetic or quadrupole spectrometers has been used by many authors ([4.87, 94–97] and others). By combining retarding potential analysis and mass spectrometry [4.98, 99], or by using more sophisticated analysers [4.100], mass and energy analyses of ions have been combined. Neutral desorption caused by electron impact has been detected from the pressure rise, monitored mass spectrometrically [4.95, 101]. The velocity of desorbed particles was measured recently by time of flight methods [4.102, 103]. The indirect methods have mostly used the work-function change coupled with adsorbate coverage, measured by field emission [4.90, 91, 104, 105] or other methods [4.106]. The change of Auger intensity [4.107], of the electron reflection coefficient [4.108], of LEED structures and intensities [4.109, 110], and the gas left on the surface, as detected by flash desorption [4.111], have also been utilized in this way.

All these investigations have shown that the data can be analysed in terms of isolated desorption processes, i.e. the rate of desorption is proportional to the number of impinging electrons per unit area and time, $\dot{n}_e$, and to the number of adsorbed particles per unit area, N_i, of the species concerned

$$-dN_i/dt = q_i \dot{n}_e N_i, \tag{4.30}$$

where the proportionality constant q_i is the total cross section of removal of species i (see Fig. 4.6). Equation (4.30) can be split up into contributions of the different desorbing species (various ions and neutrals) and other processes like conversion to another adsorbed state or decomposition, so that

$$-dN_i/dt = -\sum_k (dN_i/dt)_k = \dot{n}_e N_i \sum_k q_{ik}, \tag{4.31}$$

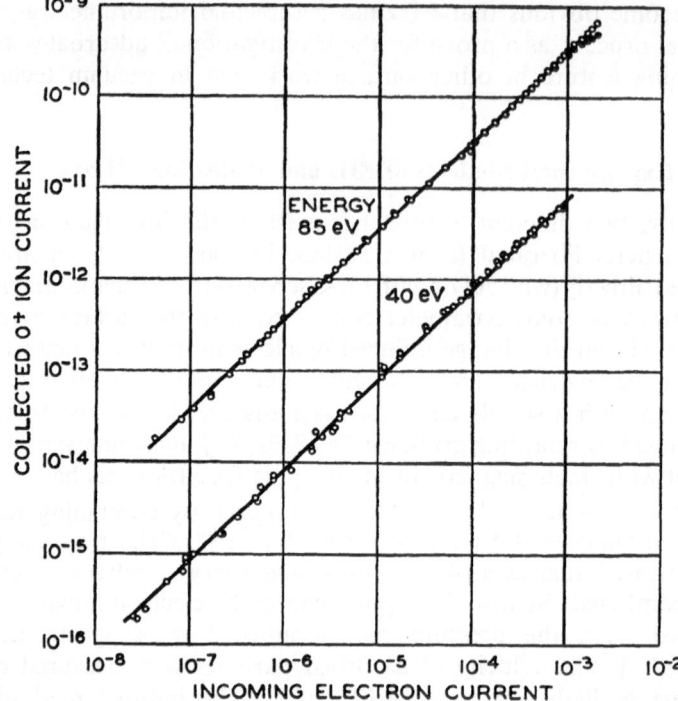

Fig. 4.6. Proportionality between electron current and EID ion signal for CO on Mo. (After [4.87])

where the q_{ik} are the partial cross sections for process k from state i. Equations (4.30) and (4.31) immediately suggest two basic uses:

a) If $\dot{n}_e$ is made large enough to change N_i measurably by EID ("high current mode"), then the time constant of this change can be used to measure the total cross section q_i. Because of the nature of (4.31), any quantity X related linearly to N_i can be used directly, for instance, the EID ion current or the work function:

$$N_i(t)/N_i(0) = X(t)/X(0) = \exp(-\dot{n}_e q t) = \exp(-i_e q t/\varepsilon), \qquad (4.32)$$

where ε is the electronic charge. The q_i-values given below have all been measured in this way (see Fig. 4.7).

b) If $\dot{n}_e$ is so small that N_i is essentially unchanged by EID ("low current mode"), the EID signal can be used to monitor the surface coverage during other processes (adsorption, thermal desorption etc.). This use of EID as a probe depends on the constancy of the q_{ik} within

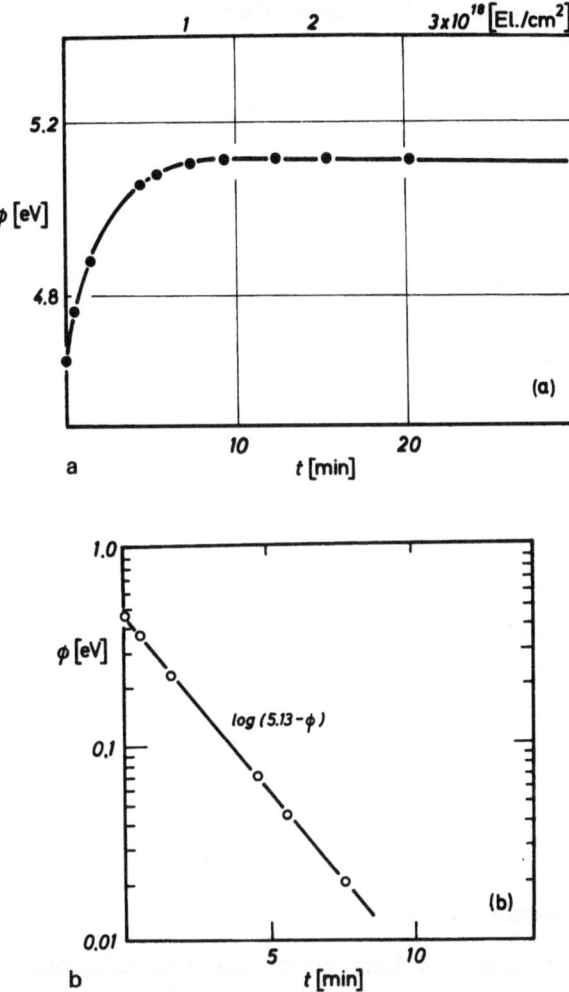

Fig. 4.7. (a) EID-induced work function change, measured in a field emission microscope, for removal of α-CO from W, and (b) plot according to (4.32). (After [4.90])

a given adsorbate state. q_{ik}-values cannot be measured as directly as q_i-values. The directly measured quantity is the desorption probability p_{ik} (number of particles per impinging electron), which is connected to q_{ik} via the coverage:

$$p_{ik} = \dot{n}_{ik}/\dot{n}_e = q_{ik} N_i .$$
(4.33)

Table 4.4. Some examples for EID products and cross sections

System	State	Product	Ionic cross section [cm^2]	Total cross section [cm^2]	Ref.
			(for 100 eV electrons)		
CO/W (poly)	virgin			2–$5 \cdot 10^{-19}$	[4.90]
	β			5–$8 \cdot 10^{-21}$	
	α			$3 \cdot 10^{-18}$	
CO/W (poly)	α_1/virgin	CO, CO$^+$	4–$8 \cdot 10^{-20}$	4–$8 \cdot 10^{-18}$	[4.95]
[mostly (100)]	α_2/β-prec.	CO, O$^+$	$5 \cdot 10^{-20}$	$1 \cdot 10^{-18}$	
	β	O$^+$		$< 10^{-21}$	
CO/W (110)	physisorbed			10^{-16}	[4.114]
	virgin			$7 \cdot 10^{-17}$	
	α			10^{-16}	
	β-prec.			$2 \cdot 10^{-17}$	
				$-2 \cdot 10^{-18}$	
	β			$< 10^{-21}$	
H$_2$/W (poly)	1			$4 \cdot 10^{-20}$	[4.89]
	2			$5 \cdot 10^{-21}$	
H$_2$/W (poly)	β_2	H$^+$	2–$6 \cdot 10^{-23}$	$5 \cdot 10^{-19}$	[4.115, 116]
	β_1	H$^+$	$< 10^{-25}$		[4.116]
	κ (on CO)	H$^+$, H$_2$	$8 \cdot 10^{-20}$	$1.2 \cdot 10^{-16}$	[4.8]
H$_2$/W (110)	β_2	H$^+$	$5 \cdot 10^{-22}$	$< 1.4 \cdot 10^{-18}$	[4.117]
O$_2$/W (poly)	1			$4.5 \cdot 10^{-19}$	[4.89]
	2			$< 2 \cdot 10^{-21}$	
O$_2$/W (110)	1	O$^+$	$3 \cdot 10^{-20}$	$3 \cdot 10^{-18}$	[4.97]
				$-7 \cdot 10^{-19}$	
O$_2$/W (100)		O$^+$		$2 \cdot 10^{-18}$	[4.118]
				$-4 \cdot 10^{-19}$	

4.2.2. Experimental Results

Most work to date has been concerned with adsorption layers on metals, although some papers on other solids have also been published (see the compilation in [4.85]). The main findings which have been obtained using the methods described are:

1) Impact of electrons with energy above about 10 eV often liberates ions and neutrals from adsorbed layers. The ionic contribution is only a few percent or less. Emission of excited neutrals has also been observed [4.92, 103].

2) The total and partial cross sections are usually much smaller than those of similar processes in molecules (excitation, ionization, dissociation). While the latter are typically between 10^{-16} and 10^{-15} cm^2 for 100 eV electrons [4.112], EID cross sections mostly lie between 10^{-17} and 10^{-22} cm^2. Exceptions to both limits are known.

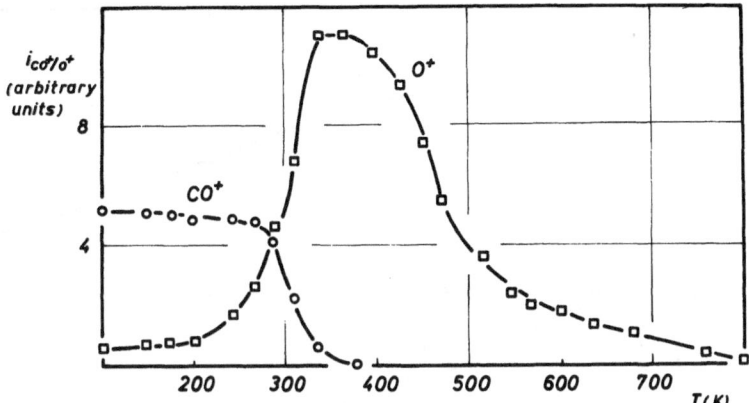

Fig. 4.8. EID of CO$^+$ and O$^+$ from CO on W as a function of temperature (low current mode). The two ionic species have been ascribed to emission from two substates of α-CO, whose populations can thus be probed. (After [4.95])

3) For different binding states of the same particle on the same surface, the q_i can be very different (e.g. for different states of CO on W, values between 10^{-17} and 10^{-21} cm^2, and for hydrogen on W, values between 10^{-23} and 10^{-16} cm^2 have been measured; see Table 4.4). In general they tend to be large for weakly adsorbed states and smaller for more strongly bound ones, although there are exceptions. For metallic adlayers, the q_i are immeasurably small ($<10^{-23}$ cm^2). Co-adsorption also has a strong influence [4.8, 105, 119–121].

The different binding states can also be characterized by EID of different particles. For instance, the weakly bound α-state of CO on W (see Chapter 3 and Section 4.1) was demonstrated by EID to consist of two substates, one of which emits CO$^+$ ions and the other O$^+$ ions ([4.95]; for a different interpretation see [4.114]), which can be interconverted by heating, as shown by EID in the low current mode (Fig. 4.8). This behavior was originally found on polycrystalline ribbons and has recently been reproduced on W (100) [4.122] and also (110) [4.114]; weakly adsorbed CO on other metals shows similar behavior [4.123, 124].

4) Besides desorption, electron impact can also cause a break-up of molecular adsorbates and conversions from one adsorbate state to another. For instance, EID of adsorbed CO leads to carbon or oxygen deposition, as has been shown by field emission microscopy [4.90], LEED [4.125] and Auger spectroscopy [4.126]. The two α-substates of CO on W can be interconverted by electron impact [4.95, 114, 122] (see Fig. 4.9); the minimum energy required for this conversion seems to be equal to that for neutral desorption [4.95].

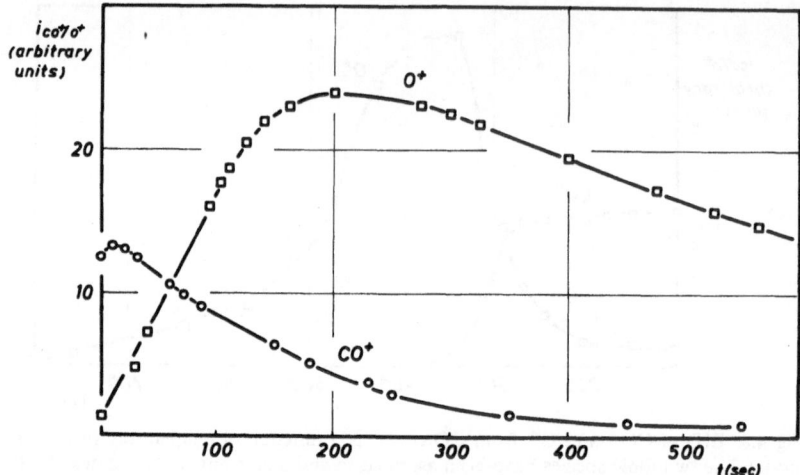

Fig. 4.9. EID of CO⁺ and O⁺ from CO on W at 100 K as a function of bombardment time (high current mode). The conversion is seen to be induced by electron impact also. (After 4.95])

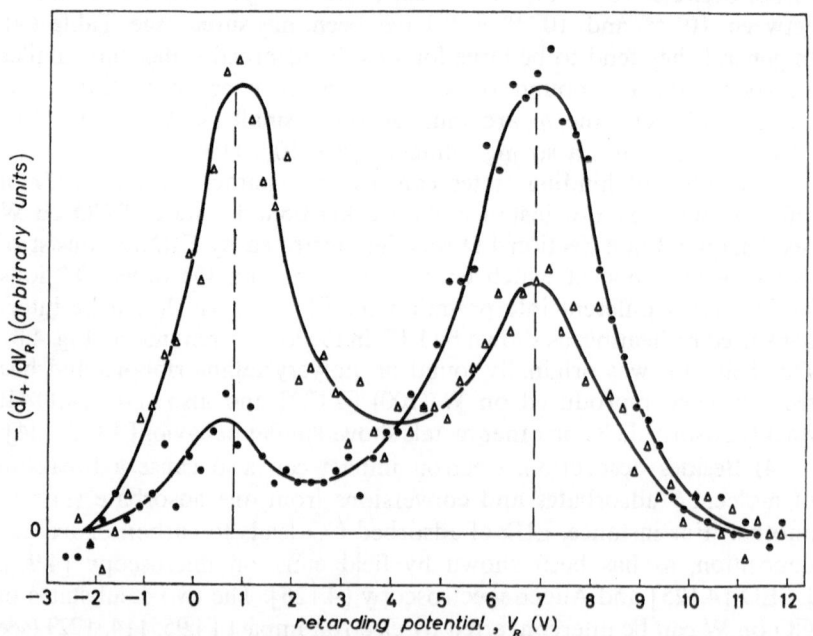

Fig. 4.10. Energy distribution of EID ions for CO on W at two different substrate temperatures (● 300 K; △ 195 K). (After [4.93])

5) EID ions have energies between 0 and $\sim 10\,\text{eV}$. The energy distribution can usually be represented approximately by Gaussian peaks with half widths between 2 and 4 eV. Multiple peak structures have been observed, which were explained by contributions from different coexisting adsorption states [4.92, 93]. For instance, the energy distributions shown in Fig. 4.10 have been found for CO on W for different conditions, which can be correlated to the substates mentioned in 3) and 4) and have been shown to correspond to CO^+ (low kinetic energy peak) and O^+ (high kinetic energy peak) emission [4.93].

6) There are strong isotope effects in the q-values in the sense that higher mass isotopes have smaller q; the effect increases with decreasing q [4.97, 127]. Temperature effects have also been reported [4.128–130].

7) The threshold energies, i.e. the minimum electron energies required for EID, seem to be lower for neutral than for ionic desorption [4.131], although some disagreement exists here [4.132]. For a discussion of the values of ionic thresholds, see e.g. [4.132]. The cross sections have no maxima around 100 eV as for isolated molecules, but keep rising slowly even above a few hundred eV [4.132]; this is probably due to the contribution of secondary electrons to EID [4.133].

These results can be understood in terms of the mechanism developed by REDHEAD [4.88] and MENZEL and GOMER [4.89], which will be discussed next.

4.2.3. Theory of EID

It is easy to show that elastic energy transfer by collisions between electrons and nuclei cannot be responsible for EID [4.89], so that electronic excitations must be the cause. Starting from the normal concepts of electron impact excitation of molecules and of essentially localized adsorbate bonds, the following two-step mechanism has been developed:

1) The adsorbate complex can be characterized by potential curves, as shown in Fig. 4.11. Electron impact causes a Franck-Condon transition from the ground state G to an excited state (repulsive or excited neutral, or ionic state). The cross section for this primary excitation is about that of comparable transitions in molecules. After this transition, the particle (ion or neutral) starts to move away from the surface.

2) During this movement, a recapture process can occur, which transfers the excitation energy into the solid and brings the particle back into this or another ground state (if multiple binding states exist). In the case of ionic desorption this recapturing transition can be approximated by a simple tunneling process (resonance or Auger, depending on the relative situation of the energy levels) of an electron from the

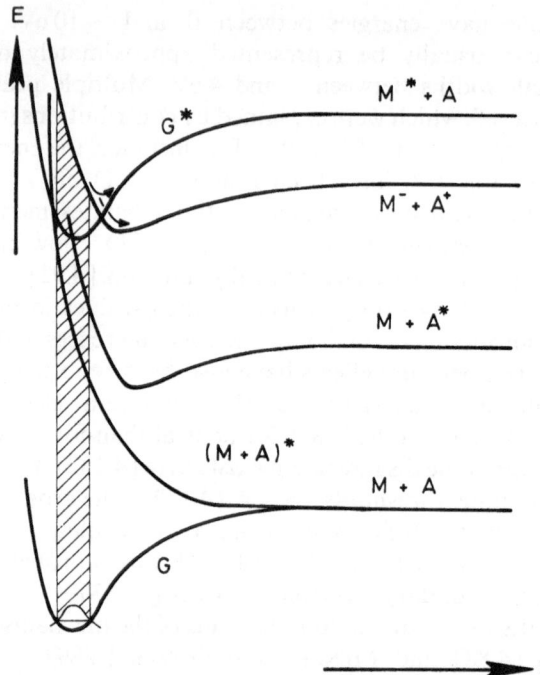

Fig. 4.11. Potential energy diagrams for an adsorbate system. [G: adsorbed ground state. $M^- + A^+$: ionic state. $(M + A)^*$: antibonding state. $M + A^*$: excited state of the adsorbate. $M^* + A$: Adsorbate ground state with excitation energy in the metal (vertically shifted replica of G).] The vibrational distribution in G and resulting ESD ion energy distribution is indicated.

solid into the hole created in the adsorbate complex. If this retunneling occurs after the moving ion has acquired a higher kinetic energy than necessary to surmount the ground state binding energy, neutralization without recapture will occur, so that a neutral particle desorbs. For neutral excitation the process is somewhat more complicated.

Qualitatively this model explains all the results mentioned above. The energy distribution of desorbing ions can be understood as a reflection of the vibrational distribution of the ground state at the ionic curve, skewed by the recapture process; different ground state curves will then lead to different peak energies. The small cross sections point to a high efficiency of the recapture processes, which can be understood in terms of the high density of states in the solid, and of a high transparency of the surface barrier. The unmeasurably small q_i of metallic adsorbates are understandable from the nature of the bond which is assumed to be due to essentially delocalized electrons, i.e. a strongly decreased surface barrier between solid and adparticle, so that tunneling is especially fast. Desorbing neutrals can be created both by direct excitation to an antibonding curve and by neutralization without recapture. Since the treatment below shows that the latter alone leads to more neutral than ionic desorption, the preponderance of neutrals is explained. The first mechanism must also exist, as shown by the existence of neutral EID below the ionic threshold.

The differences of the q_{ik}-values for different adsorbate states are then due to the strong (to first-order exponential) dependence of the recapture rate on the height and width of the barrier between solid and adsorbate. In a one-dimensional model, this should lead to increasing q_{ik} for decreasing bond strength, as the bond length should increase in the same direction, and a variation of the barrier height in the opposite direction is improbable. The true process will be three-dimensional, however; a different variation of total tunneling rate and adsorption energy is then easily envisaged. Such an effect has been invoked for the reverse order of q_i for β_1 and β_2 hydrogen on tungsten [4.116]; it is one of the possible reasons for the very interesting recent finding of strongly peaked angular distributions of EID ions (CZYZEWSKI et al. [4.134]). Since the total tunneling probability depends on the time the particle spends close to the surface after excitation, i.e. on its velocity and therefore its mass, a strong isotope effect is also expected. An increase of q_i with temperature can be understood in terms of population of other vibrational levels leading to an increase of the tunneling distance [4.128].

The quantitative one-dimensional formulation of these ideas [4.88, 89] leads to the following formulae. From a consideration of the ion flux *not* recaptured between excitation and total removal from the surface, the probability of desorption of an ion after excitation is given by

$$P_I(x) = \exp\left\{- \int_{x_0}^{\infty} R(x)\,dx/v(x)\right\}$$
$$= \exp\left\{-m^{1/2} \int_{x_0}^{\infty} R(x)\,dx/[2U(x_0) - 2U(x)]^{1/2}\right\},$$
(4.34)

where x is the distance from the surface and x_0 the equilibrium distance in the ground state, $R(x)$ is the tunneling probability as a function of distance x, and $v(x)$ the particle velocity, and $U(x)$ the repulsive potential. Similarly the total probability of desorption after ionic excitation is found by integrating the same formula between x_0 and a critical distance x_c which corresponds to the point at which the kinetic energy of the moving ion $(= U(x_0) - U(x_c))$ is equal to the adsorption energy.

Evaluation of the equations using

$$R(x) = A \exp(-ax)$$
(4.35)

and

$$U(x) = B \exp(-bx),$$ (4.36)

i.e. a normal exponential tunneling law and a Born-Mayer repulsive potential, showed semi-quantitatively the correct order of magnitude of the important effects (influence of change of x_0, preponderance of neutrals) [4.89].

The best test of this mechanism is the isotope effect, which should be governed by the appearance of $m^{1/2}$ in the exponential (4.34). Indeed, the experimental values obtained for $^{16}O/^{18}O$ ($q_{16}/q_{18} = 1.5$ [4.97]) and for different states of hydrogen on W (q_1/q_2 between 5 and 150 [4.115, 117, 127]) agree very well with this prediction.

This last finding is the strongest argument against another mechanism proposed by ZINGERMAN and ISHCHUK [4.106] who found periodic maxima and minima in the dependence of q on electron energy on single crystal faces and explained this by the assumption that the primary excitation occurs inside the metal and Bragg reflection decreases the number of these inelastic events. This mechanism cannot explain most of the other points listed above and is therefore probably not applicable.

Very recently, a quantum-mechanical treatment of ionic EID has been developed by BRENIG [4.135]. The process is not taken apart into two steps, but treated as a whole, and the quantum effects in the motion of the escaping particle are considered. This is accomplished by the use of a complex potential, whose imaginary, absorptive part takes care of recapture; the distorted-wave Born approximation is used. The treatment contains the classical two-step mechanism outlined above as the limiting case for large masses, as quantum effects then disappear. For smaller masses, it leads to a decreased overall tunneling rate, as compared to the classical case, as the wave packet representing the particle is distorted by the action of the imaginary potential and has a finite starting velocity in x_0. The isotope effect is unchanged, however.

4.2.4. Practical Importance of EID

This discussion was intended to show the different aspects of EID. EID can provide interesting insights into the nature of electronic levels of adsorbates and excitation and de-excitation processes involving them, because it entails such transitions and the transfer of electrons and energy across the surface barrier. It can therefore be used to test our concepts of these entities. It can further provide a probe for the differentiation and detailed investigation of complex adsorbate systems because of the approximate constancy of cross sections within one state and the large differences between different states. This last property, on the other hand, makes EID quite useless as a general surface analysis method and a tool for cleaning surfaces. It has some other important practical aspects, however. Since low energy electrons constitute *the* favorite probe for surface investigations because of their small mean free path in solids and the consequent high surface specificity (LEED, Auger spectroscopy, electron energy loss spectroscopy), it can cause disturbances in such measurements [4.136] which must be understood in order to be avoided. Since slow electrons are used in many vacuum systems (ion gauges, mass spectrometers) and other apparatus (accelerators, plasma machines), EID can be a disturbing effect in their operation, too [4.83, 137]; this can be very troublesome and is difficult to predict because of the possible large differences of cross sections, depending on circumstances.

4.3. Photodesorption

This short chapter is only concerned with photodesorption resulting from direct excitation of the adsorbate complex by light, i.e. effects which parallel those discussed for electron impact in Section 4.2. The discussion is therefore limited to metals, since the strong effects of illumination on adsorption and desorption from semiconductors and insulators are probably due to absorption of photons by the solid and subsequent transport of the excitation to the surface proper, so that the process is largely not a surface process. For a discussion of these effects, the reader is referred to [4.138].

Photodesorption from metals is a somewhat controversial subject. Some authors have reported clear-cut effects of various magnitude for CO on Ni [4.139–144], W [4.141, 145, 146], and other metals [4.143], while others have ascribed all effects to thermal desorption caused by heating of the metal by the light beam [4.147, 148]. Exclusion of such effects is not as easy as in EID, as in all these cases the photon energy was not sufficient to cause ionic desorption, and neutrals only were desorbed. Arguments were given, however, to support the electronic nature of photodesorption, at least for CO on W [4.146] and on stainless steel [4.149]. Comparison with neutral EID for CO/W [4.131] suggests that basically the same processes take place in the two cases. For the pure metals, the yields are quite small (less than 10^{-7} molecules/photon in the range up to $6\,\mathrm{eV}$); much larger values (above 10^{-3}) were found on stainless steel [4.149]. The measurement of high energy photodesorption, ionic and neutral, by the use of synchrotron radiation [4.150] would be worth-while both in connection with EID and in view of the importance of possible PD-effects in photoelectron spectroscopy of adsorbates, for electron accelerators [4.151], plasma machines and fusion reactor design [4.83].

Another interesting aspect is the possibility that interstellar molecules are produced by photodesorption from interstellar dust grains [4.152].

4.4. Ion Impact Desorption

The impact of ions or fast neutrals (including neutrons) can either knock off adsorbates directly by momentum transfer to them, or can induce collision cascades inside the solid which, after reflection back to the surface, can push off the adsorbate particles as the last member of the collision chain. These processes are essentially the same as sputtering of substrate material and contain large contributions from the bulk properties; the reader is referred to the papers in [4.153]. Besides their use for cleaning surfaces, these processes are important in Secondary Ion Mass Spectrometry [4.154] of adsorbate layers, as a disturbing effect in Ion Surface Scattering [4.155], for ion pumps [4.83], and for the wall problem of plasma machines and fusion reactor design [4.156]. An important question in this connection is the ionisation and excitation of the sputtered particles [4.157]. Conversely, the de-excitation of impinging ions or excited neutrals causes the emission of electrons or can lead to direct excitation of surface species, which in turn can lead to processes similar to those observed in EID. Few investigations of IID as confined to well-defined adsorbate layers and separated from normal sputtering, have become known so far [4.158].

4.5. Field Desorption

Desorption of adsorbed species under the action of very high electrostatic fields (of the order of 10^8 V/cm) was first observed by MÜLLER [4.159] and then by INGHRAM and GOMER [4.160]. As shown by subsequent work, the breaking of the surface bond occurs via a transition from the ground state to the field deformed ionic state. The basic process taking place is similar to field ionization [4.161], and can be understood from Fig. 4.12, which shows schematically the situation for a covalently bound, electronegative adsorbate ($I - \phi$ large). The presence of a strong electric field with positive polarity on the metal bends down the ionic curve so far that it would intersect the ground state curve, which to first approximation (i.e. neglecting polarization effects) is unaffected. Instead of inter-

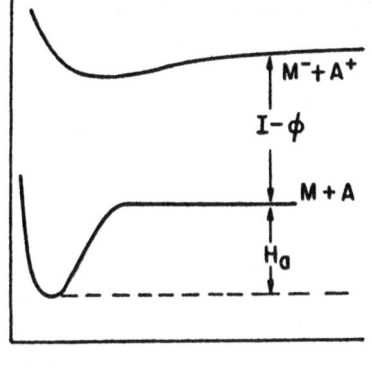
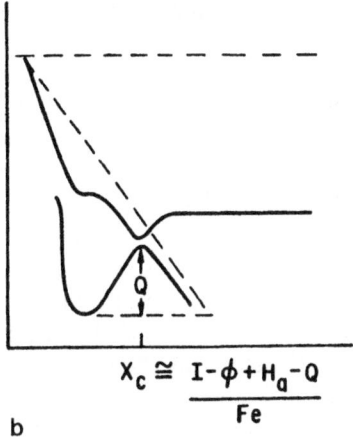

a

b

$$X_c \cong \frac{I - \phi + H_a - Q}{Fe}$$

Fig. 4.12a and b. Potential curves for field ionization of a covalently bound adsorbate. (a) Field-free case; ground state and ionic state. (b) Under high-field condition. The ionic curve is bent down; its intersection with the ground state curve produces a Schottky saddle. (After [4.167])

secting, the two curves separate, of course. An adiabatic transition from the bound (M + A) state to the free ion is then possible, if the activation energy Q (which is smaller than the thermal desorption energy) is available. If the field gets very high, the saddle can disappear altogether; tunneling through the barrier is also possible in some cases. In the case of small $I - \phi$, i.e. for ionic or polar bonding of metallic adsorbates, the applied electric field causes the ionic curve to be the lowest everywhere, and FD can again occur. In principle the same process, but at higher field because of the higher energies involved, can occur with the surface atoms of the substrate proper, and is then called field evaporation [4.161]. Both field desorption and evaporation are extensively used to clean field emitters [4.161]. Mass analysis of the desorbing species can be carried out in mass spectrometers [4.162] for the investigation of surface compounds [4.163]; the most sophisticated version is the "atom-probe", which has atomic spatial resolution by combination with a field ion microscope [4.164]. As the decisive parameters Q and x_c are clearly connected to the shape of the potential curves, measurements of FD can be used to derive the latter, e.g. [4.165]; the ground state desorption energy can be derived from determination of the conditions for $Q = 0$ [4.166]. One problem in these cases is that the products can be complex and have high charge states. The analysis then becomes difficult. The theory of field desorption has largely been developed by GOMER [4.167].

Enhancement effects on FD by the presence of field-ionizing inert gas atoms have been observed and can be explained by EID caused by the electrons liberated in the field ionization [4.168]. Adsorbed particles, on the other hand, can enhance the field evaporation of substrate atoms.

4.6. Conclusion

The field of desorption studies is seen to be interwoven with many aspects of adsorption studies in general, theoretical, experimental, and practical. While the physical mechanisms are not yet clearly understood in all cases and the experimental data still contain many uncertainties, the problems are now reasonably well defined; the experimental methods, though far from flawless, promise significant results, and the theoretical approaches become less dependent on doubtful assumptions. Important developments can, therefore, be expected in the near future, as for the general field of surface physics.

References

4.1.   See, for instance, K.J.LAIDLER: *Reaction Kinetics,* Vols. 1+2 (Pergamon Press, New York, 1963).
4.2.   R. GOMER: *Chemisorption* (to be published in the series *Solid State Physics,* Academic Press).
4.3.   D.O.HAYWARD, B.M.W.TRAPNELL: *Chemisorption* (Butterworths, London, 1964).
4.4.   G.E.MOORE, F.C.UNTERWALD: J. Chem. Phys. **40**, 2639 (1964); and references therein.
4.5.   E.A.GUGGENHEIM: Phil. Mag. **2**, 538 (1926);
       W.E.ROSEVAERE: J. Am. Chem. Soc. **53**, 1651 (1931);
       J.M.STURTEVANT: J. Am. Chem. Soc. **59**, 699 (1937).
4.6.   C. KOHRT, R. GOMER: J. Chem. Phys. **52**, 3283 (1970).
4.7.   L. A. PÉTERMANN: In *Adsorption – Desorption Phenomena,* ed. by F. RICCA (Academic Press, London, 1972), p. 227.
4.8.   W. JELEND, D.MENZEL: Surface Sci. **42**, 485 (1974).
4.9.   H. A.ENGELHARDT, D.MENZEL (to be published).
4.10. P.W.TAMM, L.D.SCHMIDT: J. Chem. Phys. **51**, 5352 (1969).
4.11. J.FRENKEL: Z. Physik **26**, 117 (1924).
4.12. F.L.HUGHES, H.LEVINSTEIN: Phys. Rev. **113**, 1029 (1959); and references therein.
4.13. J.B.HUDSON, J.S.SANDEJAS: J. Vac. Sci. Technol. **4**, 230 (1967).
4.14. J.PEREL, R.H.VERNON, H.L.DALEY: J. Appl. Phys. **36**, 2157 (1965).
4.15. M.D.SCHEER, J.FINE: J. Chem. Phys. **37**, 107 (1962); **38**, 307 (1963).
4.16. M.D.SCHEER, R.KLEIN, J.D.MCKINLEY: In *Adsorption – Desorption Phenomena,* ed. by F.RICCA (Academic Press, London, 1972), p. 169.
4.17. J.N.SMITH, JR., J.WOLLESWINKEL, J.LOS: Surface Sci. **22**, 411 (1970).
4.18. A.Y.CHO, C.D.HENDRICKS: J. Appl. Phys. **40**, 3339 (1969).
4.19. J.B.HUDSON, C.M.LO: Surface Sci. **36**, 141 (1973).
4.20. R.J.MADIX, J.M.SCHWARZ: Surface Sci. **24**, 264 (1971).
4.21. G.EHRLICH: J. Appl. Phys. **32**, 4 (1961).

4.22.   P. A. REDHEAD: Vacuum **12**, 203 (1962).
4.23.   F. URBACH: Sitzgsber. Akad. Wiss. Wien, Math.-Naturw. Kl., Abt. IIa; **139**, 363 (1930).
4.24.   R. J. CVETANOVIC, Y. AMENOMIYA: Advan. Catalysis **17**, 103 (1967).
4.25.   G. CARTER: Vacuum **12**, 245 (1962).
4.26.   G. EHRLICH: Advan. Catalysis **14**, 256 (1963).
4.27.   C. KOHRT, R. GOMER: Surface Sci. **24**, 77 (1971).
4.28.   E. LÜSCHER: Annals N.Y. Acad. Sci. **101**, 816 (1963);
        P. GODWIN, E. LÜSCHER: Surface Sci. **3**, 42 (1965).
4.29.   B. MCCARROLL: J. Appl. Phys. **40**, 1 (1969).
4.30.   J. P. HOBSON, J. W. EARNSHAW: J. Vac. Sci. Technol. **4**, 257 (1967).
4.31.   L. A. PÉTERMANN: In *Progress in Surface Science*, Vol. 3/1 (Pergamon Press, Oxford, 1973), p. 1.
4.32.   R. L. GERLACH, T. N. RHODIN: Surface Sci. **19**, 403 (1970).
4.33.   T. E. MADEY, J. T. YATES, JR.: *Structure et Propriétes des Surfaces des Solides*, Coll. CNRS No. 187, Paris 1970, p. 155.
4.34.   P. A. REDHEAD: Trans. Faraday Soc. **57**, 641 (1961).
4.35.   L. D. SCHMIDT: In *Adsorption – Desorption Phenomena*, ed. by F. RICCA (Academic Press, London, 1972), p. 341.
4.36.   C. PISANI, G. RABINO, F. RICCA: Surface Sci. **41**, 277 (1974).
4.37.   J. T. YATES, JR., T. E. MADEY: Surface Sci. **28**, 437 (1971).
4.38.   W. L. WINTERBOTTOM: J. Vac. Sci. Technol. **9**, 936 (1972).
4.39.   L. R. CLAVENNA, L. D. SCHMIDT: Surface Sci. **33**, 11 (1972).
4.40.   T. E. MADEY, J. T. YATES, JR.: J. Chem. Phys. **44**, 1675 (1966).
4.41.   L. R. CLAVENNA, L. D. SCHMIDT: Surface Sci. **22**, 365 (1970).
4.42.   P. T. DAWSON, Y. K. PENG: Surface Sci. **33**, 565 (1972).
4.43.   L. J. RIGBY: Can. J. Phys. **42**, 1256 (1964).
4.44.   C. G. GOYMOUR, D. A. KING: J. Chem. Soc. Faraday I, **69**, 736 (1973).
4.45.   M. P. HILL: Trans. Faraday Soc. **66**, 1246 (1970).
4.46.   T. TOYA: J. Vac. Sci. Technol. **9**, 890 (1972).
4.47.   C. G. GOYMOUR, D. A. KING: J. Chem. Soc. Faraday I, **69**, 749 (1973).
4.48.   D. L. ADAMS: Surface Sci. **42**, 12 (1974).
4.49.   J. R. ARTHUR, A. Y. CHO: Surface Sci. **36**, 641 (1973).
4.50.   D. A. KING, T. E. MADEY, J. T. YATES, JR.: J. Chem. Phys. **55**, 3236 (1971).
4.51.   M. J. DRESSER, T. E. MADEY, J. T. YATES, JR.: Surface Sci. **42**, 533 (1974).
4.52.   S. A. COHEN, J. G. KING: Phys. Rev. Letters **31**, 703 (1973).
4.53.   J. F. ANTONINI: Nuovo Cimento Suppl. **5**, 354 (1967).
4.54.   P. W. TAMM, L. D. SCHMIDT: J. Chem. Phys. **54**, 4775 (1971).
4.55.   V. J. MIMEAULT, R. S. HANSEN: J. Chem. Phys. **45**, 2240 (1966).
4.56.   K. F. POULTER, J. A. PRYDE: J. Phys. D **1**, 169 (1968).
4.57.   H. R. HAN, L. D. SCHMIDT: J. Phys. Chem. **75**, 227 (1971).
4.58.   J. LAPUJOULADE, K. S. NEIL: J. Chem. Phys. **57**, 3535 (1972).
4.59.   C. G. GOYMOUR, D. A. KING: J. Chem. Soc. Faraday I, **68**, 280 (1972).
4.60.   B. WEBER, J. FUSY, A. CASSUTO: J. Chim. Phys. **66**, 708 (1969).
4.61.   F. H. CONSTABLE: Proc. Roy. Soc. A **108**, 355 (1925);
        E. CREMER: Advan. Catalysis **7**, 75 (1955).
4.62.   J. LAPUJOULADE: Nuovo Cimento Suppl. **5**, 433 (1967);
        A. K. MAZUMDAR, H. W. WASSMUTH: Surface Sci. **30**, 617 (1972).
4.63.   R. MÜLLER, H. W. WASSMUTH: Surface Sci. **34**, 249 (1973).
4.64.   D. A. DEGRAS: Nuovo Cimento Suppl. **5**, 420 (1967).
4.65.   W. VAN WILLIGEN: Phys. Letters **28** A, 80 (1968).
4.66.   A. E. DABIRI, T. J. LEE, R. E. STICKNEY: Surface Sci. **26**, 522 (1971).

4.67.   T. L. BRADLEY, A. E. DABIRI, R. E. STICKNEY: Surface Sci. **29**, 590 (1972);
        T. L. BRADLEY, R. E. STICKNEY: Surface Sci. **38**, 313 (1973).
4.68.   R. L. PALMER, J. N. SMITH, JR., H. SALTSBURG, D. R. O'KEEFE: J. Chem. Phys. **53**, 1666 (1970).
4.69.   J. N. SMITH, R. L. PALMER: J. Chem. Phys. **56**, 13 (1972).
4.70.   S. GLASSTONE, K. J. LAIDLER, H. EYRING: *The Theory of Rate Processes* (McGraw-Hill, New York-London, 1941).
4.71.   S. KRUYER: Proc. K. Nederl. Akad. Wet. B **58**, 73 (1955).
4.72.   G. ARMAND, P. MASRI, L. DOBRZYNSKI: J. Vac. Sci. Technol. **9**, 705 (1972).
4.73.   F. O. GOODMAN: Surface Sci. **5**, 283 (1966).
4.74.   N. B. SLATER: Proc. Roy. Soc. A **194**, 112 (1948).
4.75.   H. A. KRAMERS: Physica **7**, 284 (1940).
4.76.   E. G. D'AGLIANO, W. L. SCHAICH, P. KUMAR, H. SUHL: In *Collective Properties of Physical Systems,* 24th Nobel Symposium 1973, p. 200 (Academic Press, New York-London, 1974).
4.77.   E. G. D'AGLIANO, P. KUMAR, W. SCHAICH, H. SUHL: To be published.
4.78.   H. SUHL, J. H. SMITH, P. KUMAR: Phys. Rev. Letters **25**, 1442 (1970);
        H. SUHL: To be published.
4.79.   P. J. PAGNI, J. C. KECK: J. Chem. Phys. **58**, 1162 (1973);
        P. J. PAGNI: J. Chem. Phys. **58**, 2940 (1973).
4.80.   J. E. LENNARD-JONES, C. STRACHAN: Proc. Roy. Soc. A **150**, 442 (1935);
        J. E. LENNARD-JONES, A. F. DEVONSHIRE: Proc. Roy. Soc. A **156**, 6, 29 (1936).
4.81.   B. BENDOW, S. C. YING: J. Vac. Sci. Technol. **9**, 804 (1972); Phys. Rev. B **7**, 622 (1973);
        S. C. YING, B. BENDOW: Phys. Rev. B **7**, 637 (1973).
4.82.   F. O. GOODMAN: Surface Sci. **24**, 667 (1971).
4.83.   P. A. REDHEAD, J. P. HOBSON, E. V. KORNELSEN: *The Physical Basis of Ultrahigh Vacuum* (Chapman and Hall, London, 1968).
4.84.   D. MENZEL: Angew. Chem. Intern. Edit. **9**, 255 (1970).
4.85.   T. E. MADEY, J. T. YATES, JR.: J. Vac. Sci. Technol. **8**, 525 (1971).
4.86.   J. H. LECK, B. P. STIMPSON: J. Vac. Sci. Technol. **9**, 293 (1972).
4.87.   G. E. MOORE: J. Appl. Phys. **32**, 1241 (1961).
4.88.   P. A. REDHEAD: Canad. J. Phys. **42**, 886 (1964).
4.89.   D. MENZEL, R. GOMER: J. Chem. Phys. **41**, 3311 (1964).
4.90.   D. MENZEL, R. GOMER: J. Chem. Phys. **41**, 3329 (1964).
4.91.   D. MENZEL: Surface Sci. **47**, 384 (1975).
4.92.   P. A. REDHEAD: Nuovo Cimento Suppl. **5**, 586 (1967).
4.93.   J. T. YATES, JR., T. E. MADEY, J. K. PAYN: Nuovo Cimento Suppl. **5**, 558 (1967).
4.94.   D. LICHTMAN, R. B. MCQUISTAN, T. R. KIRST: Surface Sci. **5**, 120 (1966).
4.95.   D. MENZEL: Ber. Bunsenges. Phys. Chem. **72**, 591 (1968).
4.96.   D. R. SANDSTROM, J. H. LECK, E. E. DONALDSON: J. Chem. Phys. **48**, 5683 (1968).
4.97.   T. E. MADEY, J. T. YATES, JR., D. A. KING, C. J. UHLANER: J. Chem. Phys. **52**, 5215 (1970).
4.98.   J. W. COBURN: Surface Sci. **11**, 61 (1968).
4.99.   T. E. MADEY, J. T. YATES, JR.: J. Vac. Sci. Technol. **8**, 39 (1971).
4.100. M. NISHIJIMA, F. M. PROPST: J. Vac. Sci. Technol. **7**, 420 (1971).
4.101. L. A. PÉTERMANN: Nuovo Cimento Suppl. **1**, 601 (1963).
4.102. R. CLAMPITT, L. GOWLAND: Nature **228**, 141 (1970).
4.103. I. G. NEWSHAM, D. R. SANDSTROM: J. Vac. Sci. Technol. **10**, 39 (1973).
4.104. W. ERMRICH: Nuovo Cimento Suppl. **5**, 582 (1967).
4.105. C. J. BENNETTE, L. W. SWANSON: J. Appl. Phys. **39**, 2749 (1968).

4.106. YA. P. ZINGERMANN, V. A. ISHCHUK: Fiz. tverdogo Tela **7**, 227 (1965); and **9**, 3347 (1967);
[Soviet Physics Solid State **7**, 173 (1965); and **9**, 2638 (1968)].

4.107. R. G. MUSKET: Surface Sci. **21**, 440 (1970).

4.108. D. A. Degras, J. LECANTE: Nuovo Cimento Suppl. **5**, 598 (1967).

4.109. J. ANDERSON, P. J. ESTRUP: Surface Sci. **9**, 463 (1968);
T. N. TAYLOR, P. J. ESTRUP: J. Vac. Sci. Techn. **10**, 26 (1973).

4.110. T. E. MADEY, D. MENZEL: Jap. J. Appl. Phys. Suppl. 2, pt. 2, 229 (1974).

4.111. J. T. YATES, JR., T. E. MADEY: In *Structure and Chemistry of Solid Surfaces,* ed. by G. A. SOMORJAI (Wiley, New York, 1969), p. 59-1.

4.112. H. S. W. MASSEY, E. H. S. BURHOP: *Electronic and Ionic Impact Phenomena*, Vols. 1 and 2. (Oxford University Press, 1969).

4.113. J. T. YATES, D. A. KING: Surface Sci. **38**, 114 (1973).

4.114. M. VASS, C. LEUNG, R. GOMER: To be published.

4.115. T. E. MADEY: Surface Sci. **36**, 281 (1973).

4.116. W. JELEND, D. MENZEL: Surface Sci. **40**, 295 (1973).

4.117. D. A. KING, D. MENZEL: Surface Sci. **40**, 399 (1973).

4.118. T. E. MADEY: Surface Sci. **33**, 355 (1972).

4.119. G. RETTINGHAUS, W. HUBER: J. Vac. Sci. Technol. **7**, 289 (1970).

4.120. W. P. GILBREATH, D. E. WILSON: J. Vac. Sci. Technol. **8**, 45 (1971).

4.121. D. P. WILLIAMS, R. P. H. GASSER: J. Vac. Sci. Technol. **8**, 49 (1971).

4.122. J. T. YATES, D. A. KING: Surface Sci. **32**, 479 (1972).

4.123. P. R. DAVIS, E. E. DONALDSON, D. R. SANDSTROM: Surface Sci. **34**, 177 (1973).

4.124. R. R. FORD, D. LICHTMAN: Surface Sci. **25**, 537 (1971).

4.125. H. H. MADDEN, J. KÜPPERS, G. ERTL: J. Chem. Phys. **58**, 3401 (1973).

4.126. J. P. COAD, H. E. BISHOP, J. C. RIVIÈRE: Surface Sci. **21**, 253 (1970);
J. M. MARTINEZ, J. B. HUDSON: J. Vac. Sci. Technol. **10**, 35 (1973).

4.127. W. JELEND, D. MENZEL: Chem. Phys. Letters **21**, 178 (1973).

4.128. D. MENZEL: Surface Sci. **14**, 340 (1969).

4.129. E. N. KUTSENKO: Zh. Tekhn. Fiz. **39**, 942 (1969).
[Sov. Phys.-Techn. Phys. **14**, 706 (1969)].

4.130. T. E. MADEY, J. T. YATES, JR.: J. Chem. Phys. **51**, 1264 (1969).

4.131. D. MENZEL, P. KRONAUER, W. JELEND: Ber. Bunsenges. Phys. Chem. **75**, 1074 (1971);
D. MENZEL: J. Vac. Sci. Technol. **4**, 810 (1972).

4.132. M. NISHIJIMA, F. M. PROPST: Phys. Rev. B **2**, 2368 (1970).

4.133. R. M. LAMBERT, C. M. COMRIE: Surface Sci. **38**, 197 (1973).

4.134. J. J. CZYZEWSKI, T. E. MADEY, J. T. YATES, JR.: Phys. Rev. Letters **32**, 777 (1974).

4.135. W. BRENIG: Z. Physik, to be published.

4.136. D. MENZEL: Surface Sci. **3**, 424 (1965).

4.137. P. A. REDHEAD: J. Vac. Sci. Technol. **7**, 182 (1970).

4.138. K. HAUFFE, S. R. MORRISON: *Adsorption* (De Gruyter, Berlin, 1974), p. 88, and references therein.

4.139. A. TERENIN, YU. SOLONITSIN: Disc. Faraday Soc. **28**, 28 (1959).

4.140. W. J. LANGE, H. RIEMERSMA: Trans. Am. Vacuum Soc. (1961), p. 167 (Pergamon, 1962).

4.141. W. J. LANGE: J. Vac. Sci. Technol. **2**, 74 (1965).

4.142. H. MOESTA, H. D. BREUER: Surface Sci. **17**, 439 (1969).

4.143. R. O. ADAMS, E. E. DONALDSON: J. Chem. Phys. **42**, 770 (1965).

4.144. P. GÉNÉQUAND: Surface Sci. **25**, 643 (1971).

4.145. H. MOESTA, N. TRAPPEN: Naturwiss. **57**, 38 (1970).

4.146. P. KRONAUER, D. MENZEL: In *Adsorption – Desorption Phenomena,* ed. by F. RICCA (Academic Press, London, 1972), p. 313.

142   D. Menzel

4.147. J. Paigne: J. Chim. Phys. **69**, 1 (1972).
4.148. J. W. McAllister, J. M. White: J. Chem. Phys. **58**, 1496 (1973).
4.149. G. W. Fabel, S. M. Cox, D. Lichtman: Surface Sci. **40**, 571 (1973).
4.150. J. Peavey, D. Lichtman: Surface Sci. **27**, 649 (1971).
4.151. G. E. Fischer, R. A. Mack: J. Vac. Sci. Technol. **2**, 123 (1965);
        M. Bernardini, L. Malter: J. Vac. Sci. Technol. **2**, 130 (1965).
4.152. L. T. Greenberg: In *Interstellar Dust and Related Topics*, ed. by J. M. Greenberg
        and H. C. Van de Hulst (D. Reidel, Dordrecht, 1973), p. 413.
4.153. *Ion-Surface Interactions, Sputtering and Related Phenomena*, ed. by R. Behrisch,
        W. Heiland, W. Poschenrieder, P. Staib, H. Verbeek (Gordon and Breach,
        London, 1973).
4.154. A. Benninghoven: Appl. Phys. **1**, 3 (1973).
4.155. D. P. Smith: Surface Sci. **25**, 171 (1971).
4.156. R. Behrisch: Nucl. Fusion **12**, 695 (1972).
4.157. A. Benninghoven: Z. Physik **220**, 159 (1969);
        J. M. Schroeer, T. N. Rhodin, R. C. Bradley: Surface Sci. **34**, 571 (1973).
4.158. A. Benninghoven, E. Loebach, N. Treitz: J. Vac. Sci. Technol. **9**, 600 (1972);
        S. K. Erents, G. M. McCracken: J. Appl. Phys. **44**, 3139 (1973);
        H. F. Winters, P. Siegmund: J. Appl. Phys. **45**, 4760 (1974);
        W. Heiland, E. Taglauer: Paper at DECHEMA-Symp. Frankfurt, 1974.
4.159. E. W. Müller: Phys. Rev. **102**, 618 (1956).
4.160. M. G. Inghram, R. Gomer: Z. Naturforsch. **10**a, 863 (1955).
4.161. R. Gomer: *Field Emission and Field Ionization* (Harvard University Press, Cam-
        bridge, Mass., 1961);
        E. W. Müller, T. T. Tsong: *Field Ion-Microscopy* (Elsevier, New York, 1969).
4.162. H. D. Beckey: *Field Ionization Mass Spectrometry* (Vieweg, Braunschweig, 1969);
        H. D. Beckey: J. Mass Spectrometry Ion Phys. **2**, 500 (1969).
4.163. J. H. Block: Adv. Mass Spectrometry **4**, 791 (1968);
        J. H. Block: J. Vac. Sci. Technol. **7**, 63 (1970).
4.164. E. W. Müller, J. A. Panitz, S. B. McLane: Rev. Sci. Instr. **39**, 83 (1968).
4.165. L. W. Swanson, R. Gomer: J. Chem. Phys. **38**, 2813 (1963).
4.166. G. Ehrlich: Disc. Faraday Soc. **41**, 7 (1968);
        E. W. Plummer, T. N. Rhodin: J. Chem. Phys. **49**, 3479 (1968).
4.167. R. Gomer: J. Chem. Phys. **31**, 341 (1959);
        R. Gomer, L. W. Swanson: J. Chem. Phys. **38**, 1613 (1963).
4.168. K. D. Rendulic, Z. Knor: Surface Sci. **7**, 204 (1967);
        D. W. Bassett: Brit. J. Appl. Phys. **18**, 1753 (1967);
        A. E. Bell, L. W. Swanson, D. Reed: Surface Sci. **17**, 418 (1969).

5. Photoemission and Field Emission Spectroscopy

E. W. PLUMMER

With 31 Figures

A surface may be viewed as a distinguishable phase of matter [5.1] with physical and chemical properties which differ from the bulk and from the gas phase molecular properties [5.2]. The termination of the bulk at a surface creates a new boundary condition with a reduced coordination number allowing for both geometrical and electronic rearrangements. To minimize the total energy the surface atoms may rearrange themselves into sites which are not characteristic of the bulk. In the simplest case this may just be a dilation or contraction of the distance between surface plane and the next plane below, while in extreme cases the two dimensional periodicity of the surface may differ from the bulk. The electronic properties of the surface must reflect more localized character than in the bulk, including the possibility of surface states or surface resonances. Clean surfaces may have very localized orbitals equivalent to the "dangling bonds" of molecular chemistry and the bonds formed when a foreign atom or molecule is adsorbed may be localized in a "surface complex". The electron spectroscopy techniques discussed in this paper attempt to characterize the electronic energy level spectrum of a clean surface which has interacted with a foreign atom or molecule. The philosophy is that all interactions of a surface with its surroundings are primarily electronic in nature and must depend upon the distribution of electronic states both in energy and space near the surface.

It should be understood that these techniques are not the whole answer. Even though the geometrical arrangement of the atoms and the electronic energy levels are inescapably intertwined, electron spectroscopies are normally not very useful in determining the geometry. Also, with the exception of X-ray induced photoemission it is very difficult to determine the atomic species present on the surface, since the valence levels depend more upon the bonding configuration than the atomic identity of the constituents. Finally the relation between what is measured and the information desired, namely the ground state configuration of the system depends on the specific spectroscopic technique used, and requires careful analysis. This will be the subject of Section 5.2 which tries to elucidate what we do or do not know about the measurement processes for field emission and photoemission spectroscopy.

There is a pragmatic approach to utilizing electron spectroscopy commonly referred to as the "finger print" technique. It assumes that every bonding configuration will have its own characteristic spectrum so that, when all possible bonding configurations of a given system have been catalogued separately, the state or states can be determined. For example, if one wanted to study the decomposition of a hydrocarbon on a given surface, the spectra of adsorbed H, C, and all decomposition products of the hydrocarbon would be separately catalogued. Then the spectrum of the adsorbed hydrocarbon would be recorded as a function of the parameters of the experiment. The latter spectrum would then be compared to the spectra of the individual components in order to ascertain the nature of the surface molecule at the various stages of the experiment. This is a powerful technique if the spectra of individual species present are additive. It avoids many of the questions of the response of the system to the probe, since one makes no pretense of making absolute measurements. In Subsection 5.5.2 this "finger print" technique will be applied to representative examples of molecular dissociation, decomposition, and multiple surface binding states.

5.1. Preliminary Discussion of Field and Photoemission

Several electron emission processes have been used to study surfaces. In all of these the emitted electrons are energy analyzed so that the energy of the final state is determined, in contrast to an absorption experiment where the energy difference between the initial and final states is measured. The objective then is to infer the initial state from the measured final state energy and the characteristics of the probe. Three of the most widely used techniques are: 1) Ion neutralization spectroscopy, [5.1, 3] where a low energy ion beam is the probe, 2) field emission spectroscopy, where an applied field is the external probe [5.4, 5], and 3) photoelectron spectroscopy, where an incident beam of photons is the probe [5.6, 7]. The energy of the incident photon can range from the ultraviolet [5.6] to the X-ray [5.7]. This article will discuss the capabilities and limitations of the latter two techniques. HAGSTRUM [5.1, 8] has described in detail the comparison of ion neutralization and photoemission. No direct experimental comparisons between ion neutralization and field emission have been made even though these two techniques should be very similar [5.1]. GOMER has also compared photoemission and field emission [5.5] and articles by BRUNDLE [5.9] and MENZEL [5.10] discuss the capabilities of ultraviolet and X-ray induced photoemission.

SCREEN WITH PROBE HOLE

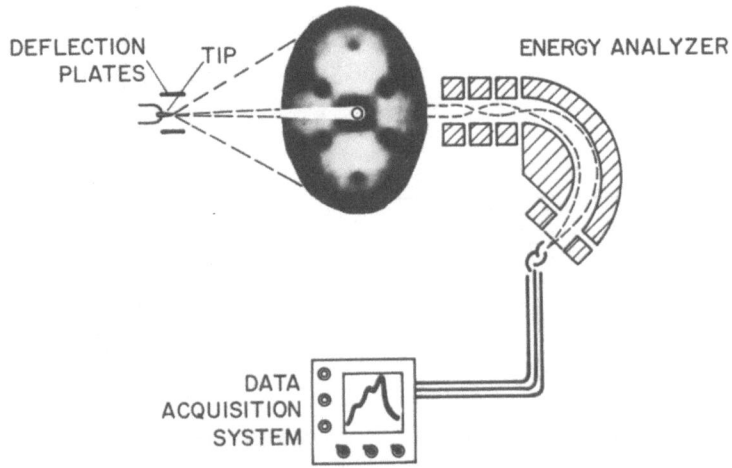

Fig. 5.1. Schematic drawing of a field emission microscope adapted for energy distribution measurement. The (110) plane is positioned over the probe hole

5.1.1. Field Emission

Field emission [5.11, 12] consists of the tunneling of electrons from a solid through the classically forbidden barrier region when the latter is deformed by the application of a strong electrostatic field, $3–6 \times 10^7$ V/cm. In order to achieve these high fields at reasonable voltages the cathode or emitter is usually etched to a sharp point (~ 1000 Å in radius) so that several kilovolts will produce the desired field. MÜLLER [5.11] realized in 1937 that a greatly enlarged ($> 10^6$) image of the spatial distribution of the tunneling electrons could be projected onto a fluorescent screen from the hemispherical emitter. Because of its small size the emitter is part of a single crystal and thus exposes all crystallographic orientations, so that individual crystal planes can be located and identified in the field emission pattern observed on the fluorescent screen. The emission characteristics of any crystal plane may be studied by placing a small "probe hole" in the screen and deflecting the field emission pattern with electrostatic deflection plates [5.13] until the plane of interest is over the probe hole. Figure 5.1 is a schematic drawing of this arrangement, showing the field emitter (tip) on a support loop and the projected field emission pattern of clean tungsten on the screen. In Fig. 5.1 the (110) plane is positioned over the probe hole and the electrons passing through it are being energy analyzed. The field dependence of the total current

a

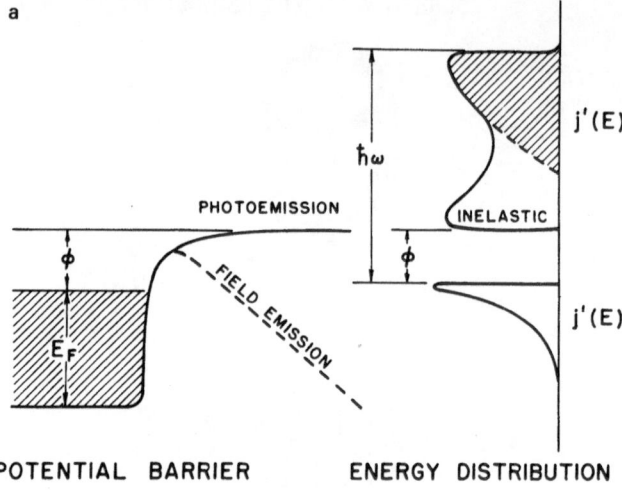

POTENTIAL BARRIER   ENERGY DISTRIBUTION

b

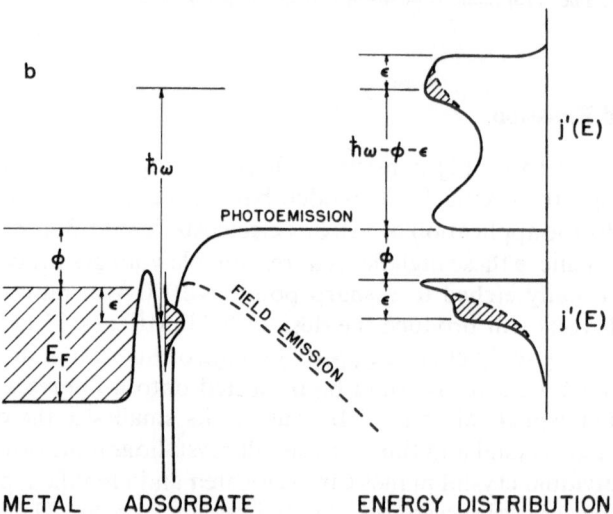

METAL   ADSORBATE   ENERGY DISTRIBUTION

Fig. 5.2. (a) On the left is a schematic representation of the surface potential of a free electron metal, with and without the application of an external field. On the right a typical energy distribution for: 1) electrons which have tunneled elastically from the metal under the application of the applied field, and 2) electrons which have been photoexcited by radiation of energy $\hbar\omega$. (b) Schematic presentation of the surface potential with an idealized adsorbed atom present, both with and without an applied field. On the right are the resultant energy distributions for field emission (bottom) and photoemission (top). The shaded areas of the energy distribution depict the increase in current coming from the "virtual level" of the adsorbate, shown at an energy ε below the Fermi energy on the left

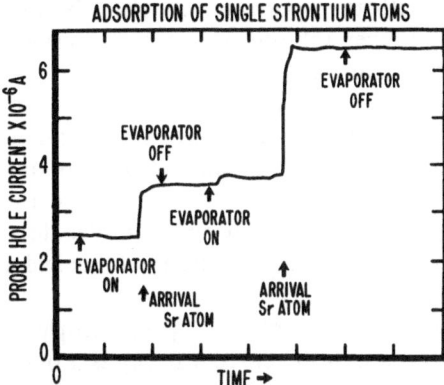

Fig. 5.3. Probe hole current (see Fig. 5.1) vs. time as a strontium source is switched on and off. The step increases in current occur when a single atom arrives on the surface being viewed [5.16]

can also be measured to obtain work function changes. These types of experiments have been reviewed recently by GADZUK   and PLUMMER [5.4], SWANSON and BELL [5.14], and GOMER [5.5].

In Fig. 5.2a we show on the left a very schematic drawing of the potential at a metal surface with an electric field applied. On the right is a typical field emission energy distribution, decreasing exponentially as the energy decreases and the barrier becomes wider and higher. The energy distribution is cut off near the Fermi energy by the Fermi-Dirac distribution function. The shape of the field emission energy distribution when viewed on a linear plot, as shown in Fig. 5.2a, is nearly impervious to any changes in the emitter or the external field, and is always ex- ponential in shape. This is a consequence of the tunneling through the field-induced barrier, which may be calculated since the barrier is almost entirely in the vacuum. The measured energy distribution with this exponential tunneling probability removed is extremely sensitive to the electronic properties of the surface [5.4]. Since all electrons which tunnel must originate at the surface, the properly analyzed field emission energy distribution will measure some function of the surface density of states. In Subsection 5.2.1 it will be shown that this function is the one- dimensional density of states evaluated at the classical turning point in Fig. 5.2a.

On the left of Fig. 5.2b we depict adsorption of a foreign atom onto the schematic surface shown in Fig. 5.2a. The bell-shaped curve in the region of the adsorbate centered at an energy ε below the Fermi energy is intended to represent the local density of states or "virtual level" on the adsorbate. This adsorbate level, sharp in the isolated atom, is broadened by interaction with the substrate, i.e., the lifetime of an electron on the adatom is no longer infinite. Duke and Alferieff [5.15] showed that tunneling through an adsorbate is equivalent to a tunneling resonance: when an electron tunneling from the solid has the same energy as a bound state in the adsorbate, resonance will occur increasing the tunneling probability by 10^2–10^4. This enhancement in tunneling probability as a function of energy can be used to measure the local density of electronic states on the adsorbed atom or molecule [5.4, 5]. Figure 5.3 illustrates quite dramatically the sensitivity of this technique for detecting single adsorbed atoms [5.16].

The limitations of field emission spectroscopy are two-fold: first, the exponential decrease in the current imposes a practical range limitation of 2–3 eV below the Fermi energy, and, secondly, there are a limited number of materials which can be prepared and cleaned for use as emitters.

5.1.2. Photoemission

While field emission is basically an elastic process, photoemission is an excitation process. An incoming photon of energy $\hbar\omega$ raises an electron to an excited state, $\hbar\omega$ eV above the initial state. If the excited electron has sufficient energy, is headed in the right direction and does not lose too much energy, it may escape from the solid to be subsequently energy analyzed. A schematic energy distribution showing both the elastically and inelastically scattered electrons is presented on the top right of Fig. 5.2a. One of the fundamental problems in interpreting photoemission data is our inability to separate the primary unscattered photoelectrons from inelastically scattered ones. The incoming photon beam has an extinction distance in the solid of ~ 100 Å in the ultraviolet region, varying with material and photon energy. The surface sensitivity of photoemission is a consequence of strong inelastic scattering in the excited state. The escape depth for an electron excited to 50 eV above the vacuum level is 2–5 Å. Figure 5.4 shows a plot of the optical absorption depth for normally incident light on tungsten [5.17] and the mean-free path before inelastic scattering of the excited electron as a function of energy. Several authors have compiled lists of the available data on electron escape depths [5.9, 18, 19] which, within experimental error, fall on the general curve shown in Fig. 5.4 for energies $> \sim 20$ eV. At energies < 20 eV the escape depth should increase depending upon the specific material, so that no generalization can be made for this range. Figure 5.4 also shows experimental escape depths for W [5.20], Ni [5.21], and Al [5.19] since these materials will be considered specifically in Section 5.4.

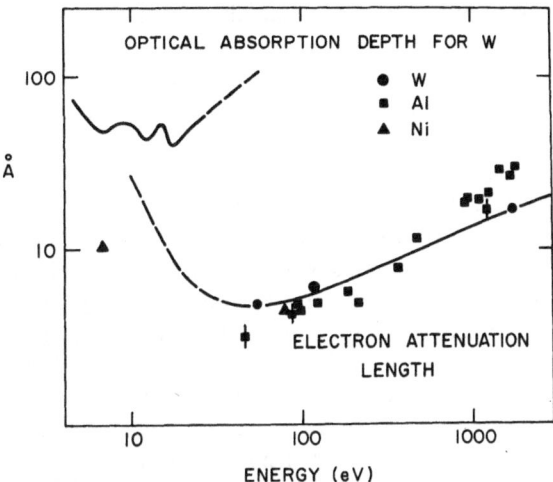

Fig. 5.4. Attenuation lengths for incoming radiation and escaping electrons as a function of energy. The top curve represents the absorption depth for tungsten from the data of JUENKER et al. [5.17]. The bottom curve indicates the excited electron attenuation length. The data for W¹ˢ is from TARNG and WEHNER [5.20], the Al data is from TRACY [5.19], and the Ni data is from EASTMAN [5.21] and a tabulation by BRUNDLE [5.9]

If the escape depth is 5 Å and the surface has the same optical excitation probability as the bulk, then ~30% of the signal will originate from the surface layer. The strong attenuation of excited electrons which is responsible for the surface sensitivity, also complicates the interpretation of the data. Even in the naive model of photoemission, which assumes that the energy distribution of primary electrons measures the density of initial states, the data can only be interpreted if the bulk and surface densities of states are identical. In the more probable case where the density of states varies layer by layer the signal will be a weighted sum of contributions for the different layers [5.22]. Inclusion of the excitation matrix element complicates matters even more by creating interference effects between the surface and bulk signals (see Subsection 5.4.5), making it nearly impossible to separate the bulk from the surface signal in a measured energy distribution.

When a foreign atom or molecule is adsorbed on the surface, an increase in signal would be expected at a kinetic energy corresponding to a level of the surface molecule, as shown in Fig. 5.2b, for an idealized case. Such bonding also causes a redistribution of energy levels in the substrate, leading to changes in the distribution not shown in the simple picture of Fig. 5.2b. When an energy level of a surface molecule or atom is below the bottom of the bulk band and hence localized the interpreta-

tion becomes somewhat easier, since there are then only weak interference effects with the bulk. A convenient way of displaying the changes
in the surface properties is to plot the difference between energy distributions taken before and after a change in surface conditions. This will be
illustrated in Section 5.3.

All atoms and molecules except hydrogen have core levels not
involved in bonding, but the kinetic energy of photoelectrons from these
levels may shift in response to the bonding of the valence electrons.
The shift in the core levels arises from charge transfer in bonding; energy
shifts of the emitted electrons can also be caused by relaxation effects
in the surface molecule during the excitation process (see Subsection
5.2.2). When X-rays are used for the excitation, the process is called
ESCA (electron spectroscopy for chemical analysis), [5.23] or XPS.
By measuring the energies of electrons excited from core levels of adsorbed
atoms one should be able to distinguish between different bonding
configurations [5.9, 24].

Section 5.2 is aimed at readers interested in details of the measurement
processes. For those readers who only want to know what can be done
with these techniques, the present section has been written in enough
detail to permit omission of Section 5.2.

5.2. The Measurement Process

This section will be devoted to an analysis of the measurement process
in field emission and photoemission. What properties of the system are
measured by an energy distribution of the emitted electrons? There is no
definite answer to this question at the present time, and much of the
following discussion will be concerned with what is unknown. In Subsection 5.5.2 the essential ingredients of any microscopic model of
surface emission will be discussed: excitation probability, relaxation
effects, and interference effects with the bulk.

5.2.1. Field Emission

Field emission was interpreted very early as quantum mechanical
tunneling. In 1928 FOWLER and NORDHEIM [5.25] calculated the emitted
current from a free electron metal with a work function φ, terminated
by a step barrier at the surface and a constant electric field outside.
NORDHEIM [5.26] modified this calculation to include the rounding of
the triangular barrier by the image potential. Their expression for field
emitted current, known as the Fowler-Nordheim equation has the form

$$\ln(i/F^2) = \ln A - 6.83 \times 10^7 \, \varphi^{\frac{3}{2}} s(y)/F \,, \tag{5.1}$$

where i is current density, F the applied field in V/cm, φ the work function in eV, $s(y)$ an image correction term close to unity, and $A = 6.2 \, 10^6 \times (E_F/\varphi)^{\frac{1}{2}}(E_F + \varphi)^{-1}$, E_F being the Fermi energy measured from the bottom of the conduction band. There has been considerable effort subsequent to their work to include effects of band structure and of adsorbed atoms or molecules [5.4, 5, 14]. Yet some fifty years after the original paper of FOWLER and NORDHEIM there still exists some controversy over what is being measured, especially for clean surfaces.

5.2.2. Clean Surfaces

The entire field of tunneling has been very carefully reviewed by DUKE [5.27], who pointed out that density of states effects in a normal metal tunnel junction are not readily observable. HARRISON [5.28] calculated the tunneling current in a non-superconducting junction, using an effective mass approximation with WKB tunneling and concluded that the tunneling current does not depend upon the density of states in any direct way. His formulation of tunneling has been adapted to field emission by most theoreticians. The work of STRATTON [5.29] and ITSKOVITCH [5.30] is the most notable using this formulation. GADZUK [5.4] has written a comprehensive review of the efforts in this area. In contrast to HARRISON [5.28], APPELBAUM and BRINKMAN [5.31] concluded from a calculation of interface effects in normal metal tunnel junctions that, "we presently know that tunneling measures the spectral function of the electrode in the vicinity of the metal-barrier interface". PENN [5.32] applied the BARDEEN [5.33] version of the OPPENHEIMER transfer Hamiltonian approximation [5.34] specifically to the case of field emission tunneling. He concluded that the field emission energy distribution can measure the "normal local density of states near the surface". Also, DUKE and FAUCHIER [5.35] calculated the field emission energy distribution for an exactly soluble one-dimensional Kronig-Penny model and found structure which can be related to the surface density of states [5.36].

The available experimental evidence supports the latter point of view. When the exponential nature of the energy distribution is divided out of a measured energy distribution from a single crystal face of the emitter, the resultant curve is usually rich in structure [5.37, 38], and this is very sensitive to the surface conditions [5.4]. In the following paragraphs the conventional application of HARRISON's [5.28] results will be described and an attempt made to show the source of the discrepancy between the two sets of calculations. Finally, the transfer Hamiltonian approach as applied by PENN [5.32, 39] will be described.

The following is based on an excellent paper by POLITZER and CUTLER [5.40] published in 1970, which assumed: 1) a one-dimensional surface potential and 2) the surface to be a perfect crystal plane with two-dimensional periodicity, so that $k_{||}$, the crystal momentum parallel to the surface, is a good quantum number and is conserved upon reflection (and transmission).

The amount of charge crossing a unit area per unit time outside the solid (to the right in Fig. 5.2a) between energy E and $E + dE$ is given by

$$j'(E)\, dE = \frac{2e}{(2\pi)^3}\, f(E) \int\int\int D(E, k_{||})\, V_z\, d^3 k . \tag{5.2}$$

Spin degeneracy which is assumed accounts for the factor of 2. The question of spin polarization will be addressed later. $f(E)$ is the equilibrium Fermi-Dirac distribution. V_z is the group velocity perpendicular to the surface and $D(E, k_{||})$ is the transmission probability for an electron of energy E and reduced transverse wave vector $k_{||}$. The volume integral in k space is over the region between the constant energy surfaces defined by E and $E + dE$. Equation (5.2) is quite easy to understand; it says: integrate over all states deep in the solid between energy E and $E + dE$ the product of the group velocity (to obtain the supply of electrons) and the transmission probability to obtain the external current.

The transmission probability D is defined as the ratio of the current density of the outgoing wave as $z \to + \infty$ (far to the right in Fig. 5.2a) to the current density of the incoming wave as $z \to - \infty$ (far to the left in Fig. 5.2a), and must be evaluated by solving the wave equation across the barrier. The approximations used for D lead to the apparent discrepancy in the different theories discussed at the beginning of this section.

By using the identity

$$V_z = \frac{1}{\hbar}\, \frac{\partial E(k)}{\partial k_z} \tag{5.3}$$

for the group velocity the integral in (5.2) can be converted to a surface integral over the constant energy surface E, and one obtains for the density in energy

$$j'(E) = \frac{2e\, f(E)}{\hbar (2\pi)^3} \int\int \frac{D(E_1 k_{||}) \frac{\partial E}{\partial k_z} dS}{|V_k E|} . \tag{5.4}$$

Let θ be the angle between the surface normal for the incremental surface area dS and the z axis (normal to the interface), and ds its projection on the plane in k space parallel to the interface, $ds = dS \cos \theta$. Since $\hat{z} \cdot \nabla_k E = |\nabla_k E| \cos \theta = \partial E / \partial k_z$ we have

$$j'(E) = \frac{2e f(E)}{\hbar (2\pi)^3} \int \int D(E_1 \boldsymbol{k}_{||}) \, ds \,, \tag{5.5}$$

where the integral is over the projection of the constant energy surface E onto the k_x, k_y plane, i.e. the "shadow" on the k_x, k_y plane of the surface of constant energy E.

Since we have assumed that $k_{||}$ is conserved in crossing the interface the integral in (5.5) can be evaluated using the values of $k_{||}$ outside the barrier where $E_{||} = \hbar^2 k_{||}^2 / (2m)$. Converting to polar coordinates we have either

$$j'(E) = \frac{2e f(E)}{\hbar (2\pi)^3} \int_0^{2\pi} d\varphi \int_{k_{||}^{\min}(E, \varphi)}^{k_{||}^{\max}(E, \varphi)} D(E, \boldsymbol{k}_{||}) k_{||} \, dk_{||} \tag{5.6a}$$

or

$$j'(E) = \frac{2em f(E)}{\hbar^3 (2\pi)^3} \int_0^{2\pi} d\varphi \int_{E_{||}^{\min}}^{E_{||}^{\max}} D(E, E_{||}) \, dE_{||} \,, \tag{5.6b}$$

where the limits on the second integral are the extremes of the projected energy surface on the k_x, k_y plane in k space. The limits on (5.6b) are the extremes in parallel (or transverse) energy outside the barrier. For elastic tunneling both E and $k_{||}$ are conserved, but $E_{||}$ may be different inside compared to outside.

POLITZER and CUTLER [5.40] pointed out that (5.5) or (5.6) are quite general and valid for any electron dispersion relation $E = E(\boldsymbol{k})$. The mistake many authors make is approximating $D(E, \boldsymbol{k}_{||})$ by the usual form of the WKB transmission coefficient for an image potential barrier [5.29], based on the assumption that far from the barrier region the WKB asymptotic solutions are the eigenstates of a slowly varying potential; the periodic potential of the solid and solid surface does not satisfy this condition. If a simple WKB exponential tunneling factor is used for D, then all the information concerning the wave function matching at the interface is thrown out and leads immediately to the result that all density of states terms in (5.5) disappear since $\partial E / \partial k_z$ cancels with $|\nabla_k E|$. *The important information about the surface is contained in the transmission factor and much is obviously lost in a calculation which assumes that D does not depend upon the nature of the wave function at the interface.*

There are now several calculations which evaluate D correctly in (5.5). All of them predict structure in $j'(E)$ due to the electronic properties of the emitter. The simplest example is furnished by a model calculation of PLUMMER and YOUNG [5.13] for a free electron emitter with a tri-angular barrier, and an attractive square-well potential at the surface. Equation (5.5) or (5.6) is very simple for this one-dimensional free electron case,

$$j'(E) = \frac{2em\,f(E)}{\hbar^3(2\pi)^2} \int_0^E D(W)\,dW, \tag{5.7}$$

where W is the normal energy $W = E - \hbar^2 k_{\parallel}^2/(2m)$.

First let us assume that $D(W)$ for the triangular barrier is given by the WKB form (no wave matching)

$$D_{\mathrm{WKB}}(W) \equiv \exp\left(-2\int_{z_0}^{z_1} \varkappa\,dx\right), \tag{5.8}$$

where

$$\varkappa = \left(\frac{2m}{\hbar^2}\right)^{\frac{1}{2}} \sqrt{V(z) - W}$$

$$= \left(\frac{2m}{\hbar^2}\right)^{\frac{1}{2}} \sqrt{E_F + \varphi - eFz - W}$$

with z_0 and z_1 the classical turning points, and E_F the Fermi energy and φ the work function; $z_0 = 0$ and $z_1 = (E_F + \varphi - W)\,eF$. This gives

$$D_{\mathrm{WKB}}(W) \equiv \exp\left[-\tfrac{4}{3}\left(\frac{2m}{\hbar^2}\right)^{\frac{1}{2}} (E_F + \varphi - W)^{\frac{3}{2}}\right]. \tag{5.8}$$

Since this transmission factor does not account for the properties of the wave functions at the surface, caused by the attractive square well it predicts a free electron energy distribution $j_0(E)$, obtained by a first-order expansion of W about E

$$(E_F + \varphi - W)^{\frac{3}{2}} \simeq (E_F + \varphi - E)^{\frac{3}{2}}\left(1 - \tfrac{3}{2}\,\frac{W - E}{E_F + \varphi - E}\right),$$

which gives

$$j_0''(E) \simeq \frac{e\pi\sqrt{2m\,f(E)}}{\hbar^3\,[E_F + \varphi - E]^{\frac{1}{2}}} \exp\left[-\tfrac{4}{3}\left(\frac{2m}{\hbar^2}\right)^{\frac{1}{2}} (E_F + \varphi - E)^{\frac{3}{2}}\right], \tag{5.9}$$

the free electron energy distribution derived by YOUNG [5.41]. On the other hand, if D is calculated correctly by matching across the barrier

[5.27], the energy distribution $j'(E)$ contains structure due to the nature of the wave function near the surface. PLUMMER and YOUNG's [5.13] model calculation showed that

$$\frac{j'(E)}{j_0'(E)} \propto |\psi(z=0)|^2 ,$$

where $z=0$ is the classical turning point in this model.

POLITZER and CUTLER [5.40, 42] calculated the field emission from ferromagnetic Ni in an attempt to explain spin polarization measurements [5.43]. They showed more rigorously that it is essential to include the matching conditions properly when calculating the transmission factor. The WKB transmission factor given by (5.8) is always multiplied by a pre-exponential term $P(E, k_{||})$ so that

$$D(E, k_{||}) = P(E, k_{||}) \, D_{WKB}(E, k_{||}) . \tag{5.10}$$

$P(E, k_{||})$ must be evaluated from the wave functions at the interface. The ratio of the actual energy distribution to the "free electron WKB" energy distribution, $j_0'(E)$ will be called the enhancement

$$R(E) = \frac{j'(E)}{j_0'(E)} . \tag{5.11}$$

$R(E)$ reflects the properties of the pre-exponential term $P(E, k_{||})$.

POLITZER and CUTLER [5.42] used this approach to calculate $P(E, k_{||})$ for the d-bands in nickel relative to the free electron bands; they found as a consequence of the localization of the d wave functions at the surface, that P is 10^{-1}–10^{-2}. GADZUK [5.44] had previously estimated a decrease of 10^{-3} for d-band tunneling based on a localization argument.

The above should make it clear that the use of WKB tunneling probabilities will, in general, lead to incorrect energy distributions, which can be very misleading. There are exceptions, for example, when there are pronounced extremes in the projected energy surfaces of (5.6), such as gaps or necks [5.45].

Although (5.5) is very useful for treating tunneling from clean metal surfaces its limitations become apparent when attempts are made to apply it to tunneling from adsorbed impurities [5.44, 46]. A more appealing method for handling this case consists of perturbation theory, or more porperly the transfer Hamiltonian method [5.33] whose final result looks exactly like Fermi's golden rule formula for time dependent perturbation theory

$$j'(E) = (2\pi/\hbar) f(E) \sum_k \sum_R |\langle \Phi_k | \tau | \Phi_R \rangle|^2 \, \delta(E - E_R) \, \delta(E - E_k), \qquad (5.12)$$

where

$$\tau = -eFz, \qquad (5.13)$$

and Φ_k and Φ_R refer to wave functions in the metal and beyond the barrier, respectively.

PENN et al. [5.32, 47] have used this approach to evaluate energy distributions from clean as well as adsorbated covered surfaces. They evaluated (5.12) by using Airy functions for Φ_R and insert a normalization factor N_m for the metal wave functions Φ_k at the classical turning point. The final result obtained by PENN [5.32, 48] is the following

$$j'(E) = \frac{2\hbar}{m} f(E) \sum_k D_{WKB}(E, k_{||}) \, N^2(E, k_{||}) \, \delta(E - E_k). \qquad (5.14)$$

If WKB wave functions are used for Φ_k, (5.14) obviously reduces to the WKB result (5.5) and (5.8) with no dependence on the wave function at the surface, i.e. no density of states information. The potential near the classical turning point on the metal side of the barrier changes quite rapidly and the WKB approximation for the wave function is not appropriate.

PENN [5.32] has shown that the normalization constant N_m^2 can be related to the wave function amplitude at the turning point z_0 so that (5.14) can be written as

$$j'(E) \simeq \lambda^{-2}(E) \frac{2\hbar}{m} f(E) \sum_k D(E, k_{||}) \, |\psi_k(z_0)|^2 \, \delta(E - E_k), \qquad (5.15)$$

where D is now the WKB tunneling probability of (5.8), and $\lambda(E)$ is a slowly varying function of energy. In this form $|\psi_k(z_0)|^2 | \lambda_2(E)$ is equivalent to POLITZER and CUTLER's [5.40] pre-exponential factor $P(E, k_{||})$. Since the tunneling probability $D(E, k_{||})$ falls off exponentially as $k_{||}$ increases from zero, the $|\psi_k(z_0)|^2$ term may be extracted from the sum if it does not change drastically for small changes in $k_{||}$ about $k_{||} = 0$. That is if $\partial|\psi_k|^2/\partial k_{||}|_{k_{||}=0}$ is small compared to the magnitude of $|\psi_k(k_{||}=0)|$.

This gives

$$R(E) = \frac{j'(E)}{j'_0(E)} \propto \varrho_m^\perp(E)|_{z=z_0}, \qquad (5.16)$$

where $\varrho_m^\perp|_{z=z_0}$ is the one-dimensional density of states evaluated at the classical turning point z_0.

Equation (5.16) immediately shows the origin of the reduction in tunneling from localized d states [5.42, 44]. The amplitude of the wave function will be considerably smaller at the turning point for a localized state than for a free electron state.

CAROLI et al. [5.49] have derived the field emission energy distribution applying an out-of-equilibrium formalism using KJELDYSH Green's functions [5.50]. SOVEN [5.51] has shown for the one-dimensional case that this formalism is equivalent to (5.5) and gives $R(E)$ of (5.16) as the one-dimensional density of states at the turning point to order of the tunneling probability.

The electrons which are field emitted may have a spin polarization. PENN [5.52] has shown, using the transfer Hamiltonian approach outlined above, that a spin polarization measurement will measure the one-dimensional "surface" density of states for a given spin band. The problem then reverts back to calculating the one-dimensional density of states for each spin band at the classical turning point. This is in fact what POLITZER and CUTLER calculated for Ni [5.42].

5.3. Adsorbate Covered Surfaces

When an atom or molecule is chemisorbed on a surface the energy levels of both metal and adsorbate will shift, and field emission energy distributions can reveal the characteristics of the local density of states on the adsorbate. Qualitatively, this is obvious from Fig. 5.2b, where electrons with appreciable amplitude on the adsorbate have a considerably smaller barrier to tunnel through than electrons in the metal. It is fairly easy to quantify this statement.

5.3.1. Tunneling Resonance

Without doubt the paper by DUKE and ALFERIEFF [5.15] pioneered in this field. They calculated the transmission through a triangular barrier with an adsorbate potential within the barrier, represented by an attractive square well plus a δ-function repulsive core. The wave functions and the current were calculated exactly for various positions and strengths of the adsorbate potential. For energies corresponding to that of a standing wave in the adsorbate potential well penetration of the barrier is much easier, i.e. tunneling resonance occurs. This approach does not show explicitly how the energy distribution is related to the "local density of states" on the adsorbate. This problem can be overcome by calculating for each energy the amplitude squared $|\psi_a|^2$ of the wave function on the adsorbate.

Subsequent to DUKE and ALFERIEFF's computer calculation several theoretical treatments have appeared which express the energy distribution from the adsorbate, $j_a'(E)$, analytically. All of these calculations [5.47, 49, 53] yield nearly the same result, i.e.

$$\frac{\Delta j'}{j_{cl}'} \equiv \frac{j_a'(E) - j_{cl}'(E)}{j_{cl}'} \equiv U^2(E)\, \varrho_a(E), \tag{5.17}$$

where ϱ_a is the local density of states on the adsorbate (see Section 2.1), and $U^2(E)$ a term which compensates for the difference in tunneling from the adsorbate relative to tunneling from the metal. The function $j_{cl}'(E)$ denotes the assumed energy distribution from the clean surface which would be obtained if the work function and field were those corresponding to the adsorbate covered surface. This quantity is of course impossible to measure experimentally. Further PENN (Ref. [5.47], Eqs. (14b), and (15a)) has shown that (5.17) is only valid when there is no structure in the clean energy distribution, i.e. when $j_{cl}'(E) = j_0'(E)$. In general, (5.17) should be written as

$$\frac{\Delta j'(E)}{j_{cl}'(E)} = U^2(E)\, \Delta\varrho_a(E). \tag{5.18}$$

$\Delta\varrho_a$ is the change in the density of states at the adsorbate site caused by the presence of the adsorbate. PLUMMER and YOUNG [5.13] observed an adsorbate induced change in the substrate local density of states for adsorption on (100)W, and DUKE and FAUCHIER [5.35] demonstrated theoretically that $\Delta j'/j_{cl}'$ can display structure which is not present in $\varrho_a(E)$. In practice [5.4, 37] the data are usually plotted in the form of $R(E)$ curves defined by (5.11), with $j_0(E)$ calculated for the field and work function of the specific adsorption system.

Only a few calculations have considered the effect of structure in clean energy distributions. BAGCHI and YOUNG [5.54] calculated the energy distribution from an adsorbate covered surface, representing the

---

Fig. 5.5. (a) Field emission energy distribution $j'(\varepsilon)$ [Δ] from clean (112) tungsten at 78 K, $j_0(\varepsilon)$ [$-$] is the calculated free electron energy distribution for a work function of $\varphi = 4.90\,\text{eV}$ and a field measured from the slope of the Fowler-Nordheim plot [5.4], $R(\varepsilon)$ [$\langle\bigcirc\rangle$] on the right hand scale is calculated by dividing j' by j_0 for each energy. R. was arbitrarily normalized to 10 at $\varepsilon = 1.0\,\text{eV}$. (b) Measured energy distribution $j'(\varepsilon)$, calculated free-electron energy distribution j_0' ($\varphi = 6.14\,\text{eV}$) and R factor for $\sim 10^{-6}$ Torr $\cdot$ sec oxygen exposure on (112) tungsten at 78 K

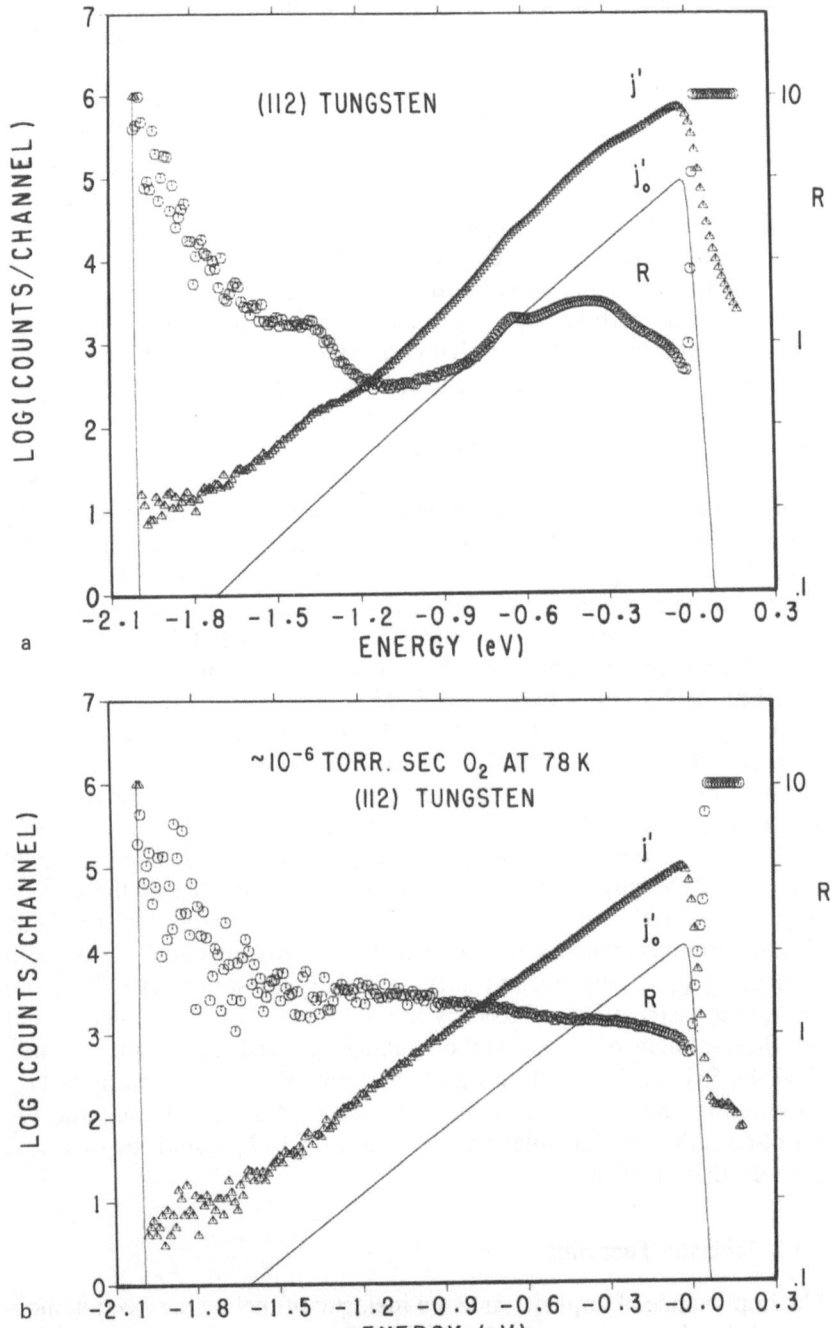

density of states of the metal by two Lorentzian peaks at 2 eV and 3.5 eV below E_F and the adsorbate density by a Lorentzian peak some 6–9 eV below E_F. Since ϱ_a included contributions from the density of states in the substrate, peaks in the latter appeared in ϱ_a and could thus show up in $j'_a(E)$ with enhanced amplitude because of the increased tunneling probability for electrons on the adsorbate. Thus they concluded that the presence of an adsorbate can produce structure in the energy distribution which reflects the density of states of the substrate, at energies far from any resonant levels in the adsorbate.

In concluding this subsection, two examples will be presented to show first that (5.16) is basically correct for clean surfaces and secondly that problems still exist in applying (5.18) for adsorbates. Figure 5.5a is a plot of the measured energy distribution $j'(E)$ for clean (112) tungsten at 78 K. The free electron energy distribution [5.4] $j'_0(E)$ and the $R(E)$ factor are defined by (5.11), and calculated for $\varphi = 4.90$ eV and a field measured from the Fowler-Nordheim plot. Since the pre-exponential terms involving the area sampled were not measured, $j'_0(E)$ is arbitrarily normalized at $\varepsilon = -1.0$ eV where $\varepsilon = E - E_F$. This curve shows the type of structure predicted by (5.16) and in Subsection 5.5.1 it will be compared with actual band structure calculations to show that it is a reasonable surface density of states. The message is that there is structure in the clean energy distribution which is very visible when the exponential tunneling probabilities have been divided out.

In Fig. 5.5b the same (112) plane of tungsten was exposed to $\sim 10^{-6}$ Torr · sec of oxygen at 78 K. The R curve is almost completely flat. Not only have no new levels due to the adsorbate appeared but all the structure due to the substrate has disappeared. When this oxygen layer is heated to 1000 K it orders and a single level appears in the R curve, peaked at $\varepsilon = -1.2$ eV with a half width 0.45 eV [5.4]. At present there is no definite explanation of this penomenon. It could be the result of scattering of tunneling electrons by the adsorbed O atoms, or could result from changes in ϱ_a or the normal surface density of states resulting from surface reconstruction, induced by the presence of O. This suggests that the three-dimensional aspects of tunneling need to be investigated [5.40, 49, 55] in a formulation which does not assume conservation of $k_{||}$. Presumably, the scattering formalism used by MODINOS [5.56] and the out-of-equilibrium formulation of CAROLI [5.49] could be adapted to study this problem.

5.3.2. Inelastic Tunneling

The importance of impurity assisted inelastic tunneling has been demonstrated by LAMBE and JAKLEVIC [5.57] for tunnel junctions. They

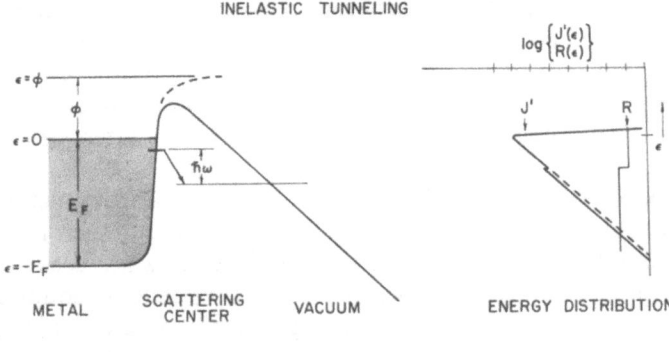

$R(\epsilon) = J'(\epsilon)/J'_o(\epsilon)$

$J'_o \equiv$ FREE ELECTRON ENERGY DISTRIBUTION

Fig. 5.6. Schematic drawing of the inelastic tunneling process. A tunneling electron loses energy $\hbar\omega$ by inelastically scattering at the scattering center. The sharp Fermi edge produces a cut-off on the energy of the tunneling electrons so that the resultant R curve has a step at an energy $\hbar\omega$ below the Fermi energy

deposited various molecules such as H_2O, CO, and C_2H_2 at one interface of a metal-insulator-metal junction and measured the current voltage characteristics of the junction. The second derivative of the junction current with respect to the bias voltage revealed sharp line structure at voltages near the energies of the infrared active vibrational modes of the molecules or hydroxyl groups. WEINBERG's group [5.58] is now using this technique to identify molecules formed at one of the interfaces after some prescribed chemical reaction has been carried out.

Corresponding processes will occur in field emission, as shown schematically in Fig. 5.6. A tunneling electron may undergo an inelastic collision with an adsorbed molecule or a surface complex. The inelastically scattered electron will appear in the energy distribution with its energy reduced by the energy of the vibrational mode. Since the cross section will usually be small for such a process the change in the signal will be small and appear as a threshold effect (Fig. 5.6), so that the R curve will have a small step at $\varepsilon = -\hbar\omega$. The width of the step will be related to the temperature of the crystal or the sharpness of the Fermi edge, while the height will depend upon the cross section and density of scattering centers. FLOOD [5.59] has presented a theory of this process which assumes equal tunneling probability for inelastically scattered and corresponding elastic electrons. Quantitatively an energy distribution $j'_{in}(E)$ can be written as

$$j'_{in}(E) = j'_{el}(E) + j'_{el}(E + \hbar\omega) \, N \, T(\hbar\omega), \qquad (5.19)$$

where N is the density of scattering centers and T the scattering cross section. $T(\hbar\omega)$ contains both the probability that an electron of energy $E + \hbar\omega$ will lose energy $\hbar\omega$ and the ratio of the escape probabilities at energy E and $E + \hbar\omega$. This makes it relatively hard to calculate $T(\hbar\omega)$, since one cannot assume a priori that an electron scatters before or after tunneling.

Experimentally the signals observed both in tunnel junctions [5.57] and in field emission [5.37] for small molecules are much less than the background current. Therefore second derivatives are used to analyze tunnel junctions and computer processing of data is required in field emission. By contrast the inelastic signal from large organic molecules adsorbed on a field emitter may be of the same order of magnitude as the primary signal [5.60].

5.3.3. Many-Body Effects and Photoassisted Field Emission

Neither of these topics will be discussed at any length. Many-body effects have been reviewed by GADZUK [5.4] and photoassisted field emission by BAGCHI [5.61] and LEE [5.62]. Many-body effects will be defined as any electron-electron interaction which modifies the one-electron tunneling picture presented in the previous two sections. Experimentally, these effects give rise to wings on the energy distribution not predicted by the one-electron theory [5.63, 64]. They may be caused by multi-particle tunneling [5.63, 65], by hot-hole electron cascade [5.64], by uncertainty-principle arguments [5.66] or by electron-electron scattering outside the emitter [5.67]. Whatever the mechanism for these tails, there is probably a fundamental limit imposed by the interaction of electrons on the current range over which the one-electron picture is useful. Experimental data [5.63, 64] indicate that this is approximately six orders of magnitude.

Photoassisted field emission [5.61] is an excellent topic to end the discussion of field emission, since it will point out some of the problems we will encounter in the section on photoemission. Photoassisted field emission refers to the process of exciting an electron with an incident photon beam to an energy below the top of the barrier in Fig. 5.2a, so that the electron must still tunnel through the barrier in order to escape. The simplest theories are those of NEUMANN [5.68], and LUNDQUIST et al. [5.69], who view photofield emission as a two-step process in which an electron is first excited by a photon and subsequently tunnels out. This model suggests that the energy of the final state in an optically allowed transition can be mapped as a function of direction in k-space collected by the probe hole. The mean-free path of excited electrons is quite large near the Fermi energy (Fig. 5.4) so that this model should be quite realistic. Unfortunately, experimental observations [5.62, 70] in several directions from a tungsten emitter do not substantiate this theory. All the energy distributions show a peak which is nearly a replica of the initial energy distribution moved up by the photon energy and decreased in amplitude. When the direct transition model of photo-excitation is discussed in the next section it will become obvious that this behavior cannot be reconciled with the band structure of the solid and k conservation in the transition.

BAGCHI [5.61] has developed a theory to explain the data of LEE [5.62], in terms of the surface photoelectric effect. The surface destroys the periodicity in the direction normal to the surface so that the three-dimensional crystal momentum k does not have to be conserved; therefore, the restriction on the allowed transitions in the bulk is removed. This leads to an excited distribution which looks like the ordinary Fermi distribution displaced by the photon energy. If BAGCHI is correct and the surface photoelectric effect is dominant at these low photon energies, where the electron mean-free path must be ~ 100 Å, then it could be even more important at the higher photon energies used in photoemission where the mean-free path of an electron may be as short as 5 Å.

5.4. Photoemission

A rigorous theoretical treatment of photoemission including surface effects has not been developed [5.71]. In this section the pertinent physical parameters of the measurement process will be discussed.

5.4.1. Photoexcitation; Primarily Atoms and Molecules

Photoionization in gases is much simpler than the analogous process in solids [5.72, 73]. Since gases adsorbed on a surface may retain many of their gas phase properties it is appropriate to begin the discussion of photoemission with a description of gas phase photoionization. The (non-relativistic) Hamiltonian of an electron in a electromagnetic field is [5.74]

$$\mathcal{H} = \frac{1}{2m}(\boldsymbol{P} + e\boldsymbol{A})^2 - e\Phi + V_0 = \boldsymbol{P}^2/2m + \left(\frac{e}{m}\right)(\boldsymbol{A} \cdot \boldsymbol{P} + \boldsymbol{P} \cdot \boldsymbol{A})$$
$$+ V_0 - e\Phi + \frac{e^2}{2m}\boldsymbol{A}^2,$$

(5.20)

where A and Φ are the vector and scalar potentials, respectively, and P is the momentum operator. By virtue of gauge invariance, we may demand that either Φ or $\nabla \cdot A$ (but not both) vanish. Here we chose $\Phi = 0$, which means that $\nabla \cdot A$ does not, in general, equal zero [5.74], a point we will use later. Neglecting the last term (quadratic in A) and treating the terms linear in A as a perturbation of the rest of $\mathcal{H}$ leads to a transition rate between states ψ_i and ψ_f of

$$P_{if} = \left(\frac{4\pi^2 e^2}{\hbar^2 m^2} \right) |\langle \psi_f | A(r) \cdot P + P \cdot A(r) | \psi_i \rangle|^2 \, \delta(E_f - E_i - \hbar\omega), \quad (5.21)$$

(using lowest-order time-dependent perturbation theory). Here $E_i(E_f)$ is the energy of $\psi_i(\psi_f)$ (an eigenstate of the unperturbed system), and $\hbar\omega$ is the photon energy.

There are several approximations applied to (5.21) to make it easier to handle. In the next few paragraphs each of these steps will be spelled out, indicating why it is usually valid for atomic or molecular calculations and whether it is a valid step when applied to a solid.

In nearly all calculations using (5.21) it is assumed that $\nabla \cdot A = 0$. This is based not on gauge invariance (which was invoked to eliminate the scalar potential) but on the approximation of A by a plane wave

$$A(r, t) = A_0 \exp(-i\omega t + i q \cdot r), \quad (5.22)$$

where $A \cdot q = 0$ so that $\nabla \cdot A = 0$. This reduces (5.21) to

$$P_{if} = (4\pi^2 e^2 / \hbar^2 m^2) |\langle \psi_f | A \cdot P | \psi_i \rangle|^2 \, \delta(E_f - E_i - \hbar\omega), \quad (5.23)$$

where $A(r)$ in (5.21) is $A(r) = A_0 \exp(i q \cdot r)$. This is the usual form for the matrix element seen in the solid state [5.21] or gas phase literature. In the solid and especially near the surface where the A vector is changing rapidly, $\nabla \cdot A$ does not equal zero and may in fact be very large [5.75]. On physical grounds, it is easy to see why this term is important in UV photoemission. The escape depth of the excited electron is $\sim 5\,\text{Å}$ (Fig. 5.4) and the region where the gradient in the perpendicular component of the field occurs is something like 2–3 Å, so in the region from which emission occurs there may be a large gradient in A_0, i.e., charge imbalance.

In (5.22) or (5.23) the $\exp(i q \cdot r)$ term coming from A, see (5.23), is usually expanded in a power series, with often only the first term, unity, being retained. This approximation is based on the large wavelength of the light compared to atomic dimensions. When this approximation is applied to (5.23), the "electric dipole" approximation is obtained

$$P_{if}(\text{dipole}) = \frac{4\pi^2 e^2}{\hbar^2 m^2} |\langle \psi_f|P|\psi_i\rangle \cdot A_0|^2 \, \delta(E_f - E_i - \hbar\omega). \qquad (5.24)$$

This approximation also ignores any retardation effects across the atom or solid.

In the "electric dipole" approximation the momentum matrix element $\langle \psi_f|P|\psi_i\rangle$ must be evaluated. This can be and is frequently converted to a spatial operator r matrix element or to a gradient in potential ∇V operator by using simple commutation relations [5.76]. If, as we have assumed, ψ_i and ψ_f are eigenstates of H_0 then we can write

$$\langle \psi_f|P|\psi_i\rangle = im\omega \langle \psi_f|r|\psi_i\rangle, \qquad (5.25a)$$

or

$$\langle \psi_f|P|\psi_i\rangle = \frac{-i\hbar}{\omega} \langle \psi_p|\nabla V|\psi_i\rangle. \qquad (5.25b)$$

Using these equalities, (5.24) can be written in the alternate forms

$$P_{if}(\text{dipole}) = \frac{4\pi^2 e^2 \omega^2}{\hbar^2} |\langle \psi_f|r|\psi_i\rangle \cdot A_0|^2 \, \partial(E_f - E_i - \hbar\omega), \quad (5.26a)$$

or

$$P_{if}(\text{grad } V) = \frac{4\pi^2 e^2}{m^2 \omega^2} |\langle \psi_f|\nabla V|\psi_i\rangle \cdot A_0|^2 \, \partial(E_f - E_i - \hbar\omega). \quad (5.26b)$$

In calculations (except those for hydrogen) ψ_f and ψ_i are usually approximate wavefunctions which are not true eigenstates of $\mathcal{H}_0$. Thus, (5.25a) and (5.25b), and consequently (5.26a) and (5.26b) are approximations which should be used with some caution. BETHE and SALPETER [5.76] pointed out that (5.26a) with the r matrix elements will accentuate the importance of the wavefunctions at large r, while the ∇V matrix element (5.26b) will accentuate the small r terms (for an atom). For the p operator in (5.25), one finds that intermediate values of r are most important. In many atomic calculations of the photoionization probabilities, the cross section is calculated using all three methods as a check on the wavefunctions.

Equation (5.26b) is usually applied to calculating the photoemission from the potential step at the surface. It appears that the $1/\omega^2$ term in

(5.25b) has led people to believe that surface photoemission will dominate at low photon energies over bulk emission. SCHAICH and ASHCROFT [5.77] have shown that this is not true.

To this point we have considered transitions of the whole system, but in most cases a one electron approximation is used. Often this helps in visualizing the process, but it should always be remembered that photoemission is not a one-electron process. If the total wave functions ψ_i and ψ_f are written as properly anti-symmetrized determinants of N one-electron wave functions Φ_j, then by virtue of Koopman's theorem [5.78] the matrix elements in (5.24) or (5.26) are just integrals over one-electron wave functions where $N-1$ of these functions are exactly the same. Note that this approximation assumes an unrelaxed ionic core. This picture is, of course, not exact and the difference between the ionization potential and one-electron orbital energies is called the relaxation energy in this paper. This relaxation energy may be as large as 70 eV for photo-ionization from the $1s$ state of Xe [5.79]. On the other hand, however, COOPER [5.80] has argued that even though the energies of the photo-emitted electrons may be badly estimated in the one-electron approxima-tion, the probability for excitation is fairly accurate.

5.4.2. Cross Section for Photoemission

The relative cross section for photoionizing electrons from different orbits of an atom or molecule has long been known to depend upon photon energy. PRICE [5.81] presented general physical arguments which show the trends in the relative cross sections for photoionization. Everything else being equal, the more localized in space the initial state is, the better it will couple with a higher energy final state, so the maximum in the cross section moves to higher energy as the mean radius becomes smaller.

These same sorts of effects are quite common in spectra from solids [5.82]. A very nice example is shown in Fig. 5.7 for EuS from the work of EASTMAN and KUZNIETZ [5.83]. This figure shows the photoelectron energy distribution from EuS as a function of photon energy from $\hbar\omega = 10.2$ eV up to 40.8 eV. The peak near -2 eV has been identified as the $4f$ level in EuS and the two lower lying peaks as originating from the p bands [5.83]. The insert in the top left of Fig. 5.7 shows the ratio of the photoionization probability of the $4f$ to p-band peak. It increases with the third power of photon energy. The UV data of BRODEN et al. [5.84] on Yb, Ba, and Eu show the same behavior of the $4f$ levels. At photon energies below 12 eV there is little sign of the $4f$ levels in Yb or Eu, and the photoelectron spectra of these metals look much like Ba.

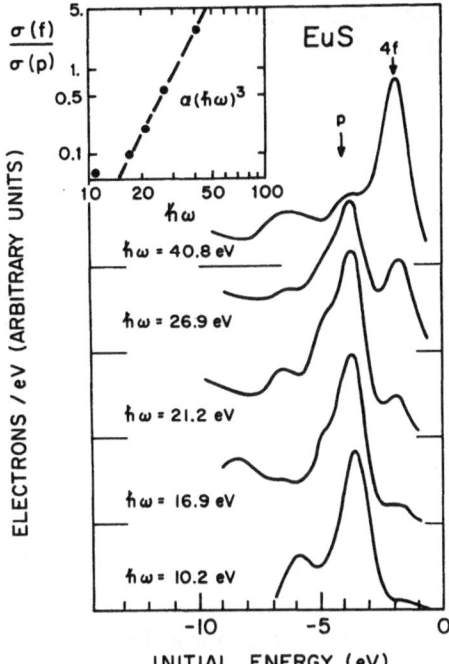

Fig. 5.7. Photoelectron energy distributions of EuS as a function of the energy of the incident photon. The energy scale is the energy of the initial state with zero being the Fermi energy [5.83]

On the other hand, the X-ray photoelectron spectrum [5.82] from Yb and Eu shows only the $4f$ levels with little if any sign of the $6s$ band.

If a photoelectron experiment is to be designed to study the adsorption of molecule X on the surface of metal M the choice of the appropriate photon energy is very important. For example, to study Eu or Ba, one surely would not use a photon energy less than 10 or 15 eV, because the Eu $4f$ levels have a very low photoionization cross section at these low energies [5.84]. These qualitative statements can be made more quantitative by actually calculating the photoionization cross section using one of the forms of (5.24, 26).

The cross sections can be estimated by using a one electron hydrogenic wavefunction for ψ_i and a plane wave approximation for ψ_f, permitting the matrix elements to be calculated in closed form. Then (5.24) becomes

$$P_{if} = \left(\frac{4\pi^2 e^2 B^2}{m^2}\right) |A_0 \cdot k|^2 |\int \exp(ik \cdot r)\, \psi_i d\tau|^2\, \delta(E_f - E_i - \hbar\omega), \quad (5.27)$$

where k is now the wave vector of the outgoing electron. If ψ_i is a hydrogenic wave function [5.76]

$$\psi_i = R_{nl}(r)\,Y_{lm}(\theta, \varphi) \tag{5.28}$$

then [5.76, 85]

$$\int \exp(i\boldsymbol{k}\cdot\boldsymbol{r})\,\psi_f\,d\tau = F_{nl}(\varrho)\,Y_{lm}(\theta, \varphi), \tag{5.29}$$

where θ, φ are the angles of k relative to the orientation of the initial state in momentum space and $F_{nl}(p)$ is the Fourier transform of the radial part of the wave function in (5.28) normalized so that

$$\int_0^\infty dp\,p^2\,|F_{nl}(p)|^2 = 1, \quad \text{with} \quad \hbar k = p.$$

Hence

$$P_{if} \propto p^2\,|F_{nl}(p)|^2 \times \left|\frac{A_0\cdot k}{|k|}\right|^2 Y_{lm}(\theta, \varphi)\,\delta(E_f - E_i - \hbar\omega), \tag{5.30}$$

and if we average over all exit directions k then

$$\langle P_{if}\rangle \propto P^2\,|F_{nl}(p)|^2. \tag{5.31}$$

The $F_{nl}(p)$ functions are given explicitly by [5.76] and [5.85]. The effective radius of the wavefunction in (5.28) can be adjusted to fit experimental values by using an effective nuclear charge Z leaving n and l unchanged. The number of nodes in the wave function is important [5.80] so SLATER type wavefunctions [5.86] cannot be used. Figure 5.8 is a plot of (5.31) for the oxygen $1s$ and oxygen $2p$ levels as a function of the energy of the photoemitted electron. For an excitation energy of $\hbar\omega = 1250$ eV the electron from the $2p$ level would have an energy of ~ 1220 eV and a probability of $\sim 1.6 \times 10^{-3}$, while the electron from the $1s$ level would have a final energy of ~ 700 eV with a probability of 1.5×10^{-1}. This ratio must be multiplied by the number of electrons in each subshell to give a ratio of 70:1 between emission probabilities for $1s$ and $2p$ electrons. This is in good agreement with the ratio of 100:1 obtained by SCOFIELD [5.87]. If the oxygen atom is adsorbed on a transition metal, the excitation probability from a d state will peak at a higher photon energy than the $2p$ state of oxygen. SCOFIELD's [5.87] calculation indicated the following ratio for the cross section σ

$$\left.\frac{\sigma(O2p)}{\sigma(W5d)}\right|_{\hbar\omega = 1000\text{eV}} \simeq 0.1$$

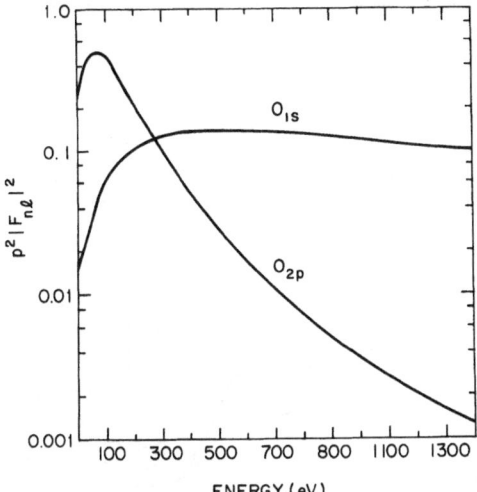

Fig. 5.8. Calculated photoionization probability for the $1s$ and $2p$ levels of oxygen, plotted as a function of the energy of the excited electron

and

$$\left.\frac{\sigma(\mathrm{O}2p)}{\sigma(\mathrm{Ni}3d)}\right|_{\hbar\omega=1000\,\mathrm{eV}} \simeq 0.05 \; .$$

Therefore, one would not expect to see the oxygen $2p$ levels in X-ray induced photoelectron spectra from oxygen on tungsten or nickel, while they should be very pronounced at low photon energies (Fig. 5.8).

Figure 5.9 shows the photoelectron spectra from clean tungsten before and after exposure of $\sim 10^{-5}\,\mathrm{Torr\cdot sec}$ oxygen at room temperature. The top set of curves are for (100) tungsten at a photon energy of 21.2 eV [5.88]. The bottom set of curves are from X-ray [$\hbar\omega = 1254\,\mathrm{eV}$] photoemission data of MADEY et al. [5.89]. The UV spectrum at the top shows that the levels derived from the oxygen p-levels are very pronounced. The X-ray spectrum show no signs of an oxygen p-band. The dotted line on the bottom is what would have been expected solely from the change in escape depth (Fig. 5.4). [The peak near the Fermi energy in the clean UV data is a surface state (see Subsection 5.5.1.) which probably was not present on the polycrystalline ribbon used for the X-ray experiment. If it had been present, it would not have been seen since WACLAWSKI and PLUMMER [5.90] showed that its photoexcitation probability relative to the d-band peak near -2 eV peaked at very low photon energies.] Figures 5.7–5.9 reveal that it is very important to

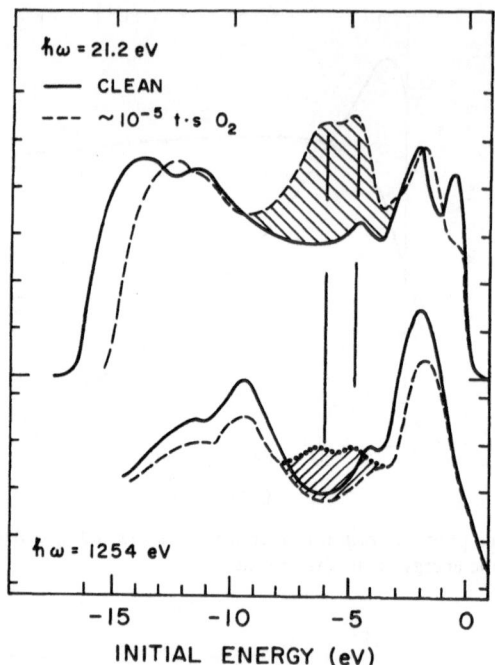

Fig. 5.9. Comparison of the effect of oxygen exposure on the photoelectron spectra from clean tungsten. The top set of curves are for $\hbar\omega = 21.2$ eV [5.88]. The bottom set of curves is for X-rays, $\hbar\omega = 1254$ eV [5.89]. The structure near — 11 eV is due to the W(4d) levels, excited by an impurity X-ray line

choose a photon energy appropriate to the system. The relative changes in the photoexcitation probability as a function of photon energy are usually a much more important consideration in a surface experiment than the electron escape depth as a function of photon energy.

Cooper [5.80] has shown theoretically that if the initial wave-function has nodes, there are nodes in the photoionization cross section. Initial states of the form 1s, 2p, 3d etc. should not have nodes while all others should. He calculated the photoionization cross section from the 3d levels of Cu^+ and 4d levels of Ag^+. Ag^+ had a node, and Cu^+ did not. In Figure 5.10 the photoionization probabilities of O (1p), S (2p), Se (3p), and Te (4p) are calculated using (5.31). The mean radius shown on the figure was obtained by adjusting the effective Z [5.91]. The nodes move to lower energy as the principal quantum number increases. This figure shows the same trend in excitation probability for O, S, Se, and Te at a given photon energy, as found by Hagstrum and Becker [5.92] for O, S, Se, and Te adsorbed on Ni. It also has the

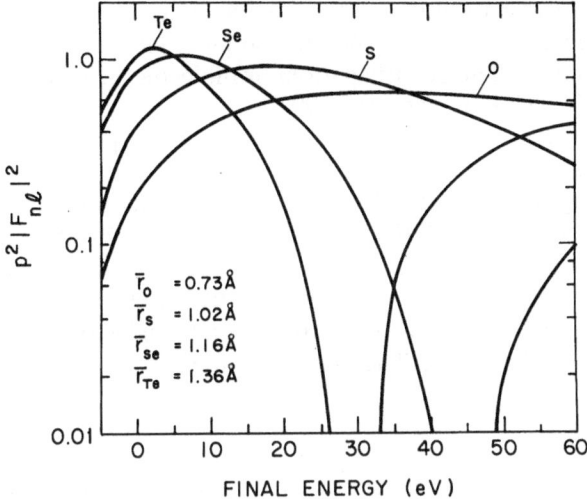

Fig. 5.10. Calculated photoionization probabilities for O, S, Se, and Te as a function of the final energy of the excited electron. Since the ionization potentials are nearly the same the final state energy is just the photon energy minus the ionization potential

same photon energy dependence, with the Te signal dropping faster than that from O as $\hbar\omega$ increased. It would be very interesting to find out whether the Te signal increases for $\hbar\omega > 30$ or 40 eV.

5.4.3. Angular Dependence of Emission

The angular distribution of electrons photoemitted from atoms or molecules in the gas phase has received considerable attention both theoretically [5.93–95] and experimentally [5.96–97]. The objective of these measurements is to determine which molecular orbital is associated with a given photoionization peak in the spectrum of a molecule or, for that matter, an atom. The variation in the angular dependence of one orbital compared to another is never very dramatic because the atoms or molecules in the gas phase have random orientation. On the other hand, a surface can orient all the adsorbed atoms or molecules so that the angular distribution of the emitted electrons has much more structure than the corresponding gas phase spectra [5.98]. A proper analysis of the angular dependence can be used to determine the symmetry of the bonding site [5.98, 99] and potentially the symmetry of each orbital [5.98]. In this section, the theory of the angular distribution from random atoms or molecules will be briefly compared with that from oriented atoms or molecules to illustrate the wealth of new structures produced by orientation. In Subsection 5.46 these angular effects will be discussed within the context of an adsorbed atom or molecule, using the formalism of GADZUK [5.98] and LIEBSCH [5.99].

For simplicity consider an incident beam of polarized photons. If a plane wave were an appropriate description of the final state, then (5.30) would give the probability of observing a photoemitted electron from an oriented atom with an initial state given by the hydrogenic wave function (5.28). θ and φ are the angles between the k vector of the outgoing electron and the orientation of the initial orbital. For a gas phase atom (5.30) must be averaged over all angles θ and φ, i.e., all orientations of the atom. The resulting intensity I as a function of angle is

$$I(\gamma) \propto \cos^2 \gamma ,$$

where γ is the angle between the polarization vector A_0 and the direction of emission k. More general calculations for gas phase atoms or molecules [5.94, 95] show that the intensity I can be expressed as

$$I(\gamma) \propto \left[1 + \frac{\beta}{2} (3 \cos^2 \gamma - 1) \right], \tag{5.32}$$

where β is an asymmetry parameter which depends upon the initial and wave final state. The parameter β can range from -1 to $+2$; $\beta = 2$ for a plane.

The form of the continuum function for the final state should be expressed as a sum of spherical harmonics, not a plane wave [5.93, 94]. If the initial state has an angular momentum l_i then the electric dipole selection rules restrict the angular momentum of the out-going photo-electron to $l_f = l_i \pm 1$ [5.82]. Therefore, for an s electron, the ejected electron has $l_f = 1$, leading to $\beta = 2$, but when the initial state is a p state, there are contributions from $l_f = 0$ and $l_f = 2$. The interference of these two out-going waves causes β to be less than 2. For example, $\beta \simeq 1.2$ for photoionization of the $4p$ levels of Kr using He(I) radiation ($\hbar\omega = 21.2$ eV) [5.97].

The effects of orientation can best be illustrated by two simple examples: 1) An atom with a single p orbital, say, boron, and 2) a simple molecule like H_2. For the atomic case assume that the final state is a plane wave. Then in the gas phase the angular intensity $I_g(\gamma)$ is

$$I_g(\gamma) \propto \cos^2 \gamma .$$

But if the atom were oriented on the surface with only a P_z orbital (z is perpendicular to the surface, then the angular intensity of the oriented atom, $I_0(\gamma)$, would be (5.30)

$$I_0(\gamma) \propto \cos^2 \gamma \cos^2 \theta$$

where θ is the angle between the emission direction and the surface normal. GADZUK [5.98] has illustrated this effect for several different oriented orbitals. In the gas phase the relevant directions are the polarization direction and the emission direction, while for an orientated atom the direction of orientation also becomes important.

The case of an oriented molecule is even more interesting. TULLY et al. [5.96] have calculated β in (5.32) for molecular H_2 in the gas phase. They found $\beta = 1.86$, or an angular dependence of nearly $\cos^2 \gamma$ for randomly oriented molecules. Now assume that the H_2 molecule is oriented with its axis along the direction of the photon beam, i.e. A_0 is perpendicular to the axis. KAPLAN and MARKIN [5.100] have calculated the angular dependence of the photoemitted electron from this configuration. They used a wavefunction for the initial state which took into account both the covalent and ionic contributions;

$$\psi_i = N_\perp \{[\varphi_a(1)\, \varphi_b(2) + \varphi_b(1)\, \varphi_a(2)]$$
$$+ \mu [\varphi_a(1)\, \varphi_a(2) + \varphi_b(1)\, \varphi_b(2)]\}\,.$$

When $\mu = 0$, ψ_i is the Heitler-London function and, when $\mu = 1$, it is the molecular orbital type wave function. Their final state is expressible by just a product of a plane wave and an atomic function

$$\psi_f = N_2 \{[\varphi_a(1) + \varphi_b(1)]\, \exp(i\mathbf{k} \cdot r(2))$$
$$+ [\varphi_a(2) + \varphi_b(2)]\, \exp(i\mathbf{k} \cdot r(1))\}$$

(N_1 and N_2 are normalization factors). When the matrix element for excitation is calculated using (5.24), the cross section contains an oscillatory factor like

$$1 + \cos(kR_0 \cos\gamma)\,,$$

where γ is the angle between the polarization vector and the angle of emission. R_0 is the internuclear spacing. This result can be understand in terms of the interference of two coherent waves emitted from the different centers. If from each center there is emitted a wave $\exp(i\mathbf{k} \cdot r)$, then in a direction θ their phase shift will be determined by the product $k \Delta r = kR_0 \cos \theta$. Maxima will be observed for those values of θ at which the path length Δr is an integral number of wave lengths.

The interference pattern from an oriented molecule will therefore give the internuclear spacing R_0 and the orientation with respect to the polarization vector. GADZUK [5.98] has used this fact to show the angular dependence of various surface molecules can be utilized to determine the structure of the molecule.

5.4.4. Relaxation Effects

The energy of a photoemitted electron is always given by

$$E_e = E_0^N - E_\alpha^{N-1} + \hbar\omega, \qquad (5.33)$$

where E_0^N and E_α^{N-1} are the total energies of the N electron initial state and the $N-1$, hole final state of the system, respectively. The subscript 0 for the N particle initial state indicates the ground state, which is the only initial state to be considered in this article. The symbol $\propto$ denotes any one of the possible excited states of the ion. The binding energy E_B of a photoemitted electron is given by

$$E_B = \hbar\omega - E_e \qquad (5.34a)$$

or

$$E_B^{(\alpha)} = E_\alpha^{N-1} - E_0^N. \qquad (5.34b)$$

In a one electron picture each binding energy $E_B(\propto)$ can be associated with an ion having a hole in the ith orbital, including the possibility for multiplet splittings.

The term relaxation energy, refers to the change in the energy of the final state E_i^{N-1}, due to the relaxation of the $N-1$ passive electrons towards the ith hole. The relaxation energy cannot be measured, being a theoretical concept, introduced to account for the difference between a binding energy calculated correctly using (5.34b) (including the final state) and a binding energy calculated entirely from the ground state properties of the neutral entity. Since the relaxation energy is a theoretical concept introduced within the one-electron approximation for an N electron system, its meaning may vary from one approximation to another. The following discussion will examine relaxation within the context of a Hartree-Fock (H-F) calculation.

In photoelectron spectroscopy the values of $E_e(i)$ are measured for a number of peaks i at the photon energy $\hbar\omega$. The binding energy of the ith peak is then calculated using (5.34a). Attempts to interpret these binding energies theoretically frequently relate them to the one-electron orbital energies of the initial state, using KOOPMANS' theorem which states that the binding energy of the ith electron in a Hartree-Fock calculation is the orbital energy ε_i of that one electron state, if the $N-1$ orbitals of the ion are taken to be the same as for the neutral atom. This is in fact the meaning of the orbital energy. Therefore, the binding energy using KOOPMANS' theorem with frozen orbitals in the ion, is

$$E_B^{KT}(i) = -\varepsilon_i .$$

In fact the $N-1$ passive orbitals of the ion relax towards the positive hole in the final state, and consequently are not the same as the $N-1$ orbitals of the neutral atom. Within the framework of H-F theory we can define a relaxation energy $E_R(i)$ for the ith orbital as the difference in energy between the binding energy calculated using KOOPMANS' theorem and (5.34b). E_i^{N-1} in (5.34b) is the total energy of the relaxed ion with a hole in the ith orbit,

$$E_R = E_B^{KT}(i) - E_B(i) = -\varepsilon_i - E_B(i) . \tag{5.35a}$$

The binding energy in H-F theory $E_B(i)$ is then given by the orbital energy minus the relaxation energy

$$E_B(i) = -\varepsilon_i - E_R . \tag{5.35b}$$

This subsection will discuss the magnitude of E_R(H-F) for different orbitals in a gas phase molecule, [5.101, 102] using calculations for CO as an example. At the end of this subsection the magnitude of the relaxation energy of an atom or molecule (specifically CO) adsorbed on a metal surface will be discussed. When an atom or molcule is adsorbed on the surface of a metal the electrons of the metal atoms will relax towards the positive hole left on the adsorbate so that the relaxation energy will be greater than in the gas phase. This surface induced relaxation will cause a shift in the binding energies of an adsorbate compared to the gas phase binding energies, which can be measured in contrast to E_R defined by (5.35b). The important question is how this surface induced relaxation depends upon the properties of the hole state.

The simplest example of relaxation effects in a H-F calculation is the binding energy of the $1s$ electron in He. The H-F orbital energy is -1.79 Ry while the difference between the total energies of He and He$^+$ is -1.70 Ry, so that $E_R = 1.22$ eV. This means that there is a 5% error in using KOOPMANS' theorem to calculate the binding energy of the $1s$ electron in He.

Extensive H-F calculations have been completed for the CO molecule, and for the different related states of CO$^+$ [5.101, 103]. A ground state wavefunction ψ_i is determined in a Hartree-Fock self-consistent field

Table 5.1. Comparison of relaxation energies [eV] in various environments

Configuration	Level	Relaxation energy	Binding energy	Binding energy relative to E_F	Work function φ	Surface shift
Gas phase						
O	O_{1s}	18.0 [a]	544.3 [a]			
CO	$O_{1s}(1\sigma)$	20.24 [a]	542.6 [b]			
NO	$C_{1s}(2\sigma)$	11.42 [a]	296.2 [b]			
CO	$O_{1s}(1\sigma)$	20.9 [a]	543.3 [b]			
CO	1π	1.90 [a]	16.54 [c]			
CO	5σ	1.41 [a]	14.0 [c]			
Adsorbed CO						
Molecular CO on polycrystalline W						
α_1 state	O_{1s}			534.2 [d]		
	C_{1s}			287.2 [d]		
α_2 state Virgin state	O_{1s}			532.8 [d]		
	O_{1s}			531.5 [d]		
	C_{1s}			285.4 [d]		
β state	O_{1s}			530.5 [d]		
Adsorbed atomic oxygen	O_{1s}			530.3 [d]		
	O_{2s}		28	22 [k]	~6.0 [k]	
Molecular CO on W(100)						
α_1 state	1π		13.7	8.7 [e]	4.98 [e]	2.8
	O_{1s}		538.0	533.0 [d]	4.98 [e]	4.6
	C_{1s}		291.2	286.2 [d]	4.98 [e]	5.0
α_2 state	1π		13.3	8.3 [e]	5.04 [e]	3.2
	4σ		16.5	11.4 [e]	4.05 [e]	3.2
Molecular CO on W(110)						
Virgin state	1π		13.1	7.3 [e]	5.75 [e]	3.4
	4σ		16.5	10.7 [e]	5.75 [e]	3.2
α state	1π		13.2	7.8 [e]	5.50 [e]	3.3

Table 5.1. (continued)

Configuration	Level	Relaxation energy	Binding energy	Binding energy relative to E_F	Work function φ	Surface shift
Molecular CO on Ni(111)						
	1π		13.6	7.6[f]	6.0[f]	2.9
	4σ		16.7	10.7[f]	6.0[f]	3.0
Molecular CO on Ni(100)						
	1π		13.7	7.8[g]	5.86[g]	2.8
	4σ		17.0	11.1[g]	5.86[g]	2.7
Molecular CO on Fe						
	1π			7.3[h]		
Organics						
C_2H_4 on Ni(111)	levels		10.9	6.4[f]	4.5	2.1
			12.7	8.2[f]	4.5	2.1
			14.1	9.6[f]	4.5	2.1
C_2H_2 on Ni(111)	3		13.7	9.1[f]	4.6	2.7–3.2[i]
	2		15.8	11.1[f]	4.6	2.7–3.2[i]
C_2H_2 on W(110)	3		13.3	9.1[j]	4.1	3.1
	2		15.2	11.1[j]	4.1	3.2

a [5.101].
b [5.104].
c [5.105].
d [5.106].
e [5.88].
f [5.107].
g [5.108].
h [5.107].
i [5.109] DEMUTH and EASTMAN calculated a relaxation of 3.2 eV. Comparison of their binding energies relative to E_f, including the work function to TURNER [5.105] gives 2.7 eV.
j [5.110].
k [5.163].

calculation so that the energy is minimized with respect to variations of the orbitals. The ground state wavefunction can be written as [5.101]

$$\psi_0 = [1\sigma_0^2, 2\sigma_0^2, 3\sigma_0^2, 4\sigma_0^2, 5\sigma_0^2, 1\pi_0^4],$$

where the orbitals $1\sigma_0$ and $2\sigma_0$ are essentially the atomic $1s$ orbitals on O and C, respectively. The subscript 0 denotes the orbitals with lowest total energy for the neutral CO molecule. Now we ask what the binding energy of the O_{1s} electron ($1\sigma_0$ level) is. Using KOOPMANS' theorem, the final state energy is calculated from the frozen orbitals (FO) of neutral CO with an electron missing from the $1\sigma_0$ level

$$E_{1\sigma_0}^{N-1}(FO) = \langle \psi_0^+ | H | \psi_0^+ \rangle$$

where

$$\psi_0^+ = [1\sigma_0^1, 2\sigma_0^2, 3\sigma_0^3, 4\sigma_0^2, 5\sigma_0^2, 1\pi^4]$$

and

$$E_0^N - E_{1\sigma}^{N-1}(FO) = +\varepsilon(1\sigma).$$

On the other hand, BAGUS [5.101] calculated the variational "best" orbitals for the O_{1s} hole configuration, i.e., solved the Fock equation for the configuration of an electron missing in the O_{1s} shell. This new relaxed hole-state wavefunction ψ_R^+ was then used to calculate the relaxed final state energy

$$E_{1\sigma}^{N-1}(R) = \langle \psi_R^+ | H | \psi_R^+ \rangle.$$

Using (5.34b) and (5.35b), the relaxation energy is

$$E_R = E_{1\sigma}^{N-1}(FO) - E_{1\sigma}^{N-1}(R). \tag{5.35c}$$

This definition of the relaxation energy is, in principle, more general than the H-F definition of (5.35a). The difference in energy of the ion with frozen and relaxed orbitals can, in principle, be calculated in any one-electron scheme.

The calculated binding energies for CO using (5.34b) with the relaxed final state are in very good agreement with the experimentally measured binding energies [5.103]. The largest error is 4.5% for the 1π level. This means that correlation effects in the CO molecule are not very important or that for some reason they cancel when E_B is calculated. Table 5.1 lists

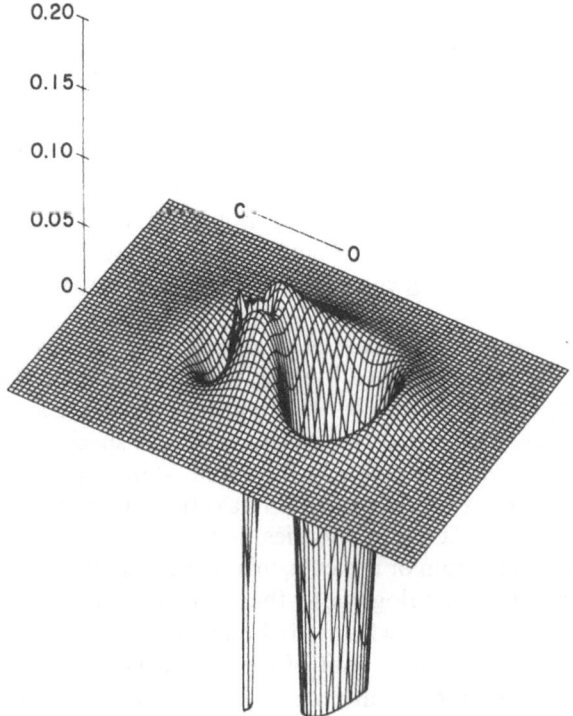

Fig. 5.11. Contour plot of the difference in charge density between frozen orbitals (KOOP-MAN's theorem) and the relaxed ion, for a hole in the $1s$ state of the oxygen. The vertical axis is in units of electrons per a.u. [5.103]

the calculated relaxation energies and the measured binding energies for several levels of CO. The binding energy and relaxation energy for the O_{1s} level in O and NO are also shown for comparison. Notice that a large fraction of the shift in the O_{1s} binding energy between O and CO or NO is a result of the change in the relaxation energy. The relaxation energy of the O_{1s} level in CO is 20.24 eV or 3.7% of the binding energy, while the relaxation energy of the 1π level is 1.90 eV or 11.5% of the binding energy. The relaxation energy or shift of the binding energy with respect to the orbital energy is not a constant or a fixed fraction of the binding energy.

The relaxation of the final state is made much more visible by the electron density difference map [5.103] shown in Fig. 5.11, where the electron density of the relaxed hole state ψ_R^+ is subtracted from that given by the "frozen" orbital—KOOPMANS' theorem state ψ_0^+, for an O_{1s} hole. The sigma orbitals (3σ and 4σ) and the 1π

orbital are both contracted. The effect of the π-orbital is discernible in the twin-humped feature between the nuclei.

The orbital energies calculated in H-F cannot be rigidly shifted or uniformly compressed in energy to obtain agreement with the binding energies. Yet the error may only be 10%, so that H-F orbital energies have served a very useful role in identification and ordering of observed binding energies in atoms and molecules. Orbital energies in other one-electron calculational schemes may not have the same significance, since analogues to Koopmans' theorem for H-F may not exist. Therefore, all calculations of binding energies should use (5.34b) and not the orbital energies.

When an atom or molecule is adsorbed on or absorbed in a metal surface, a shift in the measured binding energy (relative to the gas phase) of each energy level will result. If we view this "surface shift" in a one-electron picture, then some part of the shift of a specific level is due to bonding shifts (chemical shifts) in the initial state, while the remaining contribution to the shift comes from relaxation effects in the final state. In contrast to the relaxation energies defined by (5.35a) for the energy levels of an isolated atom or molcule, this "surface shift" can be measured. This measurement is analogous to the shift in the binding energy of a core electron in an atom when it is bound in different molecules. For example, Table 5.1 shows a shift of the O_{1s} level of 1.7 eV in CO and 1.0 eV in NO compared to atomic oxygen. If the calculated relaxation energies shown in Table 5.1 are correct then the differential relaxation accounts for 2.2 eV in CO and 2.9 eV in NO. Therefore, the "chemical shift" in the initial state must be -0.5 eV for CO and -1.9 eV for NO. This example illustrates the complexity involved in trying to extract the shift due to the initial state from the measured shift, either in the gas phase or adsorbed on the surface. Obviously, if the binding energy was calculated using (5.34b), both of these effects would automatically be taken into account, but it is not likely that these big systems (bulk plus adsorbate) will be routinely calculated. Therefore, it is useful to discuss the magnitude of the surface shift induced by relaxation in the final state.

The magnitude of this surface induced relaxation effects is important in the following situations: 1) When the photoelectron spectrum of an adsorbed molecule is being compared to gas phase spectra in an attempt to identify the chemical nature of the surface molecule [5.109, 110]; 2) when the shift in a core level of an adsorbed atom or molecule is being used to determine the "chemical shift" due to bonding, and 3) when valence band spectra of an adsorbed atom or molecule are being used to determine the bonding energy shift of a specific energy level [5.107]. The relaxation contribution to the "surface shift" could be formulated in terms of (5.35c), where the relaxation on the surface complex is calculated and the relaxation in the gas phase subtracted. This is a very cumbersome definition, since it relies on the concept of frozen orbitals and necessitates two separate calculations. A much more physical picture of the "surface shift" is obtainable by use of a simple Born-Haber cycle.

If $E_{ion}(i)$ is the bond energy of the ion to the surface, after an electron has been removed from the ith orbital, and $E_{neut.}$ is the bond energy of the atom or molecule to the surface in the ground state, then the binding energy of the ith orbital on the surface $E_B^s(i)$ can be related to the binding energy in the gas phase, $E_B(i)$ by

$$E_B^s(i) = E_B(i) - E_{ion}(i) + E_{neut.} .$$

The "surface shift", which is $E_B(i) - E_B^s(i)$, is given by the difference in the bond strength of the neutral and the ionized molecule. This energy difference will depend upon not only the nature of the ith orbit, but the degree of electronic and spatial relaxation of the ion. This equation illustrates that physically "relaxation shifts" and "bonding shift" are inseparable, unless one or the other can be calculated.

The limiting case of photoionization from a level which is not involved in the bonding or shifted as a consequence of bonding can be discussed within the context of the above equation. This means that we are considering the "surface shift" of an energy level whose H-F orbital energy is the same in the gas phase as it is when bonded to the surface. In addition, we assume that: 1) The relaxation of the outer electrons of the molecule on the surface is the same as in the gas phase and, 2) that the relaxation of the electrons about the hole does not change the bond strength of the neutral molecule. In other words, we will consider only effects due to screening of the hole by the electrons in the metal. This means that the bond energy of the ion is the neutral bond energy plus the change in the energy of the metal due to screening the hole. SMITH et al. [5.111] have calculated the change in energy of a jellium metal [5.112] as a proton is brought up to the surface. The response of the electron gas depends upon the distance of the proton from the jellium surface. This energy is 5.7 eV for a proton 1.5 Å from the surface and reaches a maximum of 8.7 eV at a spacing of 0.5 Å. This calculation is for a high density electron gas so that these values for the "relaxation energy" are probably upper limits. The response of the electron gas will depend upon the density of the gas, the position of the ion relative to the surface and the spatial extent of the hole orbit. The relaxation energy due to the metal electrons screening the ion will decrease as the size of the hole state increases. In a one-electron model like H-F this contribution to the surface shift goes to zero for a completely delocalized hole state. This fact is the basis for assuming KOOPMANS' theorem is correct for excitation from the conduction band of a solid.

Table 5.1 tabulates the "surface shift" for several orbitals of CO adsorbed on different metals, as well as two valence orbitals of C_2H_2 adsorbed on Ni and W. There is always some uncertainty in the binding energies of an adsorbed molecule, since the measurements are made relative to the Fermi energy and it is not clear which value of the work function should be added. The work function used as well as the measured binding energies relative to the Fermi energy are shown in Table 5.1. The work functions used were measured with the adsorbate present.

The measurements for the α_1 state of CO adsorbed on (100) tungsten illustrates the effects of different orbits. The surface shift for the C_{1s} level is 5.0 eV which is larger than the shift of 4.6 eV for the O_{1s}, presumably because the CO is bonded to the surface standing up with the carbon closer to the surface. Both the C_{1s} and O_{1s} have a larger shift than the 1π level, since they are more localized. In a one-electron picture the shift should go to zero for a spatially extended orbital. The surface shift of the 1π and 4σ levels of CO shown for the α_2 state of CO on (100) W and for adsorption of CO on (110) W, Ni (111), and Ni (100), are nearly identical for all cases. The charge distribution in these states is approximately the same with a ratio of 3:1 on the oxygen compared to the carbon. The surface shift for all of the valences orbitals listed in Table 5.1 ranges from 2.7 eV to 3.4 eV, with the exception of C_2H_4 on Ni (111).

Therefore, when a gas phase photoelectron spectrum is being compared to that of an adsorbed molecule, the former should be shifted upward by ~ 3 eV in the energy range of the valence orbitals. If some of the orbitals do not line up properly there is no way at present, without the aid of a calculation, to tell if this is a consequence of a different relation energy or a bonding shift. The experimentalist should operate on the premise that relaxation is not uniform for all levels. There may be situations like the 1π and 4σ levels of CO where the surface shift is constant, presumably as a consequence of both orbits having a similar charge distribution relative to the surface.

5.4.5. Bulk vs. Surface Emission

In this subsection, the concepts of bulk and surface photoemission will be briefly reviewed, always from the point of view of an observer outside trying to see the bulk through the surface. There is an interference between the bulk and surface [5.71, 77] photoemission processes, and at present there is no theoretically justifiable model for extracting either the bulk contribution or the surface contribution from the total distribution. In some cases, there may be so little emission from the surface that the energy distribution is primarily bulk. Likewise, when a state at the surface is spatially localized and split off from the bulk band, it is easy to identify, but in general the surface density of states in the region of the band will be very difficult if not impossible to extract from the energy distribution.

Let us begin with the simplest solid, that is — an infinite free electron gas. The initial and final states are

$$\psi_i \propto \exp(i\mathbf{k}_i \cdot \mathbf{r}); \psi_f \propto \exp(i\mathbf{k}_f \cdot \mathbf{r}),$$

where by conservation of energy

$$\hbar^2 k_f^2/2m = \hbar^2 k_i^2/2m + \hbar\omega.$$

If we apply the momentum matrix element of (5.24), $k_f = k_i$.

This can never be true if energy is conserved and $\hbar\omega \neq 0$. Therefore, there can be no photoexcitation in an infinite free electron gas. The application of the gradient in potential matrix element of (5.25b) to a free electron gas immediately produces the above result since $\nabla V = 0$ for a free electron gas. If this gas is bounded by a surface, then $\nabla V \neq 0$ at this surface and we have what is called the "surface photoelectric effect". This fact led early investigators to conclude that photoemission occurred only at the surface [5.75, 113–116]. The search for this effect has spanned three decades, from the work of MITCHELL in 1934 [5.113] up to the present [5.116]. The theory of the "surface photoelectric effect" has been worked out in detail for a free electron metal by ENDRIZ [5.116]. (A free electron metal allows no bulk excitations so in this case there is no possibility of interference between bulk and "surface emission" [5.77].)

In a real metal, however, the concept of surface emission is not clearly defined. In a non-free electron metal, there are gradients in the potential throughout the bulk caused by the ion cores. "Bulk photoemission" by definition, originates from this three-dimensional, periodic array of the potentials. Near the surface, the magnitude, spacing and periodicity of the potentials may change. The consequence of these changes (see the introduction of this chapter) is that the electronic properties in the vicinity of a surface atom are different from that of an atom in the bulk. The term "surface emission" can be taken to mean either the emission from the potential step at the surface or from the surface layer or layers of atoms. In a real crystal with ion cores, there does not seem to be any theoretical way of defining and separating the "surface" from the "bulk" effect [5.71, 77]. The only cases where a separation can be justified is when either one of the components becomes very small. When one considers a free electron metal, the bulk contribution is zero and the functional form of the surface emission may be predicted. Likewise, when the incident light is normal (so that $A \cdot \nabla V = 0$) and the electron escape depth is much larger than the region of the bulk which has been perturbed by the surface, then the surface contribution should be small and the properties of the emission predicted by the bulk properties. In any intermediate region, the two effects are so mingled that they may be inseparable. For UV photoemission, anything from 10% to 90% of the signal may come from the surface.

In Subsection 5.4.2 where photoexcitation cross sections were introduced it was pointed out that the $P \cdot A$ term in (5.21) was usually dropped assuming that any gradients in A were small. The surface-polarization charge induced by the component of A perpendicular to the surface (A_z) causes a rapid change in A_z over a region of ~ 2 Å. ENDRIZ [5.116] has shown how important this terms is in calculating the "surface photoelectric effect".

Let us now describe a more sophisticated model for bulk photo-emission, usually called the "direct transition" model. Here we write the initial and final states as 3-D Bloch states.

$$\psi_f \propto \exp(i \boldsymbol{k}_f \cdot \boldsymbol{r}) \, U_{\boldsymbol{k}_f}^n(\boldsymbol{r}) \qquad \psi_i \propto \exp(i \boldsymbol{k}_i \cdot \boldsymbol{r}) \, U_{\boldsymbol{k}_i}^m(\boldsymbol{r}),$$

where n and m are band indices. If we apply the momentum matrix operator of (5.24), we have

$$
\begin{aligned}
P_{if} &\propto |A_0 \cdot \langle \psi_f | \boldsymbol{P} | \psi_i \rangle|^2 \\
&\propto |A_0 \cdot [\int \exp(i(\boldsymbol{k}_f - \boldsymbol{k}_i) \cdot r) \, U_{\boldsymbol{k}_f}^n \, \nabla U_{\boldsymbol{k}_i}^m \\
&\quad - i \boldsymbol{k}_i \exp(i(\boldsymbol{k}_f - \boldsymbol{k}_i) \cdot r) \, U_{\boldsymbol{k}_f}^n U_{\boldsymbol{k}_i}^m \, d\tau]|^2 \, .
\end{aligned}
$$

The second term is zero unless $n = m$, which cannot be true if the energy of the final state is to be larger by $\hbar \omega$ than the energy of the initial state. The first term gives the criterion that $\boldsymbol{k}_f = \boldsymbol{k}_i$, where $\boldsymbol{k}_f$ and $\boldsymbol{k}_i$ are the crystal momentum in the reduced zone scheme. The $\boldsymbol{K}$ of the final state in the extended zone scheme is $\boldsymbol{K}_f = \boldsymbol{k}_f + \boldsymbol{G}_n$, where n denotes the reciprocal lattice vector associated with the nth band. The energy distribution of excited electrons $P_0(E, \omega)$ is given by summing all bands and integrating over all k in the reduced zone [5.21]

$$P_0(E, \omega) \propto \sum_{n,m} \int d^3 k \, P_{if} \, \delta[E_n(k) - E_m(k) - \hbar \omega] \, \delta(E - E_n), \qquad (5.37)$$

where

$$P_{if} \propto \int U_{\boldsymbol{k}}^n \, A_0 \cdot \nabla U_{\boldsymbol{k}}^m \, d\tau \, .$$

If we want to calculate the number of electrons excited in a direction K, then the reduced zone scheme of (5.37) is not very useful, since the reduced k_f does not give the direction of propagation of the final state. For a free electron band structure, each band m has a specific G_m which was used to fold the extended zone back to the reduced zone. For example, consider the case of a band structure in the (001) direction of a simple cubic. The first band obviously has $G_1 = 0$. The second band is folded back with $G_2(00\bar{1})$. The next set of bands are composed of reciprocal vectors of the form $G_3(010)$. An excitation from an initial state given by k_i in the first band to a final state k_f (reduced zone) in the second band occurs when

$$\frac{\hbar^2}{2m}\left[k_f + \frac{2\pi}{a}(0, 0, \bar{1})\right]^2 = \hbar\omega + \frac{\hbar^2}{2m}k_i^2 \qquad .$$

or since $k_f = k_i$

$$\hbar\omega = -\frac{4\pi}{a}|k_f| + E_G \quad \text{where} \quad E_G = \frac{\hbar^2}{2m}\left(\frac{2\pi}{a}\right)^2 .$$

The direction of this excited electron is

$$K = K_f + (2\pi/a)(0, 0, 1) = \left(0, 0, \frac{-2\pi}{a} + k_f\right).$$

If positive k_z is out of the crystal, then this direction is back into the crystal since $k_f < \pi/a$. But the initial state of $-k_i$ excited to $-k_i + (2\pi/a)$ (0, 0, 1) comes out. In contrast, excitation from $k_i(G = 0)$ to k_f in the band associated with $G = (2\pi/a)(0, 1, 0)$ occurs when

$$\hbar\omega = E_G = (\hbar^2/2m)(4\pi^2/a^2).$$

The direction of the final K is implicitly given by

$$K = (2\pi/a)(0, 1, 0) + k_i$$

or if θ is the angle from the z axis,

$$\tan\theta = 2\pi/a|k_i| \qquad \theta = \tan^{-1}\left(\frac{2\pi}{a|k_i|}\right) \geq 63°.$$

This excitation would not be seen by a detector normal to the surface. In general, a plot of the bulk energy bands in a reduced zone scheme will indicate which transitions are allowed for a given photon energy, but it will not yield information about the direction of propagation of the excited electron, unless the reciprocal lattice vectors of the initial and final state bands are known. The direction of propagation of the excited electron need not be in the same direction as the k of the initial state.

In a complicated band structure, there are many G's for a given k_f in a band and the weight associated with a given G changes as k_f changes. Therefore, it is not possible to make a unique identification of k_f in the reduced scheme with a single K_f in the extended zone. In the extended zone scheme, the energy distribution of electrons with final energy E and final direction K_f is given by

$$P_0(K, E, \omega) \propto \sum_{i,f,G} P_{if} \delta(E_f - E_i - \hbar\omega)\, \delta(K_f - k_i - G)$$
$$\cdot \delta(K_f - K)\, \delta(E - E_f), \tag{5.38}$$

where

$$P_{if} \propto |\int U_{K_f,G} A_0 \nabla U_{ki,G} d\tau|^2.$$

P_0 in (5.37) or (5.38) is usually multiplied by an escape function, assuming a three step process of light penetration, excitation, and transport and escape [5.117]. This term, even if it is appropriate does not shed much light on the physics of the process; consequently, we will leave it out.

If the matrix element P_{if} of (5.37) is assumed to be a constant for all i or f, then the calculated energy distribution is called the energy distribution of the joint density of states (EDJDOS) [5.118]. If the delta function on crystal momentum is removed from (5.37), the newly allowed transitions are called "non-direct" since momentum is not conserved. If P_{if} is set equal to a constant in this case, the energy distribution is proportional to the product of the initial and final density of states.

Assuming that the excitation process may be separated from the escape function, let us investigate the effect of the short attenuation length on the direct transition model. To begin with, let us ignore the fact that the short mean-free path localizes the excitation near the surface; consider for the moment the excitation process deep in the bulk. The scattering effects can be taken into account in a phenomenological fashion by assigning an uncertainty ΔK to the wave vector of the final state. ΔK can be estimated by using the uncertainty principle in terms of the electron attenuation length λ

$$\Delta K = 1/\lambda.$$

The smear in k is over ~20% of a reduced zone for a lattice spacing $a = 3$ Å and an escape depth of 5 Å. This washes out the sharp structure predicted by a "direct transition model" for one electron bands; the energy distribution begins to look like the "non-direct" model.

Now consider the effect of localizing the excitation near the surface. In this region, the component of the K vector perpendicular to the surface (K_z) is strictly speaking not a good quantum number. There will still be conservation of the crystal momentum parallel to the surface

$$k_f^{||} = k_i^{||} + G_{||} ,$$

where G is a surface reciprocal lattice vector. MAHAN [5.119] has shown that if $a/\lambda \ll 1$ the excitation process is described by conservation of the 3-dimensional k (a is a characteristic spacing in the crystal). In the region of interest to us $\dfrac{a}{\lambda} \sim 1$ so we would expect only a small remnant of k_z conservation. This again will have the effect of washing out both angular dependences and final state structure predicted by a bulk model [5.119]. (A problem which is related to this is the change in the density of both initial and final states near the surface [5.120–123].)

One of the simplest examples in the literature which illustrates the problem of separating bulk and surface effects is the Kronig-Penney model used by SCHAICH and ASHCROFT [5.77]. This is basically the same model used by DUKE and FAUCHIER [5.35] for field emission calculations consisting of two dimensional δ function sheets spaced an equal distance apart and parallel to the surface. The surface is a potential step. SCHAICH and ASHCROFT concluded that, to the extent that the "volume" and "surface" effects may be separated, they *must* interfere since the matrix element given by (5.25b) has a term containing the gradient in potential at the surface and a term containing the periodic bulk potential. The current is given by the square of this probability, and either constructive or destructive interference may occur. In Ref. [5.77], Fig. 1, the effect of this interference on the energy distribution as a function of the electron escape depth is shown. For an escape depth of 5 Å, the energy distribution does not look like what would have been predicted from the bulk, the surface, or any linear combination of the two. Figure 5.12 taken from [5.77] shows the angular dependence of emission for a final energy 1.4 eV above the vacuum level, as a function of the electron escape depth. This illustrates many of the points we want to make. If the escape depth is very large, the curve shown in Fig. 5.12 would be a delta function as MAHAN's theory would have predicted [5.119]. The effect of the inelastic scattering on the final state is to wash out this sharp structure, as can be seen even for an escape depth of 30 Å. The dashed line at the bottom is the surface photoelectric effect from the step potential at the surface [5.113]. As the electron escape depth is decreased, two effects occur — the amplitude decreases with an increased half width and there

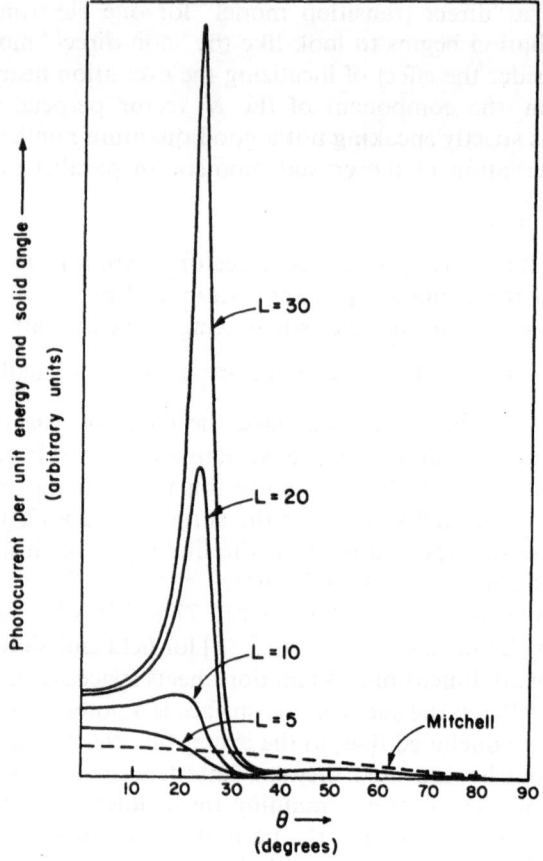

Fig. 5.12. Angular distribution of photoemitted electrons from a modified Kronig-Penney model [5.77]. The emitted electrons have an energy of 1.4 eV above the vacuum. L is the parametrized electron escape depth λ. The dashed curve is the angular dependence of the surface photoelectric effect for the step potential at the surface, first calculated by MITCHELL [5.113]

is both constructive and destructive interference between the bulk and surface effects. The effects seem to be constructive for $\theta < 15°$ and destructive for $\theta > 20°$. This model could be extended by changing the strength and position of the first delta function potential to represent a surface atom, but the effect of the interference is clearly illustrated by Fig. 5.12.

One final observation in this section is that if for any reason the hole state becomes localized, the direct transition model will break down [5.124]. In this case, the final state is not a Bloch state, and there will be

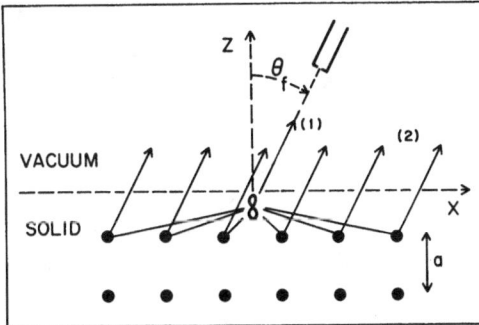

Fig. 5.13. Illustration of two processes contributing to photoemission from an adsorbate p_z orbital 1) direct emission into a plane-wave final state, 2) indirect emission via backscattering from the substrate. Only single scattering from the first layer is indicated [5.99]

no selection rules on crystal momentum. This was suggested by SPICER in 1967 to explain the fact that the energy distributions from many metals appeared as if they were caused by non-direct transitions.

5.4.6. Angular Resolved Surface Emission

The possibility of orienting molecules by adsorbing them on a surface presents great opportunities for angular and energy resolved photoemission. Subsection 5.4.3 already pointed out the expected effects for an oriented gas phase molecule. GADZUK [5.98] calculated the angular dependence for a variety of atomic orbitals and molecular surface complexes, and showed that information about the bond geometry may be obtained from the angular measurements. If one knew the direction of the A vector the symmetry and possibly the direction of the orbital being observed could be measured.

The calculations described in Subsection 5.4.3 and those of GADZUK [5.98] assume a plane wave final state. When the molecule or atom is adsorbed the final state must include the scattering from the crystal. This situation is best described by Fig. 5.13 where a hypothetical adsorbate is pictured above a periodic lattice [5.99]. If the detector is positioned at a polar angle θ_f and an azimuthal angle φ, two coherent waves are detected: 1) The direct excitation into a plane wave (GADZUK's process) and 2) Backscatter waves from the surface. LIEBSCH [5.99] has developed a "one-step" model for photoemission which uses a multiple-scattering theory such as has been applied to LEED to treat the final state. By applying this formalism to the case of localized adsorbate energy levels he could show that the symmetry of the adsorption site was as important

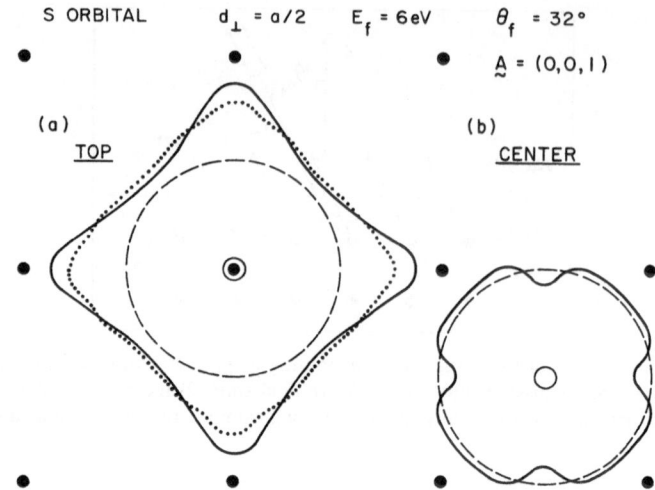

Fig. 5.14a and b. Photoemission intensity (arbitrary units) as a function of azimuthal angle for an s orbital adsorbed in top (a) and center (b) positions on the (100) face of a cubic crystal: Single scattering (solid curve), multiple scattering (dotted curve), shown only in panel (a), and no scattering (dashed curves) [5.99]

in determining the angular dependence as the symmetry of the bonding orbital.

Figure 5.14 shows an azimuthal plot of the intensity from an s orbital on a simple cubic substrate. Scattering in the final state produces a significant effect, with the symmetry of the pattern reflecting the symmetry of the bonding site. In these calculations the vector potential is perpendicular to the surface.

At present very little information is available on angular resolved photoemission from adsorbate covered surfaces. Figure 5.15 displays some preliminary work of Waclawski [5.125] for oxygen adsorbed on W(100), at $\hbar\omega = 21.2$ eV. The top set of energy distributions are for clean (100)W and an exposure of 5×10^{-6} Torr sec of O_2 at a crystal temperature of 1500 K. All the emission within a polar angle of 45° was collected. The second set of curves is for the same situation except only those electrons which were emitted at a polar angle of 33° ± 2.5°, perpendicular to the plane of incidence of the light were collected. The bottom set of curves are for electrons emitted normal to the surface +3.8°. The top set of curves shows a double peak at ~ −5.0 and −6.0 eV derived from the p orbitals of oxygen. The high energy peak is accentuated at normal collection while the low energy peak is more pronounced at 33° collection. The peak at −2.0 eV in the top curves is much more pronounced at 33° than that at 0°, while the peak at −0.4 eV in the clean

ANGULAR DEPENDENCE OF O_2 ON (100)W

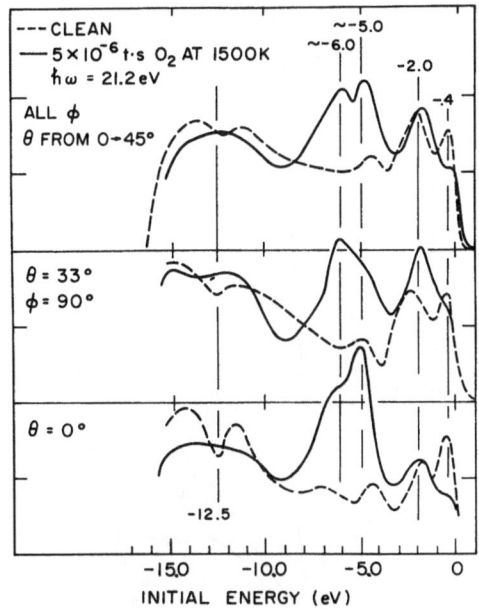

Fig. 5.15. Angular dependence of the photoemission spectra for 5×10^{-6} Torr · sec O_2 at 1500 K on (100)W [5.125]. The top set of curves are for collection of all the photoemitted electron within a polar angle of $\pm 45°$. The middle set of curves is for a collection angle of $33° \pm 2.5°$ normal to the plane of incidence of the light. The final set of curves are for normal emission ($\pm 3.3°$). The incident light was at 45° and 21.2 eV

curve which is a surface state (see Subsection 5.5.1) is more pronounced at 0° than at 33°. The dip at -12.5 eV is caused by a band gap in the final state; there are no states in the crystal for the secondary electrons. It is much more pronounced at $\theta = 0°$ than at $\theta = 33°$. If one interpreted these curves in terms of initial state effects then the lower energy oxygen-derived level would be in the plane of the surface and the higher energy level would be more in the perpendicular direction. The -2.0 eV level shown at $\theta = 33°$ could be a d-like level in the substrate mostly in the plane of the surface.

Figure 5.15 indicates a more practical problem: what one sees depends upon where one looks. For example, if the detector was fixed at large θ one would never know that the oxygen derived levels were split. Notice also that the peak position of the high energy peak is shifted by about 0.5 eV between the $\theta = 0°$ curve and the large collection angle at the top. Likewise there is about a 1 eV shift in the "d-band" peak at -2.0 eV in the clean curves. LIEBSCH and PLUMMER [5.126] have applied LIEBSCH's

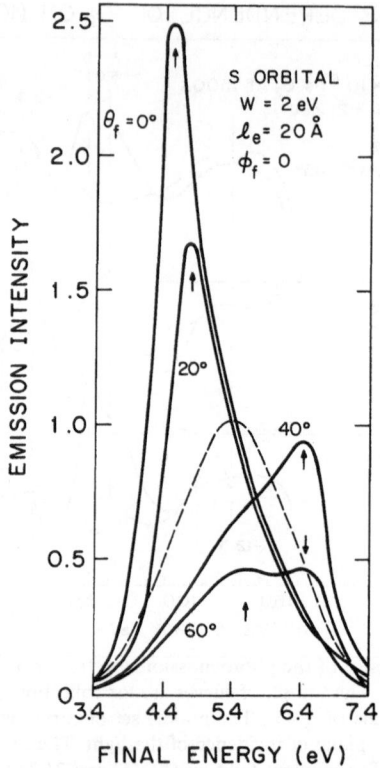

Fig. 5.16. Photoemission energy distribution as a function of final energy for a 2 eV wide p_z orbital adsorbed in the top position on a cubic lattice. The solid curves represent the intensity along four different detector angles. The dashed curve shows the intensity in the absence of scattering. The arrows indicate the effective peak positions caused by final state effects [5.126]

[5.99] "one-step" scattering model to investigate the effects on line shape and peak position in a photoemission spectrum from an adsorbate covered surface as a function of the angle of detection and the photon energy. The initial state of the adsorbate was assumed to be Gaussian in shape centered around a binding energy E_i (relative to the vacuum) with a full width at half maximum given by W. Then for a photon energy $\hbar\omega$ we would expect to see electrons emitted with a final energy in the range

$$-E_i + \hbar\omega - W \leqq E_f \leqq -E_i + \hbar\omega + W$$

peaked at $E_f = -E_i + \hbar\omega$. Because of the coherent scattering in the final state, the photoemission intensities exhibit a considerable amount of structure over the energy range shown above, thus causing the observed peak to deviate in shape and position from the Gaussian distribution centered around $-E_i + \hbar\omega$. This effect is illustrated in Fig. 5.16 where the calculated energy distributions are shown for an adsorbed s orbital sitting in the "top" position of a cubic crystal. The solid curves are for various detector directions θ_f and the dashed curve gives the distribution corresponding to no scattering. Over the 60° range of detector angles the peak is seen to vary considerably in shape and peak position (~ 2 eV). At $\theta_f = 60°$ a weak splitting of the peak takes place. These effects depend upon the final energy E_f (photon energy for a fixed adsorbate level), the electron attenuation length λ and the width W of the adsorbate level.

Averaging over large collection angles has the effect of averaging out the final state scattering effects. Therefore it seems advisable to use as large a collection angle as possible if the objective of the experiment is to look only at the energy level spectra of adsorbates. Photoemission used for the "Fingerprint" technique should certainly use a large collection angle.

5.5. Experimental Results

The last two sections will be devoted to presenting experimental data from photoemission and field emission experiments. In general, the emphasis will be placed on systems where both field emission and photoemission data are available.

5.5.1. Clean Surfaces

It was shown in Subsection 5.2.1 that field emission can measure the "one-dimensional density of states" at the turning point. There are band structure calculations [5.127] and photoemission measurements [5.128], for several directions of tungsten which we will be compared to the field emission measurements. Our objective will be twofold: 1) to determine if the "one-dimensional surface density of states" differs from what a bulk calculation would predict and 2) compare field emission and photoemission to determine the sensitivity of photoemission to the surface density of states. Figure 5.17 shows a comparison of the calculated energy bands in the direction perpendicular to the surface [5.127], the measured field emission R curves (see (5.11)), the smoothed one-dimensional bulk density of states and the photoemission energy distributions of FEUERBACHER [5.128, 129]. The photoemission energy distributions were measured normal to the surface within $\pm 6°$ and the photon energy was chosen to avoid predicted bulk direct excitations [5.127]. CHRISTEN-

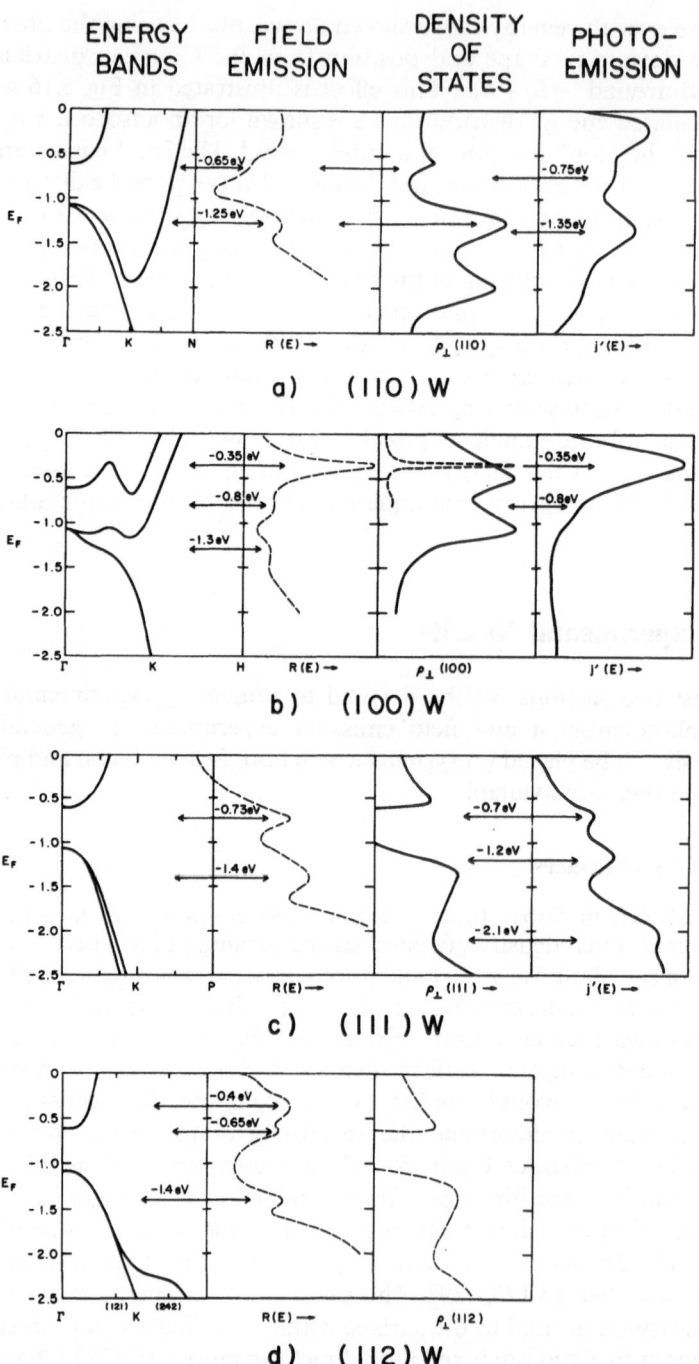

a)   (110) W

b)   (100) W

c)   (111) W

d)   (112) W

SEN has calculated that in the (110) direction there is a gap in the final state extending from approximately 7.8 eV to 9.8 eV above the Fermi energy. The photoemission curve in Fig. 5.17 is for $\hbar\omega = 9.5$ eV, putting the excited state right in this gap.

At first glance the resemblance of the field emission, photoemission and one-dimensional density of states is fairly good, with the exception of the peak at -0.35 eV on (100) W. For all four faces the peak at -1.3 to -1.4 eV seems to line up fairly well with the bulk calculation (third column) while there is a noticeable difference on (111) W between the -1.4 eV level in field emission at the -1.2 eV level in photoemission. The peaks in both the photoemission and field emission distributions at the top of the gap near -0.7 eV seem to be lower than the bulk calculations would predict. Four specific features in the field emission curves cannot be explained by the bulk band structure: 1) the large peak at -0.35 eV on (100) W which is a surface resonance in the spin-orbit split gap in that direction [5.129–131]; 2) a peak and or shoulder on (110) W near -0.4 eV; 3) an additional peak at -0.4 eV in the (112) direction and 4) the current which appears in the total gaps in the (111) and (112) directions.

The first three items are indications of density of states changes near the surface. The most pronounced is the surface resonance on the (100) face. The dashed curve in the third column is the surface density of states calculated by STURM and FEDER [5.133], who used a 3 band tight binding model which takes into account the surface resonance in the spin-orbit induced gap. Their calculation predicts a slightly narrower peak than found in field emission which itself is narrower than the photoemission peak. It has been argued that the photoemission curve is broader because it samples a larger region of k-space. In these measurements FEUERBACHER [5.129] used a 12° collection angle. The 5° collection angle shown in Fig. 5.15 does not produce a peak which is noticeably narrower.

WACLAWSKI and PLUMMER [5.131] examined this surface resonance on (100) W as a function of exposure to hydrogen, in field and photoemission. This comparison is shown in Fig. 5.18. The measured work function changes indicated that < 0.2 of a monolayer of hydrogen was

Fig. 5.17a–d. Comparison of energy bands [5.127], field emission R curves, one-dimensional bulk density of states [5.127] and normal emission photoemission energy distributions [5.128] for four low index faces of tungsten. The energy is $\hbar\omega = 9.5$ eV for (110) W; 10.2 eV for (100)-W and 10.2 eV for (111) W. The density of states shown in Column 3 is the one-dimensional density of states calculated by CHRISTENSEN [5.127] from energy bands (shown in the first column), which have been smoothed by hand to remove the singularities. The dashed curve in the third column for (100) W is from a surface resonance calculation by STURM and FEDER [5.133]

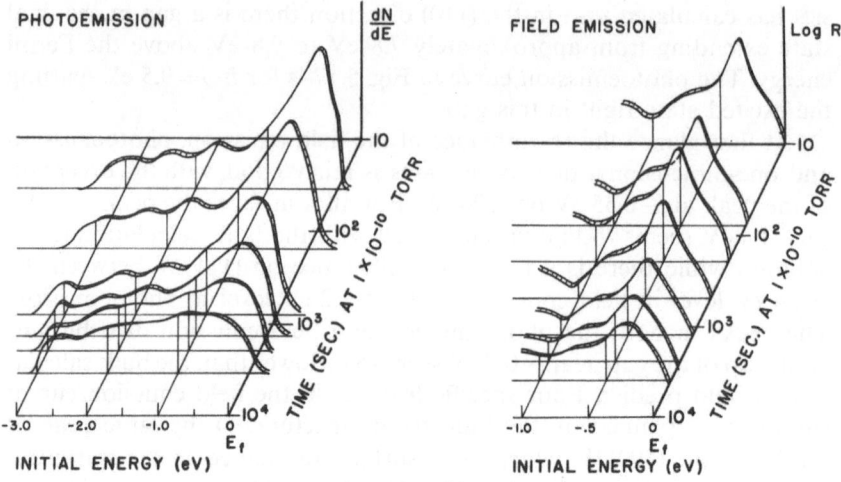

Fig. 5.18. Comparison of the sensitivity of the surface state on (100)W to hydrogen adsorption for photoemission at $\hbar\omega = 7.7$ eV (left) and field emission (right) [5.131]

adsorbed in the final curves. Within the accuracy of two different vacuum gauges these two sets of curves show the surface resonance disappearing in the same manner as hydrogen is adsorbed. Thus photoemission is dominated in this case by the signal from the surface. The latter decreases relative to the bulk signal as the photon energy is increased (see Fig. 5.15) in contradiction to what an escape depth argument would predict. WACLAWSKI and PLUMMER [5.131] measured the optical excitation probability and showed that it peaked at low energies, i.e. the variation in the photoexcitation cross section as a function of energy is more important than the variation of the electron escape depth with energy.

The fourth point of discrepancy between the field emission—photoemission curves and the bulk calculations is the current which is present in the gap regions on (111) and (112) tungsten. It would be fairly easy to attribute this either to electron scattering or to breakdown of conservation of k across the interface. But before jumping to that conclusion one must consider that is measured on these high-index faces. Consider first the case of photoemission, and assume for the moment that the surface is merely a window through which the bulk may be viewed. We now require

$$K_f = k_i + G, \quad K_{f_\parallel} = 0, \quad \text{and} \quad E_f = E_i + \hbar\omega .$$

There are two ways to satisfy the $K_\parallel = 0$ criterion; the first is obvious and corresponds to the slice of k-space shown in Fig. 5.17, that is if z

is the normal direction then

$$k_{ix} = k_{iy} = 0$$

and

$$G_x = G_y = 0 .$$

The other solution is more interesting, i.e.

$$\left.\begin{array}{l} k_{ix} = - G_x \\ k_{iy} = - G_y \end{array}\right\} k_{i_\parallel} = - G_i^\parallel$$

so that transitions other than along the cut in k-space shown for each plane in Fig. 5.17 may occur. The criterion for this condition is the existence of a reciprocal lattice vector G such that the vector

$$Q = (n \times G) \times n$$

lies within the first Brillouin zone. The unit vector n lies normal to the surface. For a low-index face this criterion can never be satisfied so that the E vs k curves for (110) and (100) W are the only possibilities for direct transition. For the (111) face there is another slice in k space where $k_{i_\parallel} = - G_\parallel$ besides the Γ to P cut shown in Fig. 5.17. This is along the zone face between the points H and P. Comparing the points H and P it is obvious that there will not be a gap in this cut in the energy range -0.6 to -1.1 eV. How much these transitions contribute to the energy distribution is uncertain, but the band structure from Γ to P does not tell all the story even if the surface contributions are neglected. All planes of higher index than (111) have other lines of allowed transitions besides the line from Γ in k space. Therefore, we can now understand why the intensity in the photoemission spectra should not go to zero in the gaps shown in Fig. 5.17c and d, without involving surface effects.

Even though there are no excitations in field emission the same type of process may occur, if we require that the $k_\parallel$ of the lattice potential be conserved across the barrier then

$$k_\parallel \text{ (outside)} = k_\parallel \text{ (inside)} + G_\parallel .$$

If we ignore the slight spread in $k_\parallel$ (outside) then $k_\parallel$ (outside) $= 0$ and

$$k_\parallel \text{ (inside)} = - G_\parallel .$$

Thus bulk states with $k_\parallel \neq 0$ can be diffracted by the surface into $k_\parallel = 0$ states. This would allow one to look at the bulk states along the P to H cut in a *bcc* crystal when looking at the (111) surface. Therefore we would not expect to have zero current in the gap in the (111) and (112) directions of tungsten. For (112) W the line in k space where $k_\parallel$ (inside) $= -G_\parallel$, enters the Brillouin zone at N [(110) direction] and leaves the Brillouin zone at a point where another [112] vector intersects. Therefore the peak at -0.4 eV on (112) W and -0.4 eV on (110) W might originate from structure near the K point in the Brillouin zone. For example, the band rising above E_F near N in Fig. 5.17a might intersect the zone face below E_F.

The photoemission energy distributions at low energies, normal to the surface of a single crystal of tungsten seem to reproduce fairly accurately the "one-dimensional surface density of states" at least within 2 eV of E_F. This statement is surely not true at higher photon energies [5.87, 128], as will be seen in the next subsection, so that no general conclusions should be drawn from the work on tungsten. It would be helpul to have field emission distributions for comparison with normal photoemission distributions for several other materials but these data do not exist. How then can one find at least rough guidelines for separation of bulk and surface effects?

In certain cases such as surface states [5.134, 135] or surface resonances [5.131, 132] changes occurring upon adsorption will identify the surface contribution. For tungsten this works only for the surface resonance of the (100) face. CHRISTENSEN [5.127] and FEUERBACHER [5.128] "eliminate" bulk emission by deciding on a model for it, in this case "direct transitions". All the structure in the energy distribution which cannot be identified with the bulk model is then tentatively classified as surface emission. The photon energy dependence of the model is used to decide whether a given peak corresponds to a direction transition. What is left is by definition non-bulk and in some way represents the presence of the surface. It surely cannot be identified with the local density of states of the surface layer, except in rare cases where there is an unusually large surface excitation probability. Even when there is a gap in the final state the excitation goes from initial state to an evanescent wave which penetrates into the bulk. All of the intensity in the photo-emission curves shown in Fig. 5.17 comes from non-bulk emission.

ROWE and IBACH [5.136] used a more experimental approach to separate and bulk contributions from the (111) face of Si with different 2-*D* periodicity, by averaging the energy distributions from 3 different 2-*D* geometrical arrangements of the Si (111) plane. This averaged distribution was then subtracted from each of the original distributions of obtain the surface contribution. Justification for this procedure comes from the fact

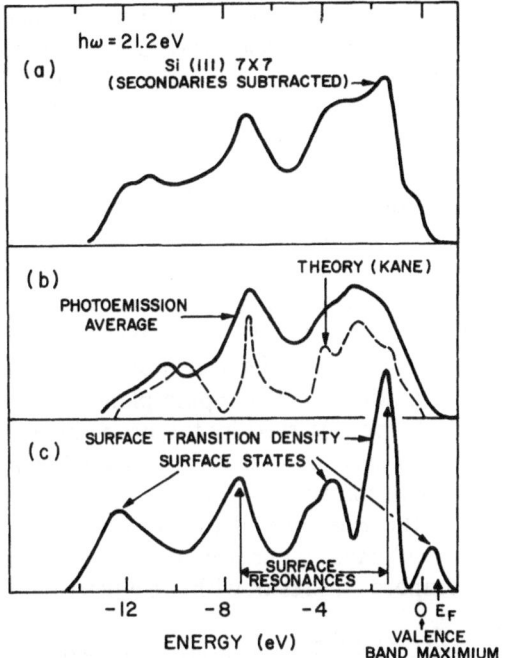

Fig. 5.19. (a) Photoemission energy distribution from Si (111) 7×7 with the estimated contribution from secondary electrons removed. (b) Comparison of "averaged" photoemission curve with the density of states calculated by KANE [5.137]. (c) Surface transition density of states obtained by ROWE and IBACH [5.136] by subtracting the assumed bulk signal [curve b] from the measured energy distribution [curve a]

that the averaged energy distribution looks very similar to the theoretical calculation of KANE [5.137] for bulk Si. There is always an uncertainty in the relative weight to be assigned to the bulk distribution in the subtraction from the measured energy distribution. In Fig. 5.19 we show the assumed bulk distribution for Si and the difference of the Si (111) 7×7 surface structure [5.138]. ROWE and IBACH [5.136] identified three states $(+0.1, -3.6,$ and -12.3 eV) as surface states and two states as surface resonances $(-1.5$ and -7.5 eV). They maintain that the peak positions of the structure shown in Fig. 5.19 were independent of the scaling factor used in subtracting out the bulk contribution. Yet it is easy to see that if the bulk curve is given more weight the "surface density of states" would become negative around -6.5 eV, -3 eV, and 0 eV. These negative regions might arise from interference of bulk and surface emission.

SWANSON and CROUSER [5.139] have measured the field emission energy distribution from (100) Mo and observed a surface state similar to that on W except narrower and shifted up in energy. Dionne and Rhodin [5.38] have measured field emission energy distributions from the low-index planes of the f.c.c. transition metals (iridium, rhodium, palladium and platinum) and compared them with the band structure calculations of Anderson [5.140]. There is systematic agreement if some band edge contraction at the surface is assumed [especially on Ir(111)].

Shepherd and Peria [5.141] have observed a surface state in the energy distribution of field emitted electrons from (100) Ge. The surface state was found to overlap the valence band and, as in the case of tungsten, to be very sensitive to contamination. This surface state which was between 0.6 and 0.7 eV below E_F was found to oscillate quite dramatically in amplitude as the Ge tip was field evaporated. They concluded that this was a result of gradual field stripping plane edge atoms away until the plane edge was being observed by the probe hole. The surface state does not exist near a plane edge so that the observed amplitude decreases. It would be extremely informative to know the minimum number of atoms required for the existence of the surface state. Modinos [5.46] has calculated the field emission energy distribution from a surface state on Ge (100) and compared his result with Shepherd and Peria's measurements [5.141].

Smith and Peria [5.141] have used Shepherd's analyzer to look at Ge (111) where they find a surface state overlapping the valence band. Eastman and Grobman [5.134] found this surface state in their photo-emission work on cleaned Ge (111).

5.5.2. Adsorption Studies

Hydrogen Adsorption

The chemisorption of hydrogen on (100) tungsten has been investigated by many experimental and theoretical techniques. The salient experimental facts are reviewed in Chapter 3. Briefly, it is known that the thermal desorption spectrum of a saturated layer has two unequal peaks at ~ 450 K and ~ 550 K, which are referred to as the β_1 and β_2 state, respectively.

Plummer and Bell [5.37] studied both the elastic and inelastic tunneling in field emission from hydrogen on (100) W as a function of coverage. Figure 5.20 shows the field emission enhancement factor as a function of the atom density on this face. The surface resonance at $E = -0.35$ eV decreases with increasing hydrogen coverage and a new level associated with β_2 hydrogen builds in at $E = -0.9$ eV with a half width ~ 0.6 eV. In the coverage region $\sim 5 \times 10^{14}$ atoms/cm^2 this level at $E = -0.9$ eV suddenly disappears. This is the same coverage range where the $C(2 \times 2)$ LEED pattern disorders. Plummer and Bell con-

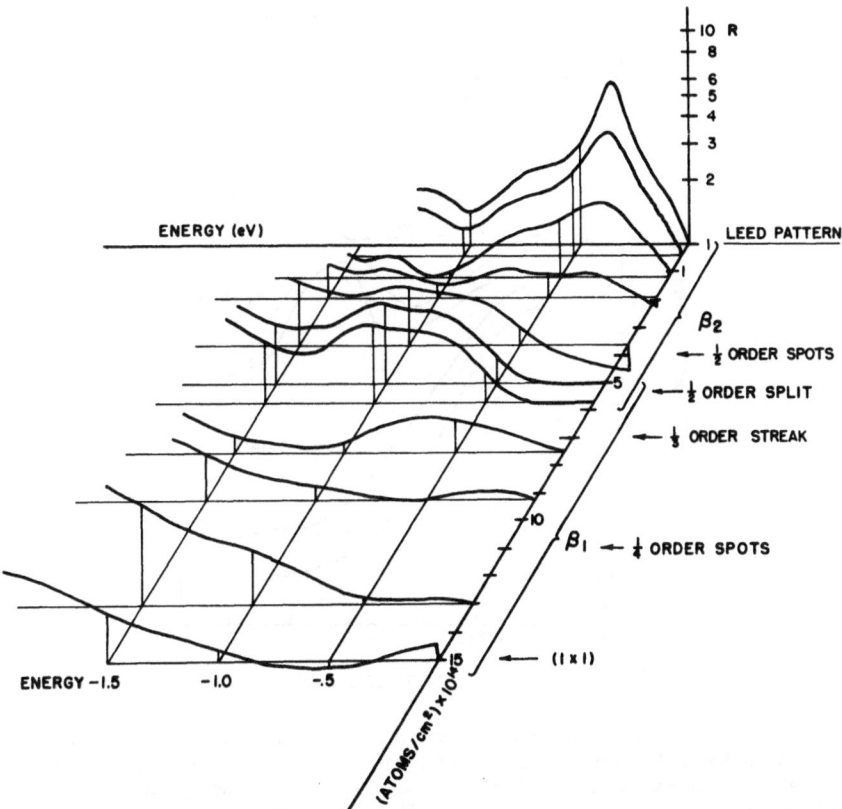

Fig. 5.20. Enhancement factor R for hydrogen and deuterium on (100) W at 300 K as a function of atom density on the surface. The saturated coverage is taken as 1.5×10^{15} atoms $\cdot$ cm^2. LEED data from [5.142]

cluded from these measurements that the β_1 state was not sequentially filling on top of the β_2 state, but that interaction between the chemisorbed hydrogen species caused a density dependent transition in the binding character, accompanied by an order-disorder transition, which is reversible with density.

In Fig. 5.21 the field emission and photoemission energy distribution are compared for clean (100) tungsten, the β_2 state, and the β_1 and β_2 states. All of the photoemission data were taken at $\hbar\omega = 10.2$ eV with collection at normal emission, $\pm 6°$ for the curves of Feuerbacher [5.143] and ± 15 degrees for those of Waclawski [5.144]. There is an arbitrary scaling factor in the field emission curves so that the amplitudes are not relevant, but the structure is almost identical. Again at these low photon energies the normal photoemission must be sampling the surface region.

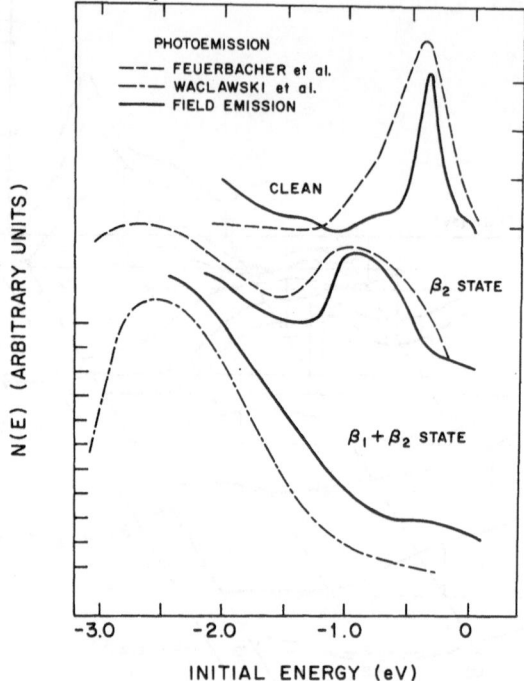

Fig. 5.21. Comparison of field emission and photoemission data for H_2 adsorption on (100) W at 300 K. The top set of curves is for clean (100) W- photoemission data are from [5.143], normal emission ($\pm 6°$). The second set of curves is for the β_2 state $\theta = .2$ and the final set of curves is for saturation coverage. The field emission data are from [5.37], with $R(E)$ curves being plotted with an arbitrary scaling factor. The bottom photoemission curve is from [5.144] for normal collection ($\pm 15°$)

If hydrogen makes localized bonds at the surface forming a "surface complex" (Chapter 2) there should be a corresponding energy level near or below the tungsten band. To observe such a level in photoemission a photon energy > 10.2 eV must be used. The resonance lines of He and Ne at 21.2 and 16.9 eV are intense and narrow enough to be used without a monochromator. The changes in the surface electron density upon adsorption can be enhanced by looking at the difference in the photoemission energy distribution as the surface conditions are changed. Figure 5.22 shows such a difference curve for adsorption of D_2 on (100) W, which should be compared to the clean energy distribution in Fig. 5.15. The scale of the difference curve is related to the clean energy distribution by normalizing the intensity of the -2.0 eV peak to 10.0. The middle set of curves represents the difference with respect to the clean energy

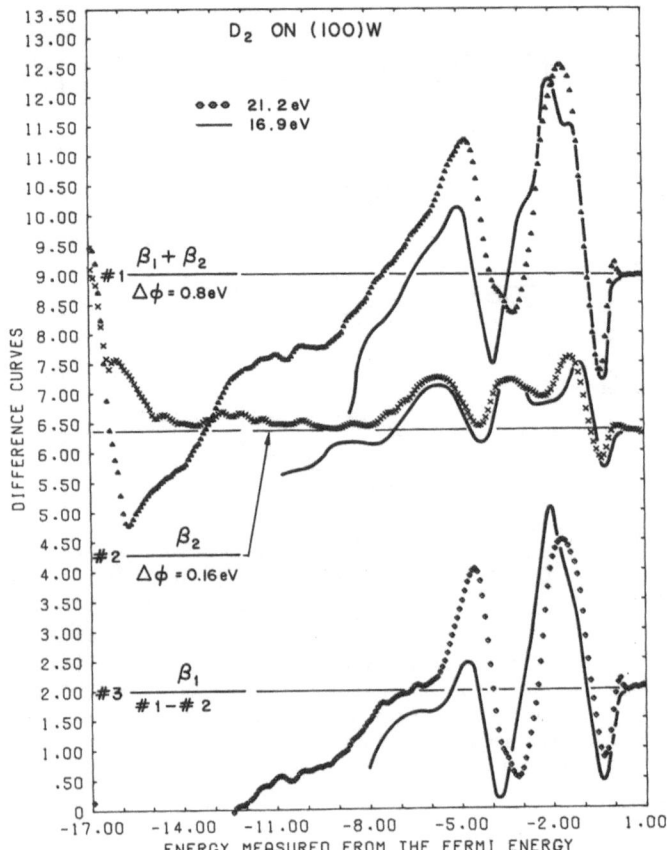

Fig. 5.22. Difference curves for β_1, and β_1 plus β_2 states of adsorbed D_2 on (100) W at 300 K. Curve 1 is saturation coverage of D_2 where both β_1 and β_2 states will desorb. Curve 2 is low coverage ($\theta \simeq 0.2$) where only the β_2 state will desorb. The bottom curves show the difference between Curves 1 and Curve 2 or the β_1 state. The solid curves are for $h\omega = 16.9$ eV and the broken (data) curve is for $h\omega = 21.2$ eV [5.88]

distribution at two photon energies for an exposure of D_2 which maximizes the β_2 state ($\Delta\varphi \cong 0.16$ eV). The clean energy distribution at $\hbar\omega = 16.9$ eV is very different from that at 21.2 eV, so it is imperative to compare difference curves for more than one photon energy. The dip near -0.3 eV is caused by the disappearance of the surface resonance while the peak near -1.2 eV is a consequence of the -1.0 eV peak shown in Figs. 5.20 and 5.21. There is no observable peak in the actual energy distribution because its intensity is too low and it is riding on the edge of the strong d-band peak at -2.0 eV. There are two additional

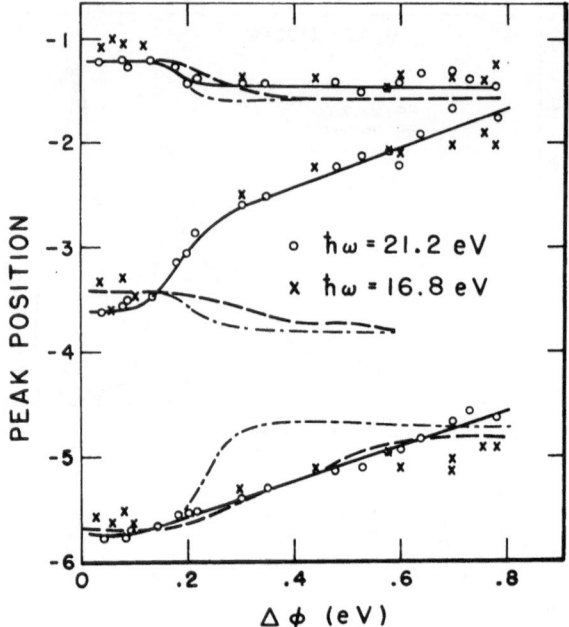

Fig. 5.23. Peak position as a function of work function change for the peaks in the difference curves for D_2 adsorption on (100) W [5.88] [(0) for $\hbar\omega = 21.2$ eV and ($\times$) for
$\hbar\omega = 16.8$ eV]. The dashed lines are the peak position variation predicted from a sequentially
filling state model, while the dot-dashed lines are predictions of a two state conversion
model

peaks in the difference curve, one at -3.6 eV and the other at -5.7 eV.
The latter is the bonding orbital mentioned earlier. The top set of curves
are difference curves for a saturated adsorption of D_2, or the β_1 and β_2
states. The two peaks visible in the β_2 spectrum are not present in these
curves and the lower lying peak has shifted up by nearly 1 eV. The
bottom set of curves are for what should have been the β_1 state if the two
states were filling sequentially; the β_2 spectrum was subtracted from that
labelled $\beta_1 + \beta_2$. The labelling here is meant to correspond to thermal
desorption and does not imply that there are distinct β_1 and β_2 states.
The large negative regions especially near -3 eV are indicative of conversion, meaning that the adsorbed atoms which were in the β_2 configuration at low coverage changed their electronic configuration, as coverage
increased.

This conversion process can be better visualized if the peak positions
shown in Fig. 5.22 are plotted as a function of coverage. Figure 5.23
shows such a plot as a function of $\Delta\varphi$ which varies linearly with coverage

[5.145] ($\Delta\varphi = 0.8$ eV at saturation). Below $\Delta\varphi = 0.1$–0.14 eV ($\theta \simeq 0.2$) the peaks are all independent of the coverage; the β_2 state is being filled. For $\Delta\varphi = 0.12$–0.22 eV ($0.16 \leq \theta \leq 0.28$) there is a marked change in the -1.2 eV and -3.6 eV peaks in the difference curves. The low lying level at -5.73V does not show any preferential shifting in this coverage range.

Many different experiments have shown this behavior. MADEY [5.146] has measured the cross section for electron impact desorption of hydrogen as a function of coverage. He found that the cross section was 1.8×10^{-23} cm^2 and constant below a coverage of 0.16 ($\Delta\varphi = 0.13$ eV) and fell very rapidly within a coverage range of 0.1 monolayers. The saturation cross section was $< 10^{-25}$ cm^2. The (1/2, 1/2) spot in the Leed diffraction pattern characteristic of the low coverage β_2 state has an amplitude proportional to coverage for $\theta < \sim 0.15$ and then decreases rapidly [5.142]. RUBLOFF et al. [5.147] have measured the change in the reflectance of a (100) W surface as a function of H coverage and wavelength of the incident light. Their data agree exactly with what is shown in Figs. 5.21–23 if the optical transition occurs to the Fermi energy. A plot of the change in the reflectance change $\Delta R(\hbar\omega = 1.4$ eV) with coverage θ is fairly constant for $\theta \leq \sim 0.18$ or $\theta > \sim 0.3$. For $0.18 < \theta < 0.3 - d\Delta R/d\theta$ peaks at $\theta \simeq 0.2$. RUBLOFF et al. conclude that a rapid conversion of β_2 to β_1 is accurring at $0.19 \leq \theta \leq 0.26$. Conversion, even though much slower, continues at $\theta > 0.3$, again in agreement with the results of Fig. 5.23. FEUERBACHER and FITTON [5.143] and EGELHOFF [5.147] have produced photoemission data at different photon energies and collection angles which show the same effects.

PLUMMER and BELL [5.37] used the inelastically tunneling electrons in field emission (Subsection 5.3.2) to measure the vibrational modes for hydrogen and deuterium adsorbed on (100) W at 77 K. In the β_2 region ($\theta < \sim 0.2$ monolayers) they observed two vibrational modes for adsorbed hydrogen, one at 0.140 eV and the other at 0.07 eV, in excellent agreement with the inelastic scattering experiments of PROBST and PIPER [5.149]. For deuterium the inelastic losses are shifted to 0.1 eV and ~ 0.05 eV; which is the shift predicted by the mass ratio. At full coverage PLUMMER and BELL could not see any inelastic losses; even the 0.14 and 0.07 eV losses disappeared at high coverage, indicating that the vibrational structure as well as the electronic properties of the hydrogen-tungsten system had changed.

The cross sections for excitation of the molecular stretching frequencies of H$_2$ (0.55 eV) and D$_2$ (0.4 eV) were measured by looking at the γ state of hydrogen on (111) W at 77 K. This state is believed to be molecular, and produced inelastic losses at the appropriate energies for H$_2$ and D$_2$ stretching. On this basis and from the signal to noise ratio for (100) W it was concluded that $< 10\%$ of the hydrogen could have been molecular on W (100) even allowing for differences in the cross sections on both faces.

We thus require a model for hydrogen adsorption consistent with the following: atomic adsorption at all coverages, into an ordered array at $\theta < \sim 0.2$: a rapid conversion in the range $0.16 < \theta < 0.3$ and a gradual change above this coverage. The LEED pattern must disorder for $\theta > \sim 0.2$, yet $\Delta \varphi$ must increase linearly with θ. We consider briefly two models for chemisorbed hydrogen [5.88]. The first is a sequentially filling state model, where the β_2 state fills for $\theta < 0.2$ ($\Delta \theta \gtrsim 0.16$ eV) and for $\theta > 0.2$ ($\Delta \theta \gtrsim 0.16$) the β_1 state fills on top of the β_2 state. This model can be tested by using curve 2 of Fig. 5.22 for the β_2 state and curve 3 for the β_1 state. The dashed curves on Fig. 5.23 are the predictions of this model. The sequential state model reproduces the upper and lower curves shown in Fig. 5.23 but not the middle curve. The second model is a two state conversion model, where for a coverage beyond the saturation of the β_2 state every newly adsorbed hydrogen atom converts one or more adsorbed atoms into a new state. The simplest two state conversion model originated with Estrup [5.142] where the β_2 state corresponded to hydrogen atoms adsorbed in every other bridge site. As the coverage increased the lateral interaction of H atoms, which is repulsive at nearest-neighbor distances [5.150] could cause the existing H atoms to be displaced. In Estrup's [5.142] model the saturation configuration, which was again ordered, corresponding to hydrogen atoms in every bridge site or two hydrogen atoms per tungsten surface atom. The dot-dashed curves in Fig. 5.25 are for a two state conversion model where every newly adsorbed hydrogen atom for $\theta \gtrsim 0.2$ ($\Delta \theta \gtrsim 0.16$) converts two existing H atoms. The conversion is over at $\theta = 0.3$. This two state conversion model fits only the upper level in Fig. 5.23. Therefore we need a more complicated picture than any two state model [5.147]. At present no obvious resolution of these puzzles is in sight!

CO *Adsorption—Multiple Binding States—Dissociation*

The adsorption of CO on tungsten is more complicated than the adsorption of hydrogen. This system is of great interest because it was considered as one of the few known cases of non-dissociative molecular adsorption [5.5, 151]. This belief was based on the fact that desorption always occurred as CO without residual C or O on the surface. It is now considered likely that the high temperature β states are dissociated (see Chapter 3). Several general features of CO adsorption seem to be preserved from plane to plane. Low temperature adsorption leads to a rather weakly bound state (binding energy ~ 1 eV) which is called virgin. In the virgin state the ratio of CO to W atoms on the surface is ~ 1.0 upon heating to

300–600 K the virgin state partially desorbs and partially converts to a more tightly bound state (called beta or beta-precursor [5.5]) for which the ratio of CO to W is ~0.5. Readsorption on a beta layer results in formation of a new state, alpha, with a ratio of the alpha state to beta state of ~1.0. Virgin adsorption results in a work function increase of 0.6–1.6 eV depending upon the plane. In the beta layer the work function change is smaller but usually of the same sign, while alpha adsorption reduces the work function relative to the beta layer [5.151]. The beta layer exhibits from 2 to 3 peaks in thermal desorption depending upon the crystal face while there may be two alpha states and usually only one virgin state.

Photoelectron spectra from CO adsorbed on tungsten can be used to separate and identify these states. We can determine: 1) whether states are being populated sequentially; 2) if they are molecular or correspond to dissociation, and 3) if there is conversion between states.

The virgin state desorbs and converts to beta below 300 K on (100) W so that adsorption at 300 K results in the filling of beta states and after prolonged exposures population of an alpha state or states [5.152]. Figure 5.24 shows the photoelectron spectra for clean (100) W and (100) W with a pressure of 5×10^{-6} Torr of CO in the chamber, recorded at $\hbar\omega = 21.2$ eV; the 16.9 eV spectra show the same features [5.88]. The peak near -8.5 eV is the 1π level of the molecularly adsorbed α state. This can be seen more easily if the difference curves are plotted as a function of heat treatment, as in Fig. 5.25. The top curve is identical to the difference curve in Fig. 5.24. The state called α_1 should desorb at 300 K [5.152], as shown in the second curve recorded after removal of CO and pumping for 15 min. φ has increased by 0.06 eV, i.e. this α state is electropositive. The difference between Curves 1 and 2 shown in Curve 5 is labeled α. It shows that the α state consists of molecular CO. More important is the fact that there is no conversion of the α_1 state to other states. The -8.7 eV peak in the α state spectrum is the 1π CO orbital. The 5σ level which is the lone pair orbital on the carbon atom has disappeared as one would expect from a donor level.

The α_2 state observed by YATES et al. [5.152] can be desorbed by heating to ~700 K. The difference curve relative to clean (100) W for the heat treatment is shown in Curve 3 of Fig. 5.25. The peak near -8.5 eV has been completely removed, indicating removal or conversion of all the CO with a 1π type level. Notice that φ decreased in contrast to the assumption that all α states are electropositive. Curve 6 shows the difference between Curves 2 and 3 which should be the α_2 state. It shows the 1π type level at -8.3 eV and a new peak at -11.4 eV (presumably either the 4σ or the bonding orbital from the 5σ bond to the substrate). Notice the negative dip in the region of -4.3 eV, which

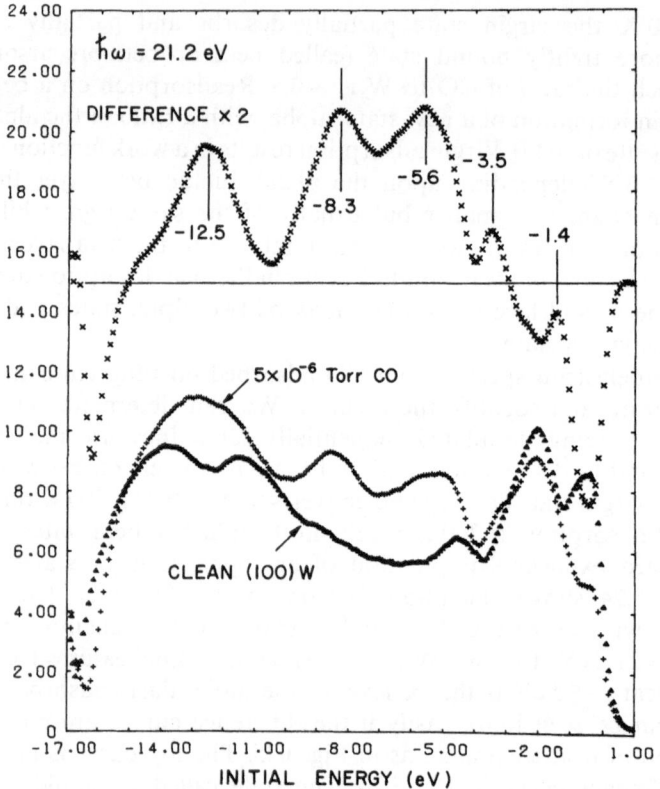

Fig. 5.24. Photoelectron spectra of adsorbed CO on (100) W at 300 K for $h\omega = 21.2$ eV. The two curves at the bottom are clean (100) W and (100) W with a pressure of 5×10^{-6} Torr of CO in the chamber. The top curve is the difference multiplied by 2 [5.88]. $\Delta\varphi = .34$ eV. The peak near -8.5 eV is the level of the molecularly adsorbed 1π state

occurs at the peak in the β state spectrum. Plummer et al. [5.88] were able to show that this peak represents conversion of what is called the α_2 state to the β state. If all the β states were saturated by prolonged exposure at a higher temperature the α_2 state did not appear upon readsorption at 300 K. This was interpreted to mean that the α_2 state is in fact a remnant of the virgin state.

Curve 4 of Fig. 5.25 was taken after heating to 1100 K. This results in a reversal in the sign of $\Delta\varphi$ and an ordered $C(2 \times 2)$ LEED pattern [5.153]. Part of the β layer is desorbed by heating to these temperatures [5.154]. The peak near -6.0 eV shows fine structure with the appearance of

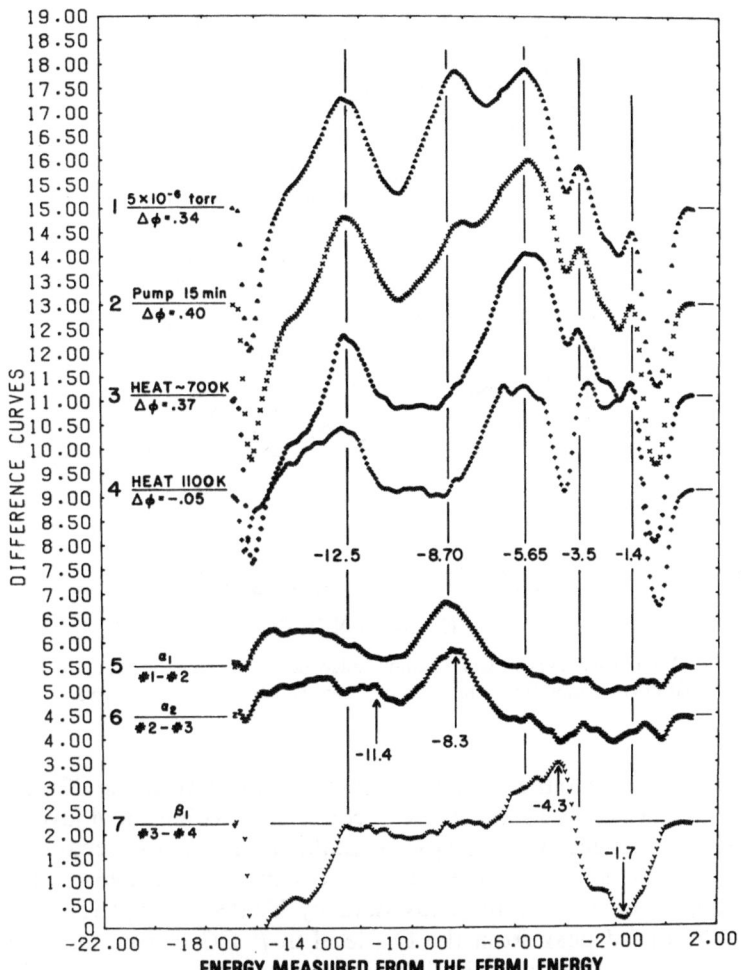

Fig. 5.25. Difference curves for CO adsorption on (100) W for various heating cycles. All of the curves are recorded at 300 K after the indicated heat treatment. The work function changes with respect to clean W are shown on the left [5.88]

three sub-peaks. This peak is very close to that seen for oxygen adsorption giving a $[C(2 \times 2)]$ LEED pattern. This led BAKER and EASTMAN to postulate that CO in this state was dissociated [5.155]. This conclusion cannot be made directly from a curve like *4*. What can safely be concluded are the following points: 1) The β states are such that both O

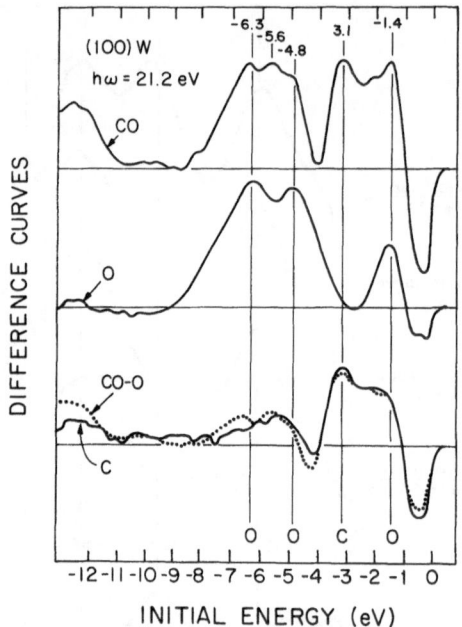

Fig. 5.26. Difference curves for CO, O, and C adsorbed in the C(2 × 2) structure on (100) W. The dashed curve is 0.69 times the CO curve minus 0.35 times the O curve

and C are bound to the substrate, 2) there is no 1π level in the CO molecule and 3) Curve 7 shows that there is conversion within the β layer as the density is changed. We [5.88] were unable to distinguish separate β_1, β_2 or β_3 states, since it appeared that a density dependent conversion process occurred. Recent X-ray data by YATES et al. [5.156] indicate that the broad peak from the O_{1s} level in β–CO can be resolved into three individual peaks. The UV data cannot be decomposed in that fashion.

We now address the question of dissociation. Our definition will be the following: If CO in the high temperature β state (1100 K) is dissociated then the photoelectron spectra should be a linear sum of the spectra from adsorbed oxygen and adsorbed carbon both in a C(2 × 2) surface geometry. This comparison is shown in Fig. 5.26. The top curve is the difference for CO, the middle for oxygen and the bottom for C [5.88]. The dotted curve is the best fit to the C curve obtained by subtracting the oxygen from the CO. The fit is very good especially above −7 eV. The −1.4 and −6.4 eV peaks are primarily O–W while the −3.10 and −5.1 eV peaks are primarily C–W. For our purposes CO in the C(2 × 2) configuration is dissociated.

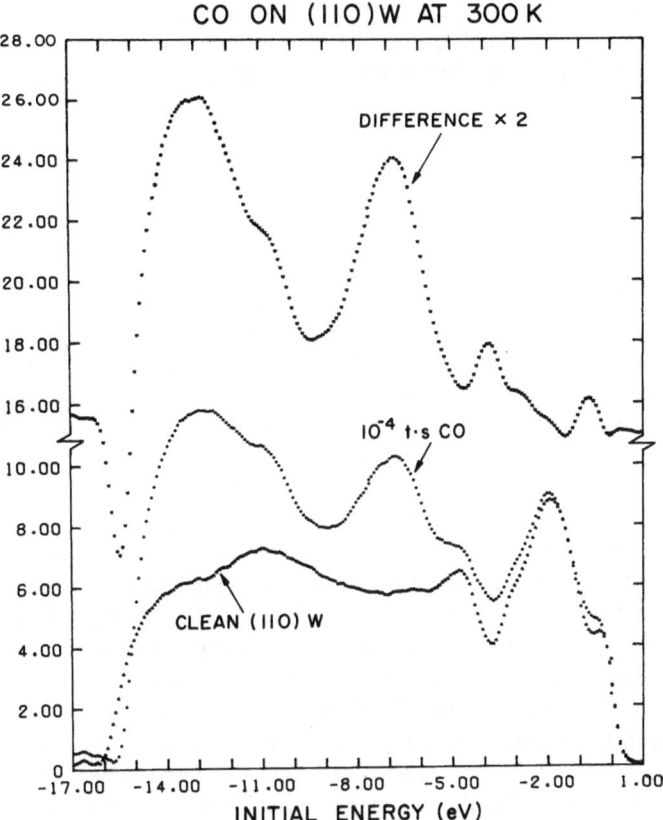

Fig. 5.27. Photoelectron spectra at $h\omega = 21.2$ eV for CO adsorbed on (110) W at 300 K. The two curves at the bottom are for clean (110) W and (110) W after an exposure to 10^{-4} Torr · sec. CO at 300 K. The top curve is the difference multiplied by 2. The peak near -7.0 eV is the 1π level of molecularly adsorbed CO in the virgin state. The shoulder near -10.7 eV is most likely the 4σ level of molecular CO [5.88]

We would now like to compare these results with those YOUNG and GOMER [5.157] for the field emission energy distributions. First observe that the -1.4 eV level which appears in all the CO difference curves of Figs. 5.25 and 5.26 is derived primarily from the O–W bond. O is bonded to W only in the β states so that one would expect an enhancement in the field emission energy distribution near -1.5 eV for 300 K adsorption. This should persist at higher temperatures until desorption begins to occur and it should also be observed with oxygen adsorption [5.88]. This is more or less what was found by YOUNG and GOMER [5.157] so that there seems to be moderately good agreement between field emission

and photoemission for the β states of CO on (100) W. YOUNG and
GOMER [5.157] pointed out from the temperature behavior of partial
layers that there is evidence for strong interaction between C and O
to at least 800 K.

Adsorption of CO on (110) W at 300 K results largely in population
of the virgin state [5.158]. This is shown in photoelectron spectra of
Fig. 5.27 for 10^{-4} Torr · sec exposure of CO on (110) W. The peaks near
-7 eV and -11 eV are the 1π and 4σ levels of the molecularly adsorbed
virgin state of CO [5.88]. The level at -3.9 eV is indicative of some form
of beta CO present at 300 K [5.88]. Figure 5.28 shows the desorption
and conversion of the virgin layer. φ increases by 0.5 eV when the virgin
state is adsorbed at 300 K. Heating the layer adsorbed at 300 to 400 K
(Curve 1, Fig. 5.28) results in desorption of CO without appreciable
evidence for conversion from photoemission data, as indicated by the
absence of appreciable amplitude in the region 0 to -5 eV. However,
there is evidence from electron impact desorption [5.159] that conversion from the low temperature CO^+ yielding state to an O^+ yielding
state occurs in this temperature region. The discrepancy may result
from the fact that a mixture of virgin, O^+ yielding beta-precursor, and
CO^+ yielding alpha-CO can coexist at 300 K. Heating from 390 to
500 K (Curve 2) converts the state remaining after heating to 390 K
without appreciable desorption. This is an agreement with the thermal
desorption measurements of KOHRT and GOMER [5.158] and the
electron impact results, which indicate disappearance of the O^+ yielding
state in this temperature range. The 300 K layer is characterized by a
1π type level at -7.0 eV and a 4σ type level at -10.8 eV. There are
two unidentified levels at -12.6 eV and -13.0 eV. The work function
difference is 0.2 eV. The -10.4 eV and -7.22 eV levels are the 4σ and
1π levels of the state being converted and the -5.3 and -2.2 eV levels
are the new levels formed by conversion. Curve 3 shows the conversion
between the two β states as the crystal is heated from 500 to 1000 K. The
β states are density dependent as on (100) W. The last curve shows the α
state which results from readsorption on a layer heated to 700 K. It
appears to be slightly electronegative [5.88] again in agreement with the
results of LEUNG et al. [5.159] on this plane.

The dissociation of the high temperature states cannot be ascertained
as on (100) W because O, C and CO do not form the same Leed patterns.
Nonetheless PLUMMER et al. [5.88] performed the same comparison
shown in Fig. 5.26 for (100) W. The agreement is not as good, but indicates
that β–CO is almost or wholly dissociated.

YOUNG and GOMER [5.5, 155] observed a peak at -2.0 eV in the field
emission energy distribution and a shoulder at $E = -0.2$ eV which could
be destroyed by heating a layer adsorbed at low temperature to ap-

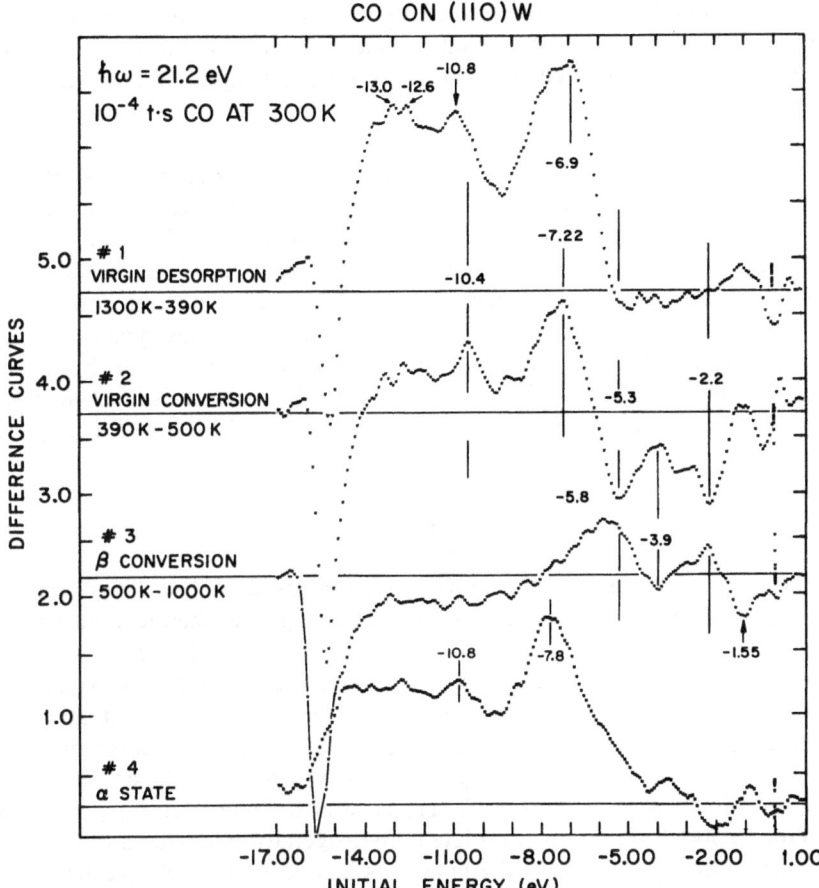

Fig. 5.28. Difference curves for CO adsorbed on (110) W. The top curve shows desorption of the virgin state without conversion. Curve 2 shows the subsequent conversion of the remaining virgin state to the β state of CO. Curve 3 shows the conversion of that occurs in the β states upon heating to 1000 K. Curve 4 shows the alpha state which results from readsorption on a saturated CO layer which had been heated to 700 K [5.88]

proximately 350 K. Readsorption on the heated virgin layer does not change the spectrum. The latter observation agrees with the photo-emission observations of the α state in Fig. 5.28. The $E = -2.0$ eV peak which should have disappeared when the virgin layer was heated to 400 K does not appear in Curve 1 of Fig. 5.28. This can easily be explained by the fact that this peak is seen in field emission only at $\theta < 0.75$ [5.157].

In concluding this discussion let us ask what we have learned about the CO on tungsten system that we did not already know. The photoelectron spectra showed that the weakly bound alpha and virgin states consist of molecularly adsorbed CO, but this was known even before the infrared studies of Yates et al. [5.152]. The UV spectra revealed that the multitude of β states observed in a thermal desorption spectrum are not sequentially filling states. Instead, as in the case of hydrogen there is a density dependent transition or conversion. The singly most important observation is the degree of dissociation which exists in the β states, with the high temperature β state on (100)W being completely dissociated as far as photoemission can determine. This type of "finger print" study used to investigate the dissociation of CO is probably the most fruitful experimental application of UV photoemission. Next we shall illustrate this technique for adsorption of C_2H_4 on Ni(111) and W(110).

Hydrocarbon Decomposition

Photoemission energy distribution studies used in the "finger print" mode will become very useful in identifying the chemical nature of adsorbed species or intermediate states. We shall illustrate here the capabilities of this technique by discussing the adsorption of C_2H_4 on Ni(111) and W(110). The photoelectron spectra of these systems have shown that C_2H_4 will adsorb as C_2H_4 at (≤ 250 K) while adsorption at 300 K or subsequent warming of a low-temperature layer results in partial dehydrogenation to a C_2H_4 type adsorbate. Heating a (110)W plane to 500 K results in complete dehydrogenation leaving a C–C bond which is not broken until the crystal is heated to 1100 K. We will now explain how this information can be obtained by comparison of the spectra with gas phase spectra or spectra obtained from adsorption of the intermediate species.

Figure 5.29 shows the difference curves relative to clean (111)Ni obtained by Demuth and Eastman [5.107, 109] for adsorption of C_2H_2 (Fig. 5.29a) and C_2H_4 (Fig. 5.29b) with $\hbar\omega = 21.2$ eV. The top curve of Fig. 5.29b is for a chemisorbed layer of C_2H_4 on Ni(111) at 100 K. The dashed curves show how the spectra were separated into four peaks. The lower three levels have binding energies of 10.9, 12.7, and 14.1 eV, respectively which, given a 2.1 eV surface induced relaxation shift, agrees with the gas phase spectra from the σ levels of C_2H_4 [5.105]. The binding energy of the peak labelled π does not agree with the gas phase binding energy because it is involved in the bonding to the substrate so that there is a shift [5.109] as well as a different relaxation shift. Demuth and Eastman confirmed that the asorbed

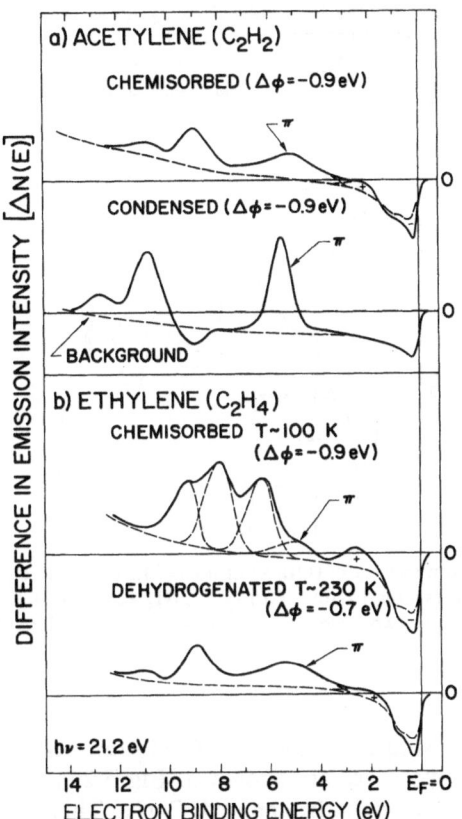

Fig. 5.29. (a) Difference curves for 1.2×10^{-6} Torr · sec exposure of Ni(111) to acetylene at $T \sim 300$ K and for condensed acetylene formed at $T \sim 100$ K with acethylene pressures of 6×10^{-8} Torr. (b) Difference curves for chemisorbed ethylene (exposure of 1.2×10^{-6} Torr · sec at $T \sim 100$ K) and for dehydrogenated ethylene (obtained by warming to $T \sim 230$ K or with initial exposure at $T \sim 300$ K) [5.107, 109]

molecule was C_2H_4 by condensing multilayers of C_2H_4 on the Ni surface. The only observed change in the three σ levels was a reduction in the surface induced relaxation energy by 0.4 eV, as expected for molecules farther from the surface. If the chemisorbed C_2H_4 is warmed to ~ 230 K the photoelectron spectrum changes quite dramatically as in the bottom difference curve of Fig. 5.29b. The adsorbed entity is no longer C_2H_4. A comparison of the two peaks at 9.1 eV and 11.0 eV with gas phase spectra [5.105] suggests that C_2H_4 has been partially dehydrogenated to a C_2H_2 complex. A surface induced relaxation shift of ~ 3.0 eV will align these peaks with the two upper σ levels of the gas phase C_2H_2. DEMUTH and EASTMAN checked this hypothesis by chemisorbing partial

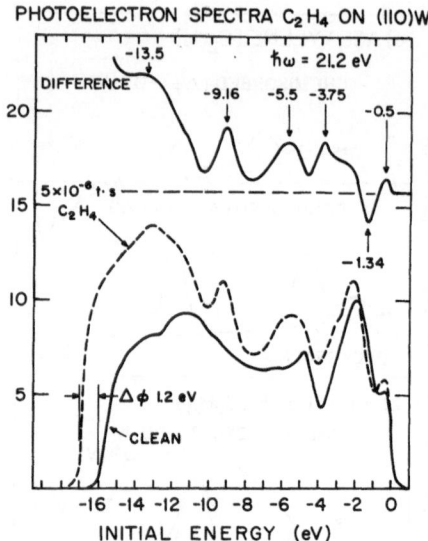

PHOTOELECTRON SPECTRA C₂H₄ ON (IIO)W

Fig. 5.30. Photoelectron spectra of (110) W at $h\omega = 21.2$ eV. Solid curve (bottom) is clean (110) W, dashed curve is after 5×10^{-6} Torr · sec C_2H_4 exposure at 300 K. At top is shown the difference between the two lower curves. Energy is plotted relative to the Fermi energy [5.110]

layers and condensing multilayers of C_2H_2 on Ni(111). The difference curves for these two cases are shown in Fig. 5.29a; chemisorbed C_2H_2 is nearly identical to dehydrogenated C_2H_4. There may be hydrogen adsorbed on the surface after dehydrogenation since the observed Leed pattern is (2×2) which indicates 0.25–0.5 layers. The condensed C_2H_2 has a reduction in the surface induced relaxation shift of 1.7 eV [5.109]. Room temperature adsorption of C_2H_4 produces adsorbed C_2H_2.

It is interesting to note in passing that SINFELT's [5.160] analysis of the rate law in C_2H_6 hydrogenolysis indicated that ethane was chemisorbed as C_2H_2 on Ni.

The situation for C_2H_4 adsorbed on (110)W seems to be very similar to Ni(111), in that room temperature adsorption of C_2H_4 results in partial dehydrogenation [5.110]. Figure 5.30 shows spectra for clean (110)W and after an exposure of 5×10^{-6} Torr·sec of C_2H_4. The two curves, taken under identical conditions of light intensity and crystal position, were subtracted to obtain the difference curve shown at the top of Fig. 5.30. The original energy distributions were arbitrarily normalized to 10 at the -2.0 eV peak in the clean energy distribution. The -5.7 eV peak is approximately the same position at the "π" peak for the Ni(111) case (Fig. 5.29b) and the -9.16 eV peak and the shoulder

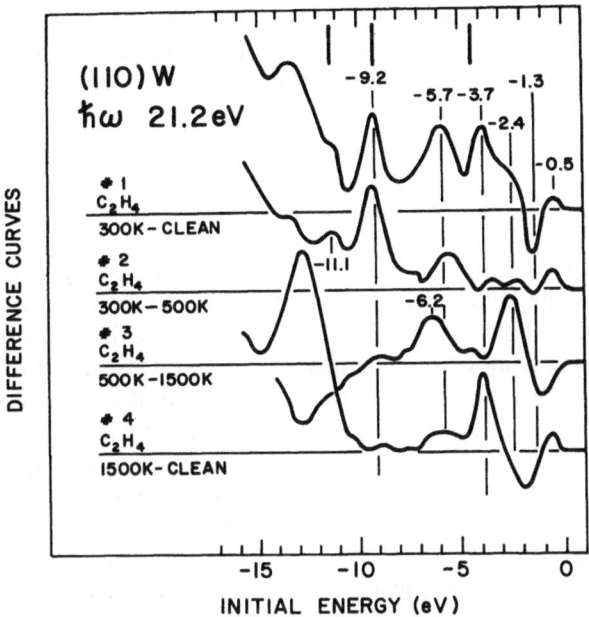

Fig. 5.31. Difference curves for C_2H_4 adsorption on (110) W at $h\omega = 21.2$ eV. Curve *1* is the difference between spectra after 5×10^{-6} Torr·sec exposure to C_2H_4 and clean (110) W. Curve *2* is the difference resulting from heating to 500 K the layer adsorbed at 300 K. Curve *3* is the difference between an adsorbed layer of C_2H_4 heated to 500 K and the 1500 K, and Curve *4* is the difference between the 1500 K heated layer and clean (110) W. The vertical lines at the top are the gas phase photoionization potentials for C_2H_2. They have been shifted up by 3.1 eV using the clean surface work function [5.110]

at -11 eV (which will show up in the subsequent figure) are the same as the two σ peaks. PLUMMER et al. [5.110] concluded that this spectrum corresponded to C_2H_2 adsorption by comparison with the gas phase spectra for C_2H_2 and with the measured photoelectron spectra for adsorbed hydrogen and carbon. The photoelectron spectra in Fig. 5.30 could not be constructed from any linear combination of a hydrogen and a carbon spectrum; therefore, the additional levels must result from C–H and C–C bonds. A 3.1 eV surface induced relaxation shift would bring the σ levels of gas phase C_2H_2 in agreement with the -9.1 eV and -11.1 eV levels.

Room temperature adsorption of C_2H_4 on (110)W results in partial dehydrogenation of C_2H_4 to C_2H_2. The two hydrogen atoms per adsorbed molecule may be adsorbed on the surface or desorbed into the gas phase. The desorption data of BARFORD and RYE [5.161] indicate that the amount of hydrogen desorbed depends on the coverage of

C_2H_2; at the high exposures used for Fig. 5.30 1/3 of the hydrogen is desorbed. Figure 5.31 shows the difference curves as the adsorbed layer is heated. A thermal desorption experiment was simulated by heating the crystal to the indicated temperature, then cooling to room temperature and recording the UV spectrum. The first difference curve is the same as shown in Fig. 5.30; the energy distribution after saturated exposure of C_2H_4 minus the clean energy distribution. The subsequent curves are differences between two different heat treatments.

If the surface molecule does not change upon heating to a prescribed temperature the difference curve comparing spectra before and after heating will be a straight line. If one or more of the energy levels are removed by decomposition or desorption upon heating the difference curve will exhibit a peak where the energy level used to be. If a new energy level is produced by rearrangement upon heating the difference curve will exhibit a negative peak at this energy. Curve 2 shows the difference between the room temperature adsorption and subsequent heating to 500 K. This treatment desorbs all the hydrogen [5.161] and completely dehydrogenates the surface. The -9.2 eV and -11.1 eV levels have been removed as a consequence of breaking the C–H bonds in the C_2H_2 molecule. The only other change in the spectrum is a peak near -5.5 eV, whose origin is not completely understood.

Before discussing Curve 3 of Fig. 5.31 let us examine the fourth curve which is the difference curve relative to clean (110)W after heating the adsorbed layer to ~ 1500 K. This difference curve is characteristic of carbon; it can be obtained by diffusion of small amounts of bulk carbon to the surface, by decomposition of CH_4, or by subtracting the oxygen spectrum from the β spectrum of adsorbed CO [5.88]. The point is that there are peaks at -6.0 eV and -2.4 eV in the original spectrum (Curve 1) which were not removed when the surface was completely dehydrogenated and which are not present in the C–W spectrum shown in Curve 4 of Fig. 5.31. Curve 3 shows that these two peaks are removed by heating above 1100 K. Our contention is that they are energy levels resulting from the C–C bond in the presence of the surface atoms, and that near 1100 K the C–C bond is broken leaving a carbon residue on the surface. The negative dip at ~ -1.0 eV in Curve 3 is a result of charge redistribution into the C–W bond. The decrease in amplitude of these two peaks, associated with the C–C bond, as a function of temperature can be used to measure the activation energy for C–C bond scission energy. This energy is $3.5 - 4.0$ eV, a reduction of > 2 eV from the gas phase.

The observed Leed patterns are consistent with this model. Room temperature adsorption produces no noticeable change in the pattern. Heating to 500 K causing complete dehydrogenation produces a diffuse background with no new spots. Heating above 1100 K breaks the C–C bond and produces an ordered Leed pattern characteristic of a carbon contaminated surface [5.110].

These studies seem to support SINFELT's [5.16] three-step model for hydrogenolysis: 1) dehydrogenative chemisorption, 2) carbon-carbon bond scission, and 3) hydrogenation and desorption. The latter step is currently being checked by performing a dynamic experiment with both C_2H_4 and H_2 beamed onto the surface at high pressure while the energy analyzer is differentially pumped.

Acknowledgements

The author is indebted to T. GUSTAFSSON, C. ALLYN, and T. EINSTEIN for critically reviewing this manuscript and to Professor R. GOMER for a heroic job in editing it. It is a pleasure to acknowledge PAUL SOVEN for the many hours he spent tutoring the author.

References

5.1.   H. D. HAGSTRUM: Science **178**, 275 (1972).
5.2.   C. B. DUKE, R. L. PARK: Physics Today **25**, 23 (1972).
5.3.   H. D. HAGSTRUM, G. E. BECKER: J. Chem. Phys. **54**, 1015 (1971).
5.4.   J. W. GADZUK, E. W. PLUMMER: Rev. Mod. Phys. **45**, 487 (1973).
5.5.   R. GOMER: Advan. Chem. Phys. **27**, 211 (1974).
5.6.   D. E. EASTMAN: *Electron Spectroscopy*, ed. by D. A. SHIRLEY (North-Holland Publishing Co., Amsterdam, 1972), p. 487.
5.7.   C. S. FADLEY: *Theoretical Aspects of X-ray Photoelectron Spectroscopy* (NATO Advanced Study Institute of Electron Emission Spectroscopy, Ghent University, Ghent, Belgium, 1972).
5.8.   H. D. HAGSTRUM, G. E. BECKER: Proc. Roy. Soc. (London) A **331**, 395 (1972).
5.9.   C. R. BRUNDLE: J. Vac. Sci. Technol. **11**, 212 (1974).
5.10. D. MENZEL: Proc. 2nd Conf. on Solid Surfaces (Madrid, September, 1973).
5.11. R. H. GOOD, E. W. MÜLLER: In *Encyclopedia of Physics*, ed. by S. FLÜGGE, Vol. 21 (Springer, Berlin-Göttingen-Heidelberg, 1956), p. 176.
5.12. R. GOMER: *Field Emission and Field Ionization* (Harvard Press, Cambridge, Mass., 1961).
5.13. E. W. PLUMMER, R. D. YOUNG: Phys. Rev. B **1**, 2088 (1970).
5.14. L. W. SWANSON, A. E. BELL: Advan. Electron. Electron Phys. **32**, 194 (1973).
5.15. C. B. DUKE, M. ALFERIEFF: J. Chem. Phys. **46**, 923 (1967).
5.16. H. E. CLARK, R. D. YOUNG: Surface Sci. **12**, 385 (1968).
5.17. D. W. JUENKER, L. J. LeBLANC, C. R. MARTIN: J. Opt. Soc. Am. **58**, 164 (1968).
5.18. C. J. POWELL: Surface Sci. **44**, 29 (1974).
5.19. J. C. TRACY: J. Vac. Sci. Technol. **11**, 280 (1974), and private communication.
5.20. M. L. TARNG, G. K. WEHNER: J. Appl. Phys. **44**, 1534 (1973).
5.21. D. E. EASTMAN: Photoemission Spectroscopy of Metals, Metals **6**, 411 (1972), ed. by BUNSHAH (Wiley, New York).
5.22. T. L. EINSTEIN: Submitted to Surface Sci.
5.23. K. SIEGBAHN: *ESCA Applied to Free Molecules* (North-Holland Publishing Co., Amsterdam, 1969).
5.24. J. T. YATES, JR., T. E. MADEY, N. E. ERICKSON: Surface Sci. **43**, 257 (1974).
5.25. R. H. FOWLER, L. NORDHEIM: Proc. Roy. Soc. (London) A **119**, 173 (1928).
5.26. L. NORDHEIM: Proc. Roy. Soc. (London) A **121**, 626 (1928).

5.27.   C. B. Duke: *Tunneling in Solids* (Academic Press, New York, 1969).
5.28.   W. A. Harrison: Phys. Rev. **123**, 85 (1961).
5.29.   R. Stratton: Phys. Rev. **135**A, 794 (1964).
5.30.   F. I. Itskovitch: Sov. Phys.-JETP **23**, 945 (1966).
5.31.   J. A. Appelbaum, W. F. Brinkman: Phys. Rev. B**2**, 907 (1970); Phys. Rev. **186**, 464 (1969).
5.32.   D. R. Penn, E. W. Plummer: Phys. Rev. B**9**, 1216 (1974).
5.33.   J. Bardeen: Phys. Rev. Letters **6**, 57 (1961).
5.34.   J. R. Oppenheimer: Phys. Rev. **31**, 66 (1928).
5.35.   E. B. Duke, J. Fauchier: Surface Sci. **32**, 175 (1972).
5.36.   Analyzing the ratio of the calculated current per unit energy from a narrow Kronig-Penny band to the free-electron tunneling probability gives an enhancement factor very similar to the surface density of states of a tight binding band.
5.37.   E. W. Plummer, A. E. Bell: J. Vac. Sci. Technol. **9**, 583 (1972).
5.38.   N. J. Dionne, T. N. Rhodin: Phys. Rev. Letters **32**, 1311 (1974).
5.39.   D. Penn, R. Gomer, M. H. Cohen: Phys. Rev. B**3**, 768 (1972); Phys. Rev. Letters **27**, 26 (1971).
5.40.   B. Politzer, P. H. Cutler: Mat. Res. Bull. **5**, 703 (1970).
5.41.   R. D. Young: Phys. Rev. **113**, 110 (1959).
5.42.   B. A. Politzer, P. H. Cutler: Surface Sci. **22**, 277 (1970).
5.43.   W. Gleich, G. Regenfus, R. Sizman: Phys. Rev. Letters **27**, 1066 (1971).
5.44.   J. W. Gadzuk: Phys. Rev. **182**, 945 (1969).
5.45.   R. D. B. Whitcutt, B. H. Blott: Phys. Rev. Letters **23**, 639 (1969).
5.46.   A. Modinos: Surface Sci. **42**, 205 (1974).
5.47.   D. Penn: Phys. Rev. B**9**, 844 (1974).
5.48.   In [5.32] Penn used Eq. (5.14), while in the article [Phys. Rev. B**5**, 768 (1972)] his Eqs. (20a) and (20b) have a slightly different form.
5.49.   C. Caroli, D. Lederer, D. Saint James: Surface Sci. **33**, 228 (1972).
5.50.   L. V. Kjeldysh: Sov. Phys.-JETP **20**, 1018 (1961).
5.51.   P. Soven: Private communication. Also T. E. Feuchtwang, Phys. Rev. B**10**, 4121, 4135 (1974) independently used the Kjeldysch formalism to show that $R(E)$ measures the one dimensional density of states at the surface.
5.52.   D. Penn: Private communication.
5.53.   J. W. Gadzuk: Phys. Rev. **61**, 2110 (1970).
5.54.   A. Bagchi, P. L. Young: Phys. Rev. B**9**, 1194 (1974).
5.55.   J. L. Politzer, T. E. Feuchtwang: Phys. Rev. B**3**, 597 (1971).
5.56.   A. Modinos: Surface Sci. **20**, 55 (1970); **22**, 473 (1970);
        A. Modinos, N. Nicolaun: Surface Sci. **17**, 359 (1969); J. Phys. C**4**, 2859 and 2875 (1971).
5.57.   R. C. Jaklevic, J. Lambe: Phys. Rev. Letters **17**, 1138 (1966);
        J. Lambe, R. C. Jaklevic: Phys. Rev. **165**, 821 (1968).
5.58.   B. F. Lewis, M. Moseman, W. H. Weinberg: Surface Sci. **41**, 142 (1974).
5.59.   D. J. Flood: Phys. Rev. Letters A**29**, 100 (1969); J. Chem. Phys. **52**, 1355 (1970).
5.60.   L. W. Swanson, L. C. Crouser: Surface Sci. **23**, 1 (1970).
5.61.   A. Bagchi: Phys. Rev. B**10**, 542 (1974).
5.62.   M. J. G. Lee: Phys. Rev. Letters **30**, 1193 (1973).
5.63.   C. Lea, R. Gomer: Phys. Rev. Letters **25**, 804 (1970).
5.64.   J. W. Gadzuk, E. W. Plummer: Phys. Rev. Letters **26**, 92 (1971).
5.65.   K. L. Ngai: phys. stat. solidi (b) **53**, 309 (1972).
5.66.   J. W. Gadzuk, A. A. Lucas: Phys. Rev. B**7**, 4770 (1973).
5.67.   J. A. Simpson, C. E. Kuyatt: J. Appl. Phys. **37**, 3805 (1966);
        B. Zimmermann: Electron and Ion Beam Tech. Conf. (1969).

5.68. H. NEUMANN: Physica **44**, 587 (1969).
5.69. B. I. LUNDQVIST, K. MOUTFIELD, J. W. WILKINS: Sol. State Commun. **10**, 383 (1972).
5.70. Two other groups have observed basically the same feature seen by LEE [5.66] in
       different crystal directions: R. POLIZZOTI, R. LIU, G. EHRLICH: 19th Field Emission
       Symposium (Urbana, Ill., 1972);
       T. VORBURGER, B. WACLAWSKI, E. W. PLUMMER: Unpublished. Also see J. WYSOCKI,
       CH. KLEINT: phys. stat. solidi (9) **20**, K57 (1973);
       H. NEUMANN, CH. KLEINT: Ann. Physik (7) **7**, 237 (1971).
5.71. C. CAROLI, D. LEDERER-ROZENBLATT, B. ROULET, D. SAINT-JAMES: Phys. Rev. B **8**,
       4552 (1973), have developed a microscopic model including inelastic scattering
       which may be quite useful when applied to specific systems.
5.72. U. FANO, J. W. COOPER: Rev. Mod. Phys. **40**, 441 (1968).
5.73. G. V. MARR: *Photoionization Processes in Gases* (Academic Press, Inc., New York,
       1967).
5.74. E. MERZBACHER: *Quantum Mechanics* (Wiley, New York, 1962), p. 447.
5.75. R. E. B. MAKINSON: Proc. Roy. Soc. (London) A **162**, 367 (1937).
5.76. H. A. BETHE, E. E. SALPETER: In *Encyclopedia of Physics*, Vol. XXXV (Springer,
       Berlin-Göttingen-Heidelberg, 1957), p. 88.
5.77. W. L. SCHAICH, N. W. ASHCROFT: Phys. Rev. B **3**, 2452 (1971); Sol. State Commun.
       8, 1959 (1970).
5.78. T. KOOPMANS: Physica **1**, 104 (1933).
5.79. D. A. SHIRLEY: Chem. Phys. Letters **16**, 220 (1972).
5.80. J. W. COOPER: Phys. Rev. **128**, 681 (1962).
5.81. W. C. PRICE, A. W. POTTS, D. G. STREETS: *Electron Spectroscopy*, ed. by D. A.
       SHIRLEY (North-Holland Publishing Co., New York, 1972), p. 182.
5.82. S. B. M. HAGSTRÜM: *Electron Spectroscopy*, ed. by D. A. SHIRLEY (North-Holland
       Publishing Co., New York, 1972), p. 515.
5.83. D. E. EASTMAN, M. KUZNIETZ: J. Appl. Phys. **42**, 1396 (1971).
5.84. G. BRODEN: Ph. D. Thesis (Chalmers University of Technology, Göteborg, Sweden,
       1972);
       G. BRODÉN, S. B. M. HAGSTRÜM, C. NORRIS: Phys. Kondens. Materie **15**, 327 (1973);
       G. BRODÉN: Phys. Kondens. Materie **15**, 171 (1972).
5.85. B. PODOLSKY, L. PAULING: Phys. Rev. **34**, 109 (1929).
5.86. J. C. SLATER: Phys. Rev. **36**, 57 (1930).
5.87. J. H. SCOFIELD: *Theoretical Photoionization Cross Sections from 1 to 1500 keV*
       (Lawrence Livermore Laboratory, TID-4500, UC-34 1973).
5.88. E. W. PLUMMER, B. J. WACLAWSKI, J. VORBURGER: in preparation.
5.89. T. E. MADEY, J. YATES, JR., N. E. ERICKSON: Chem. Phys. Letters **19**, 487 (1973).
5.90. B. J. WACLAWSKI, E. W. PLUMMER: Phys. Rev. Letters **29**, 783 (1972).
5.91. The mean radius was calculated using Eqs. (3.20)–(3.27) of Ref. [5.76].
5.92. H. D. HAGSTRÜM, G. E. BECKER: Private communication.
5.93. S. T. MANSON: Phys. Rev. Letters **26**, 219 (1971).
5.94. For a review see F. A. GRIMM: *Electron Spectroscopy*, ed. by D. A. SHIRLEY (North-
       Holland Publishing Co., New York, 1972), p. 199.
5.95. J. COOPER, R. N. ZARE: *Lectures in Theoretical Physics XI C* (Gordon & Breach,
       New York, 1969) also J. Chem. Phys. **48**, 942 (1968).
5.96. J. C. TULLY, R. S. BERRY, B. J. DALTON: Phys. Rev. **176**, 95 (1968).
5.97. T. A. CARLSON, G. E. McGUIRE, A. E. JONES, K. L. CHENG, C. P. ANDERSON, C. C. LU,
       B. P. PULLEN: *Electron Spectroscopy*, ed. by D. A. SHIRLEY (North-Holland Publish-
       ing Co., New York, 1972), p. 207.
5.98. J. W. GADZUK: Phys. Rev. B **15**, 1011 (1974); Jap. J. Appl. Phys. Suppl. 2, Pt. 2,
       851 (1974); Sol. State Commun. **15**, 1011 (1974).

5.99.   A. Liebsch: Phys. Rev. **32**, 1103 (1974).
5.100. I. G. Kaplan, A. P. Markin: Sov. Phys.-Doklady **14**, 36 (1969).
5.101. P. S. Bagus: Proc. X-ray Spectra and Electronic Structure Meeting (Munich, 1972).
5.102. D. A. Shirley: Advan. Chem. Phys. **23**, 85 (1973).
5.103. J. Cambray, J. Gasteiger, A. Streitwieser, Jr., P. Bagus: J. Am. Chem. Soc. **96**, 5978 (1974).
5.104. K. Siegbahn, C. Nordling, G. Johannson, J. Hedman, P. F. Heden, K. Hamrin, U. Gelius, T. Bergmark, L. O. Werme, R. Manne, Y. Baer: *ESCA Applied to Free Molecules* (North-Holland Publishing Co., Amsterdam, 1969).
5.105. D. W. Turner, C. Baker, A. D. Baker, C. R. Brunkle: *Molecular Photoelectron Spectroscopy* (Interscience, New York, 1970).
5.106. J. T. Yates, Jr., T. E. Madey, N. E. Erickson: Surface Sci. **43**, 257 (1974). Also see [5.156].
5.107. D. E. Eastman, J. E. Demuth: Japan. J. Appl. Phys. Suppl. 2, Pt. 2, 827 (1974).
5.108. G. E. Becker, H. D. Hagstrüm: J. Vac. Sci. Technol. **10**, 31 (1973).
5.109. J. E. Demuth, D. E. Eastman: Phys. Rev. Letters **32**, 1123 (1974).
5.110. E. W. Plummer, B. J. Waclawski, J. V. Vorburger: Chem. Phys. Letters **28**, 510 (1974).
5.111. J. R. Smith, S. C. Ying, W. Kohn: Phys. Rev. Letters **30**, 610 (1973). In this paper Fig. 3 is a plot of the screening charge density which when the original charge density is substracted will give the equivalent plot for a proton near a surface as Fig. 11 shows for CO.
5.112. N. D. Lang: Sol. State Phys. **28**, 225 (1973).
5.113. K. Mitchell: Proc. Roy. Soc. (London) A **146**, 442 (1934);
        L. I. Schiff, L. H. Thomas: Phys. Rev. B **7**, 3464 (1973);
        I. Adawi: Phys. Rev. A **788**, 134 (1964).
5.114. S. A. Flödstrom, J. G. Endriz: Phys. Rev. Letters **31**, 893 (1973).
5.115. P. O. Gartland, S. Berge, B. J. Slassvold: Phys. Rev. Letters **30**, 116 (1973).
5.116. For a recent paper see J. G. Endriz: Phys. Rev. B **7**, 3463 (1973).
5.117. C. N. Berglund, W. E. Spicer: Phys. Rev. **136** A, 1030, 1044 (1964).
5.118. N. V. Smith, M. M. Traum: Phys. Rev. B **9**, 1341, 1353, 1365 (1974).
5.119. G. D. Mahan: Phys. Rev. B **2**, 4334 (1970).
5.120. D. Kalkstein, P. Soven: Surface Sci. **26**, 85 (1971).
5.121. R. Haydock, M. J. Kelley: Surface Sci. **38**, 139 (1973).
5.122. J. A. Appelbaum, R. Hamman: Phys. Rev. B **6**, 2166 (1972).
5.123. J. W. Davenport, T. L. Einstein, J. R. Schrieffer: Japan J. Appl. Phys. Suppl. 2, Pt. 2, 691 (1974).
5.124. W. E. Spicer: Phys. Rev. **154**, 385 (1967).
5.125. B. J. Waclawski, T. V. Vorburger, J. R. Stern: J. Vac. Sci. Technol. **12**, 301 (1975).
5.126. A. Liebsch, E. W. Plummer: Paper presented at Faraday Division of the Chemical Society (Sept. 10–12, 1974).
5.127. N. E. Christensen, B. Feuerbacher: Phys. Rev. B **10**, 2349 (1974).
5.128. B. Feuerbacher, N. E. Christensen: Phys. Rev. B **10**, 2373 (1974).
5.129. B. Feuerbacher, B. Fitton: Phys. Rev. Letters **30**, 923 (1973).
5.130. E. W. Plummer, J. W. Gadzuk: Phys. Rev. Letters **25**, 1493 (1970).
5.131. B. J. Waclawski, E. W. Plummer: Phys. Rev. Letters **29**, 783 (1972).
5.132. B. Feuerbacher, B. Fitton: Phys. Rev. Letters **29**, 786 (1972).
5.133. K. Sturm, R. Feder: Sol. State Commun. **14**, 1317 (1974).
5.134. D. E. Eastman, W. D. Grobmann: Phys. Rev. Letters **28**, 1378 (1972).
5.135. L. F. Wagner, W. E. Spicer: Phys. Rev. Letters **28**, 1381 (1972).
5.136. J. W. Rowe, H. Ibach: Phys. Rev. Letters **32**, 421 (1974).
5.137. E. O. Kane: Phys. Rev. **146**, 558 (1966).

5.138. The 7 × 7 refers to the LEED pattern observed, 1 × 1 would be a bulkstructure.
5.139. L. W. SWANSON, L. C. CROUSER: Phys. Rev. Letters **19**, 1179 (1967).
5.140. O. K. ANDERSEN: Phys. Rev. B**2**, 883 (1970).
5.141. W. B. SHEPERD, W. T. PERIA: Surface Sci. **38**, 461 (1973).
5.142. P. J. ESTRUP, J. ANDERSON: J. Chem. Phys. **45**, 2254 (1966).
5.143. B. FEUERBACHER, B. FITTON: Phys. Rev. B**8**, 4890 (1973).
5.144. B. J. WACLAWSKI, E. W. PLUMMER: Unpublished.
5.145. T. E. MADEY, J. T. YATES, JR.: "Structure et Proprietes des Surfaces des Solides" (Editions du Centre National de la Recherche Scientique, Paris) **187**, 155 (1970).
5.146. T. E. MADEY: Surface Sci. **36**, 281 (1973).
5.147. G. W. RUBLOFF, J. ANDERSON, M. A. PRASSLER, P. J. STILES: Phys. Rev. Letters **32**, 667 (1974).
5.148. W. E. EGELHOFF, D. L. PERRY: Phys. Rev. Letters **34**, 93 (1975).
5.149. F. M. PROBST, T. C. PIPER: J. Vac. Sci. Technol. **4**, 53 (1967).
5.150. T. E. EINSTEIN, J. R. SCHRIEFFER: Phys. Rev. B**7**, 3629 (1973); T. B. GRIMLEY, S. M. WALKER: Surface Sci. **14**, 395 (1969).
5.151. R. GOMER: Jap. J. Appl. Phys. (in press).
5.152. J. T. YATES, JR., D. A. KING: Surface Sci. **32**, 479 (1972).
5.153. J. ANDERSON, P. J. ESTRUP: J. Chem. Phys. **46**, 563 (1967).
5.154. L. R. CLAVENNA, L. D. SCHMIDT: Surface Sci. **33**, 11 (1972).
5.155. J. M. BAKER, D. E. EASTMAN: J. Vac. Sci. Technol. **10**, 223 (1973).
5.156. J. T. YATES, JR., N. ERICKSON, S. D. WORLEY, T. E. MADEY: Bateile Colloq. "The Physical Basis for Heterogeneous Catalysis", Gstaad, Switzerland, Sept. 2–6, 1974.
5.157. P. L. YOUNG, R. GOMER: Phys. Rev. Letters **30**, 955 (1973); J. Chem. Phys. **61**, 4955 (1974).
5.158. C. KOHRT, R. GOMER: Surface Sci. **24**, 77 (1971).
5.159. C. LEUNG, M. VASS, R. GOMER: To be published.
5.160. J. H. SINFELT: Advan. Catalysis **23**, 91 (1973).
5.161. B. D. BARFORD, R. R. RYE: J. Vac. Sci. Technol. **9**, 673 (1972).
5.162. P. G. CARTIER, R. R. RYE: J. Chem. Phys. **59**, 4602 (1973).
5.163. T. GUSTAFSSON: Private communication.

6. Low Energy Electron Diffraction (LEED) and Auger Methods

E. BAUER

With 28 Figures

The two phenomena to be discussed in this chapter involve slow electrons: in LEED electrons with energies between approximately 10 eV and 200 eV are used, in Auger electron spectroscopy (AES) the useful energies of the Auger electrons range from 10 eV to 2000 eV. The basis of LEED is the interference between the electron waves scattered from the atoms which comprise the surface region of a crystalline solid, a phenomenon discovered 1927 by DAVISSON and GERMER. The basis of AES is an effect discovered 1932 by AUGER in which electrons are liberated in an atom with energies characteristic for the emitting atom. According to the basic mechanisms, LEED is governed predominantly by the periodocity, in particular by the periodicity parallel to the surface, while AES predominantly responds to the chemical composition of the material studied. The interpretation of LEED data requires the knowledge of the chemical composition of the surface. Therefore AES is complementary to LEED where LEED is applicable. But also when LEED cannot be used, e.g. with polycrystalline and amorphous surfaces, AES is a very important tool for the characterization of surfaces. Although LEED and the Auger effect were discovered more than 40 years ago, they have reached importance in surface studies only during the last ten years. The main reasons for this long delay of their wide-spread application are the past strong interest of science in the bulk of solids and the experimental difficulties of producing well-defined surfaces routinely and keeping them unchanged during observation. The recognition of the importance of surfaces for many solid state properties and many important technical processes as well as the development of ultra high vacuum (UHV) techniques in the early sixties eliminated these reasons and both techniques developed rapidly.

In LEED electrons are produced outside the crystal in form of a nearly parallel electron beam which may be described theoretically as a plane wave. In AES the electrons are produced within the crystal by electrons, X-ray photons, energetic ions or neutrals and are emitted nearly isotropically by the atoms and may be described theoretically by a spherical wave. Whatever their origin, whether incident from the outside or generated within the solid, the electrons have to propagate through the solid and the surface, and in doing so they interact with the solid and its surface. These interactions are decisive for the understanding of both LEED and AES and will therefore be discussed first (Section 6.1). By understanding elastic (Subsection 6.1.1), inelastic (Subsection 6.1.2) and quasielastic (Subsection 6.1.3) scattering processes the limitations of the kinematical and dynamical theory of the elastic LEED method (ELEED) (Subsection 6.2.2) will become evident and the occurrence of inelastic LEED (ILEED) discussed in conjunction with other special topics (Subsection 6.2.4) will become understandable. The reader not interested in the physical basis and the theory of LEED but only in the experimental aspects and results is referred to Subsections 6.2.1 and 6.2.3.

The physical foundations of the Auger methods are discussed next (Subsection 6.3.1), followed by a description of the experimental techniques (Subsection 6.3.2). The applications of AES are illustrated by the results of Subsection 6.3.3 and current basic problems and new developments in Subsection 6.3.4. The article is intended to be an introduction to the physical basis of the technique and as a synopsis of the present state of the art. Because of its brevity it cannot be comprehensive but only illustrative. For more detailed information the reader is referred to the comprehensive reviews (Appendix) which have already appeared on the subject of this article.

Before going into the theoretical discussion of Section 6.1 a few remarks should be made about the experimental limitations common to both methods. Because of the strong interaction between slow electrons and matter the experiments have to be performed in very high vacuum. Hydrocarbons in the system generally lead to strong specimen contamination. This requires special specimen chambers or the whole system must be bakeable unless desorption from the walls is suppressed by cooling. The specimens which can be studied in a meaningful manner should fulfill a number of conditions: 1) they should not be changed by the electron beam (dissociation in ionic crystals, desorption of adsorbed gases, etc.) 2) they should not charge up on electron bombardment, 3) they should have a low vapor pressure or special cooling devices have to be used and 4) they have to have long-range order over a sufficiently large portion of the surface so that diffraction can occur (for LEED only).

The basic properties of the methods outlined and the experimental limitations just mentioned determine the application range of LEED and AES. LEED is only suited for the study of surfaces with sufficient lateral periodicity, i.e., single crystal surfaces, surfaces of specimens with fiber texture (deposited films, rolled sheets, etc.), or surfaces of layered materials with disorder normal to the surface such as pyrolytic graphite. On such surfaces not only the structure but all processes can be studied in which changes in lateral periodicity or order occur such as order-disorder, decomposition, adsorption or condensation processes. In contrast to LEED, AES neither requires a periodic nor a flat surface. Thus it can be applied—and it has with great success—to such unusual samples as fracture surfaces or rocks from the moon. Surface roughness, of course, makes quantitative chemical analysis of the surface layer which is the final goal of AES very difficult, especially if the primary beam is far from normal incidence (shadow effects). But on smooth surfaces after appropriate calibration quantitative surface analysis is already possible with proper care. AES is not limited to chemical analyses, but can also give structural information via the strong attenuation of the Auger electrons on their way to the surface. Furthermore, all processes can be studied which are connected with changes in chemical composition in the surface and which can take place in vacua better than 10^{-4} Torr. The use of sputtering in conjunction with AES not only allows depth analysis but also studies in relatively dirty vacuum because the contamination can be removed continuously. AES is at present by far the most universal surface analysis method with sensitivities down to less than 1/100 of a monolayer, except for hydrogen and helium which are not detectable.

6.1. Interaction of Slow Electrons with Condensed Matter

6.1.1. Elastic Scattering

Elastic scattering is predominantly due to the ion cores which contain the nucleus and the strongly bound electrons. In order to obtain an understanding of the elastic scattering probability, in particular of the total scattering cross-section of an atom and the angular distribution of the scattered intensity it is useful to neglect initially the phase relations between the waves scattered by the atoms of a solid. This amounts to assuming a random distribution of atoms ("randium") which is approximated to a certain extent in the amorphous state. In such a system the intensities scattered by the individual atoms add up, instead of the amplitudes. Thus the angular distribution, neglecting double scattering, is simply given by that of an individual atom. Figure 6.1 [6.1] shows typical angular distributions for 50 eV electrons scattered by ion cores as used in band structure calculations. Very strong forward and strong backward scattering is evident, as well as the fact that Cu scatters much less than Al, which is quite unexpected from X-ray and high-energy electron diffraction. These deviations from the expectation that the scattering probability increases with nuclear charge Z are even more

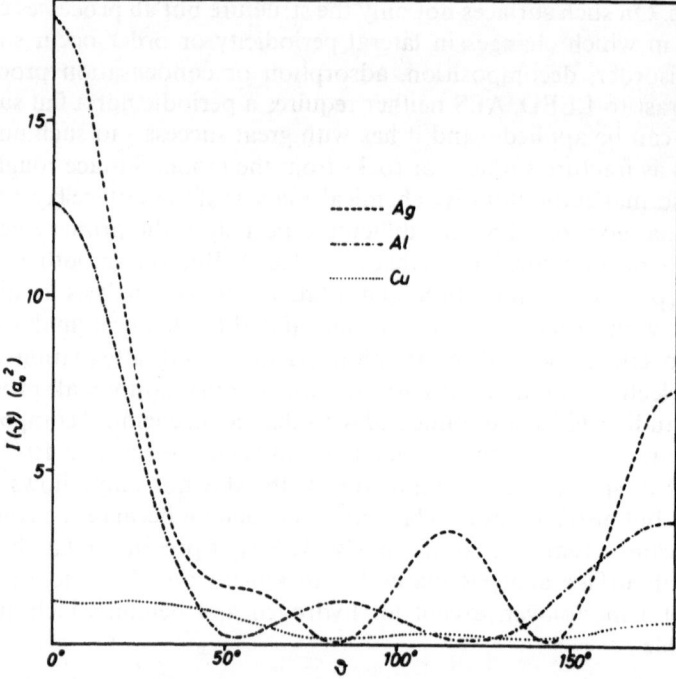

Fig. 6.1. Angular distribution of the intensity scattered by Al, Cu, and Ag ion cores as obtained from band structure calculations in atomic units ($a_0 = 0.529$ Å) [6.17]

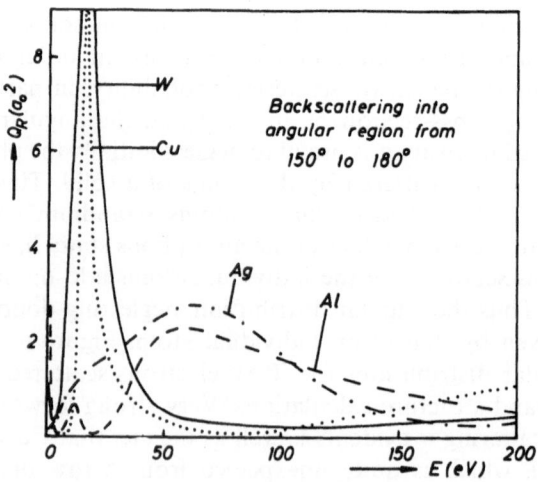

Fig. 6.2. Backscattering into a 30° cone around the backward direction: $Q_b = 2\pi \int_{150}^{180} I(\vartheta)$ · $\sin \vartheta \, d\vartheta$ for Al, Cu, Ag, and W ion cores [6.2]

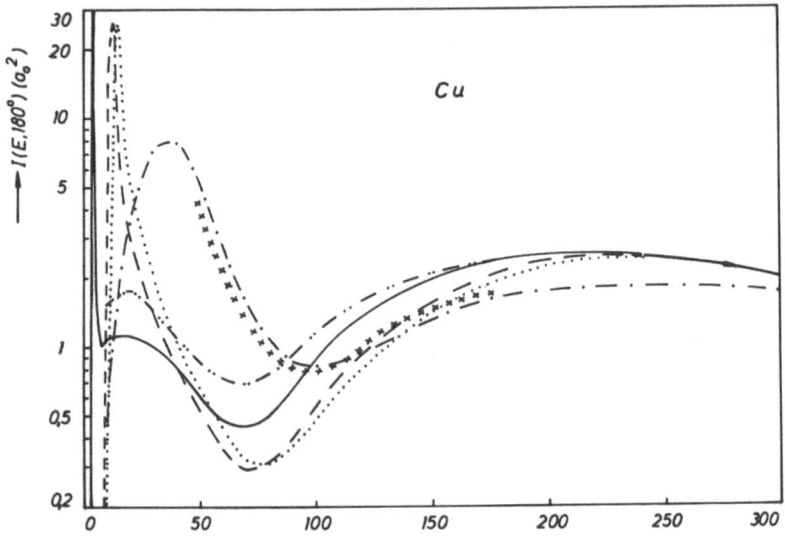

Fig. 6.3. Backscattering into $\vartheta = 180°$ by Cu described by different potentials [6.2]. See text for an explanation

evident in Fig. 6.2 [6.2] which shows the backscattering into a 30° cone around the backward direction: W $(Z = 74)$ shows less backward scattering than Al $(Z = 3)$ from about 30 eV to 200 eV! It should be noted that such calculated scattering cross-sections become increasingly unreliable with decreasing energy below about 100 eV. This is illustrated in Fig. 6.3 which shows the backscattering $(\vartheta = 180°)$ from Cu calculated with six different potentials [6.2]. The results for the two best "solid state potentials" $(---, ...)$ agree quite well with each other, as do those for free-atom potentials $(\times \times \times, -\cdot-)$, but the difference between the two pairs of curves is drastic below 100 eV, indicating that free-atom potentials become rather questionable in the low energy range. But even the results obtained with "solid state potentials" become inaccurate, if correlation (polarization of the electron cloud of the ion core by the incident electron) and exchange ("spin-spin interactions" between incident and ion-core electrons) effects are neglected. Correlation becomes important below about 100 eV, exchange below about 30 eV, but both effects are still noticeable at higher energies [6.3]. It is difficult to take these effects into account quantitatively, which represents one of the major problems of the theory of low energy electron scattering and LEED. These limitations should be kept in mind.

The attenuation $I = I_0 \exp(-vx)$ of the incident beam due to the total elastic scattering or the elastic backscattering is determined by the

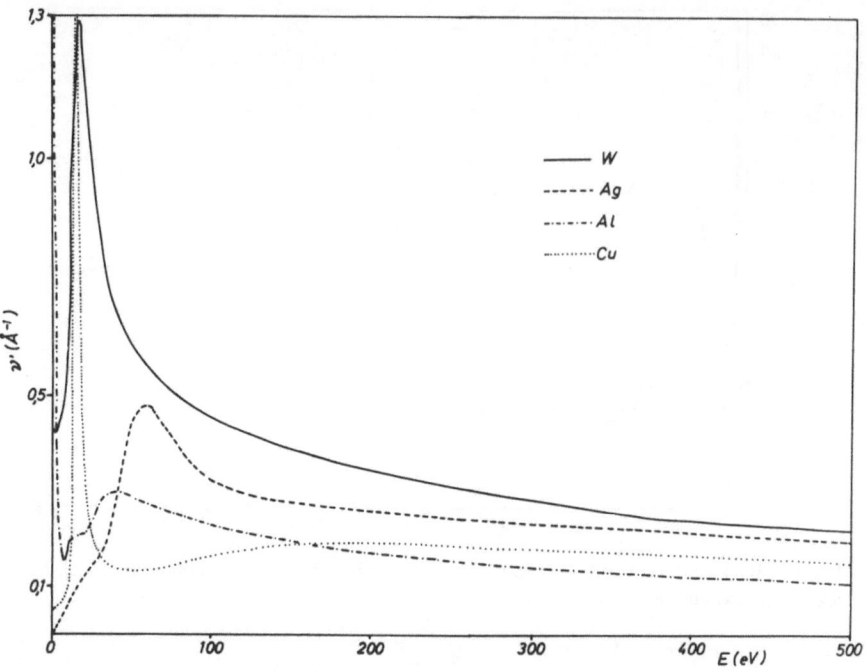

Fig. 6.4. Total attenuation coefficient (neglecting double scattering) for Al, Cu, Ag, W [6.1]

total scattering cross-section $Q_t = 2\pi \int\limits_0^\pi I(\vartheta) \sin\vartheta \, d\vartheta$ or the backscattering

cross-section $Q_b = 2\pi \int\limits_{\pi/2}^\pi I(\vartheta) \sin\vartheta \, d\vartheta$ and the density ϱ of the ion cores: $v = \varrho Q$. The primary extinction coefficient $v' = \varrho Q_t$ is shown for. Al, Cu, Ag, and W in Fig. 6.4 [6.1]. The v' values indicate that the incident intensity has already been reduced by $1/e$ after 1–10 Å which means that double and multiple elastic scattering play an important role. Electrons scattered into the forward half space are not lost from the elastic wave field and thus do not attenuate it. True attenuation due to elastic scattering occurs only by backscattering. Except below 100 eV the calculated v values are below 0.1 $Å^{-1}$, which is a weak attenuation as compared to that caused by inelastic scattering.

6.1.2. Inelastic Scattering

Inelastic scattering can occur by excitation or ionization of valence electrons or inner shell electrons. Inelastic interactions with inner shell electrons are weak compared to those with valence electrons and

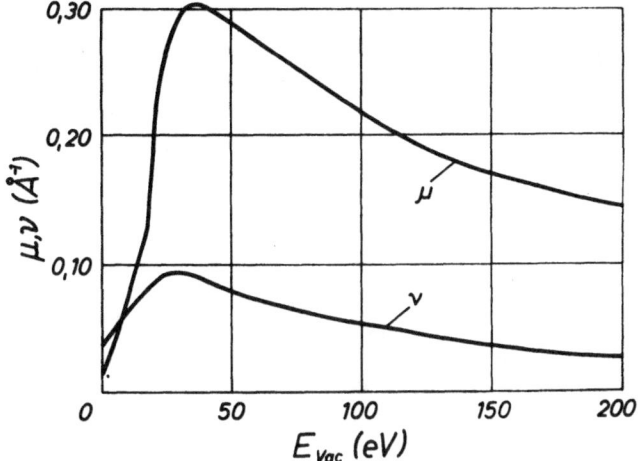

Fig. 6.5. Attenuation coefficients due to inelastic scattering by valence electrons (μ) and due to elastic backscattering (v) for Al [6.1]

important—in the context of this chapter—only for AES. They will be discussed in Subsection 6.3.1. Inelastic scattering by valence electrons is best discussed in terms of the "jellium" model: the valence electrons are considered as a free electron gas whose charge is compensated by the positive charge of the ion cores which is smeared out to form a homogeneous, isotropic background charge. This model is most applicable to free-electron-like metals such as the alkalis, Al, Be or Mg, but to a certain degree also to other metals, to semiconductors and insulators provided that the group of electrons considered is not bound too tightly ($E_i < 10$–20 eV) and that it is sufficiently separated in energy from the rest of the electrons.

Inelastic scattering in jellium occurs by one-electron excitation and by excitation of plasma oscillations, i.e. by collective excitation of the valence electrons due to the interaction with the incident electron and due to their mutual interaction. This excitation can occur only above a certain threshold, which varies with electron density from $2E_F$ to $1.6E_F$ (E_F: Fermi energy) [6.2]. Below this threshold only one-electron excitations are possible. Figure 6.5 shows the attenuation coefficient μ of slow electrons due to valence (i.e. conduction) electron excitation in Al as calculated in the jellium approximation and for comparison the attenuation coefficient v due to elastic backscattering. The dominance of inelastic scattering by valence electrons is clearly recognizable. The minimum mean-free path for inelastic scattering in Al is $\lambda_{ee}^{min} = 1/\mu_{max} \approx 3$ Å. The experimental values for Al [6.4] agree well with

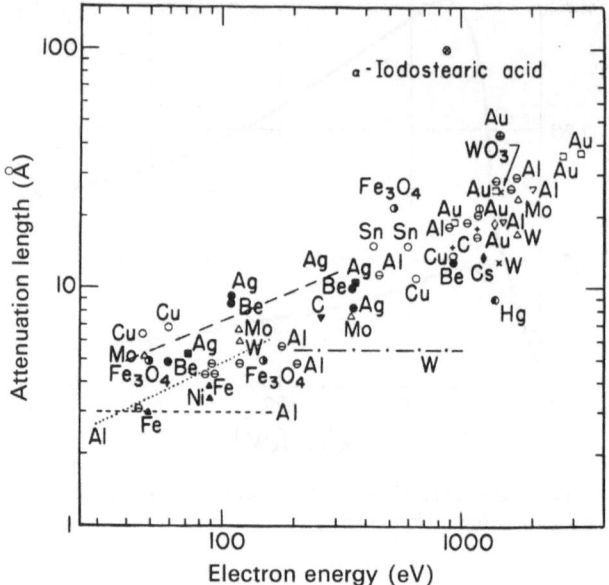

Fig. 6.6. Experimental attenuation length (inelastic mean-free path) values λ_{ee} [6.8]

the jellium predictions. This seems surprising because an important inelastic attenuation mechanism, surface plasmon excitation, has been neglected in the calculation. However, the volume plasmon excitation probability decreases towards the surface in a manner complementary to the increase of the surface plasmon excitation probability [6.5]. Therefore an approximate compensation occurs with the result that inelastic attenuation changes little right up to the surface. Curves similar to that in Fig. 6.5 are obtained for other materials [6.1] and give λ_{ee} values which lie well within the scatter of the experimental values [6.6, 6.7] of mean-free paths for inelastic scattering shown in Fig. 6.6 [6.7]. The weak dependence of λ_{ee} on the kind of material is not surprising because λ_{ee} is basically determined by the valence electron density which does not change strongly from material to material.

For the understanding of ILEED in which inelastically scattered electrons are observed, not only the total scattering cross-section, as expressed by μ, is needed but also the differential scattering cross-section (angular distribution) $I_{ee}(\vartheta)$. There are at present no explicit expressions for $I_{ee}(\vartheta)$ available for slow electrons [6.8] so that to a first approximation the formulas valid for fast electrons ($E \sim 10^4$ eV) [6.9] must be used [6.2]. For volume plasmon excitation the differential attenuation coefficient is $\mu(\vartheta) \sim \varepsilon/(\varepsilon^2 + \vartheta^2)$ which can also be written in the form $\mu(K) \sim \Delta E / K^2$. Here $\varepsilon = \Delta E / 2E$ ($\Delta E = \Delta E(K) =$ energy loss = energy of the volume plasmon excited), K is the wave number of the plasmon and E the energy of the incident electron. Thus μ is independent of ϑ for small angles and decreases like ϑ^2 at larger angles. Beyond a critical scattering angle ϑ_c corresponding to a maximal wave number K_c or minimum plasmon wavelength λ_c no plasmon excitation is possible but only one-electron excitation. For example, for 50 eV electrons in Al $\vartheta_c \approx 25°$ [6.2]; similar values are obtained for slow electrons in other materials. Thus electrons which have been inelastically scattered due to excitation of volume plasmons cannot be observed directly in the backward direction but only via *elastic* backscattering. One-electron excitation is not limited by such a critical ϑ_c but drops off so rapidly beyond ϑ_c for volume plasmons ($\sim \vartheta^{-4}$, Rutherford scattering!) that for practical purposes it can also be observed only via elastic backward scattering. Similarly electrons which have lost energy by surface plasmon excitation can generally only be observed by the same mechanism. This coupling between inelastic and elastic scattering necessary for the observation of inelastic scattering is the basis of ILEED.

6.1.3. Quasielastic Scattering

Quasielastic scattering represents the influence of thermal motion of the ion cores on the scattering process. Because of the small energy exchange (10–100 meV) occurring in this process, it cannot be energy-resolved in ordinary LEED instruments but it has a strong effect on the angular distribution of the (quasi)elastically scattered electrons. This effect is most apparent in well-ordered crystals, in which the thermal motion (lattice vibration) reduces the order and thus the intensity of the diffracted beams which depends on the same order. This phenomenon is taken into account in the usual diffraction theory by the Debye-Waller factor $\exp(-2W)$ with $2W = \langle (K \cdot u)^2 \rangle \approx 16\pi^2 \langle u^2 \rangle \sin^2(\vartheta/2)/\lambda^2$. Here $K = k - k_0$, k_0, k being the wave vectors of the incident and scattered waves, u is the instantaneous displacement of an atom from its equilibrium position, ϑ the scattering angle $\sphericalangle(k, k_0)$ and λ the wavelength of the electron. The mean square displacements of atoms in and near the surface are in general much larger than in the bulk, show strong anisotropy and depend in a complicated manner upon the distance from the surface [6.10]. To a first approximation, however, $\langle u^2 \rangle$ decreases as $\exp(-l)$ with layer number l from the surface to the bulk value $\langle u^2 \rangle_B$. For example, $\langle u^2 \rangle_l / \langle u^2 \rangle_B \approx 3.5$, 1.7, and 1.3 for $l = 1, 2, 3$. Therefore thermal motion reduces the contribution of the first layer to the intensity of diffracted beams much more than the contribution of the second layer. This fact is important for the interpretation of LEED patterns.

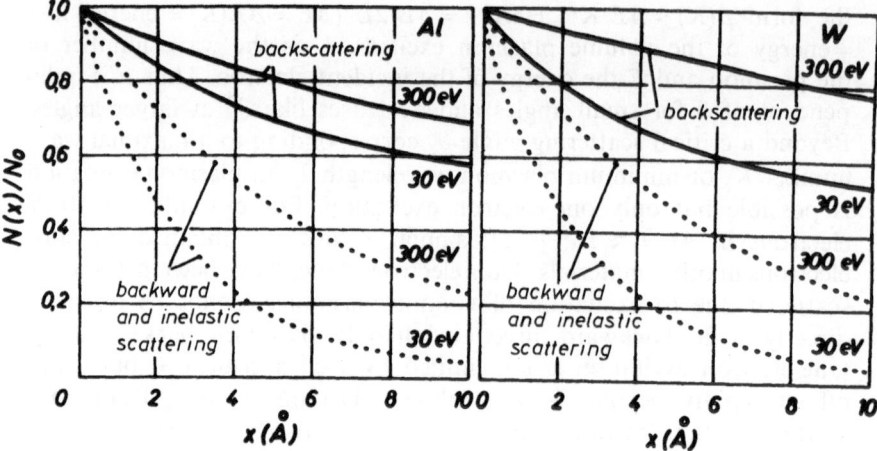

Fig. 6.7. Attenuation of the elastic electron current ($\Delta E = 0$) propagating into the foreward half-space due to backscattering only and due to inelastic scattering plus backscattering [6.1]

6.1.4. Consequences of the Scattering Processes

The important features of these scattering processes for ELEED are 1) the strong attenuation due to inelastic scattering by the valence electrons, 2) the large total cross-section for elastic scattering and 3) the strong layer dependence of the quasieleastic scattering. The first feature reduces the thickness of the surface region which contributes effectively to the LEED pattern to about 10–20 Å, depending upon energy. This is illustrated in Fig. 6.7: at a beam energy of 30 eV at which the absorption coefficient is close to its maximum, only 2–4 layers are reached by enough "elastic" electrons to contribute to the LEED pattern. Because of the high elastic scattering probability only a fraction of these elastic electrons propagate in the direction of the incident beam: the electrons form an "elastic" wave field which produces the LEED pattern in contrast to X-ray diffraction where, to a good approximation, the unattenuated incident wave produces the diffraction pattern. The third feature counteracts the first one which suppresses the contributions of deeper layers to the LEED pattern. At high energies, high temperatures or in crystals with low Debye temperatures—that is large $\langle u^2 \rangle$—the second layer may contribute as much as the first layer.

The consequences drawn for ELEED apply also to ILEED, for which still another scattering feature is important: the lack of direct back-scattering which requires combination with elastic scattering (diffraction) in order to observe inelastic scattering. The feature most important for

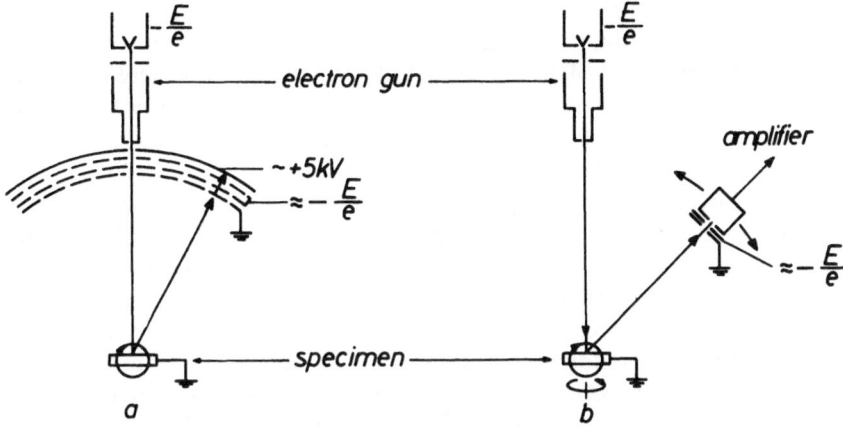

Fig. 6.8. Schematic of the basic types of LEED instruments. See text for an explanation

AES is again the strong inelastic scattering which reduces the escape depth of Auger electrons in the usual energy range to 3–20 Å (see Fig. 6.6) independent of energy, direction and kind of the exciting incident radiation (electrons, X-ray, ions).

6.2. Low Energy Electron Diffraction (LEED)

6.2.1. Experimental Methods

There are essentially two types of LEED instruments: display-type systems and diffractometers. In display-type systems (Fig. 6.8a) the electrons scattered into a large solid angle segment are observed simultaneously on a fluorescent screen. The inelastically scattered electrons and the secondary electrons generated by the incident beam are eliminated by a retarding field between the grids. In a diffractometer (Fig. 6.8b) the electrons scattered into a small solid angle are measured with a Faraday cup or channeltron after passing through a retarding field. Detector and crystal are rotated in a suitable manner so that all desired angles of incidence and diffraction may be covered. The display-type system has the advantage of speedy observation, but for quantitative intensity measurements diffractometers are far superior. There are many versions of both types of instruments, including combinations, e.g. with a channeltron behind a slit in the LEED screen.

The electron gun usually produces a beam on the specimen with 0.1–10 µA current and a diameter from 10^{-4} to 10^{-3} m depending on electron energy (10–200 eV). The coherence of the beam is determined primarily by the angular divergence and is usually such that specimen regions of the order of several 100 Å diameter scatter coherently. This means that the scattered *amplitudes* have to be added only for atoms within a coherence region and the total scattering is obtained by adding the scattered *intensity* of all coherence regions within the beam cross-section. There are many ways of specimen preparation depending upon its chemical, thermal and mechanical properties: sawing followed by mechanical, chemical or electrochemical polishing and cleaving are the most important ones. If cleaving is not done in the system after establishing UHV then the surface has to be cleaned in situ by ion bombardment or by reaction with gases such as O_2 and H_2. An example is the removal of C or S from metals by O_2, sometimes followed by reduction in H_2. The single-crystalline regions on the specimen obviously should be larger than the beam cross-section if the interpretation should not get to complicated. Utmost surface perfection (e.g. surface orientation within 0.01° of the desired plane) is essential for many quantitative studies but LEED patterns can be obtained frequently from microscopically very rough and inhomogeneous surfaces, as long as a sufficiently large fraction of the surface ($> 10\%$) produces the same LEED pattern.

In order to keep the surface clean pressures below 10^{-9} Torr have to be maintained during observation; the best systems work in the 10^{-11} Torr range. In studies of the interaction with gases pressures up to 10^{-4} Torr can be tolerated if proper electron emitters are used. Because of the interpretation difficulties it is useful if not necessary to incorporate other experimental techniques in the LEED system: AES is the most important one, but thermal desorption spectroscopy (TDS), work function measurements and ion scattering spectroscopy (ISS) are also very useful.

6.2.2. Theoretical Methods

Kinematical Theory

This is the simplest theoretical method. In its most elementary form it takes only the diffracted waves produced by the incident wave in the topmost layer into account. More refined versions include also the diffracted waves from "visible" atoms in the second layer without or with inclusion of the attenuation of the incident wave due to inelastic scattering, or the diffracted waves from many layers whose contribution rapidly diminish with increasing distance from the surface due to attenuation by inelastic scattering. All versions of the kinematical theory have in common that only the (unattenuated or attenuated) incident wave is considered to produce diffracted waves while the re-diffraction of diffracted waves (double and multiple scattering) is neglected. Because of the strong elastic scattering (Subsection 6.1.1) this is an extremely serious oversimplification, which considerably limits the application of this theory: it makes it nearly useless for the interpretation of the intensity of the diffracted beams and causes ambiguity even in the interpretation of the geometry of the diffraction pattern from non-ideal surfaces and surfaces covered with adsorbates or thin overlayers. Nevertheless, it is useful in many cases for the interpretation of the position and shape of the diffraction spots, as observed on the fluorescent screen.

A discussion of the simplest version—diffraction only by the topmost layer or in the case of adsorbates by the first two layers—suffices to demonstrate the application of the kinematical theory. The amplitude of the scattered wave is in this approximation obtained by summing over the amplitudes of the waves scattered by all atoms within the coherence region

$$\Psi(r) = \frac{e^{ikr}}{r} \, \psi(K) \quad \text{with}$$

$$\psi(K) = \sum_{lm_1m_2} f_l(K) \exp[-iK \cdot (r_l + m_1 a_1 + m_2 a_2)]$$

$$= \sum_l f_l(K) \exp(-iK \cdot r_l) \cdot \sum_{m_1 m_2} \exp[-iK \cdot (m_1 a_1 + m_2 a_2)]$$

$$= F \cdot G,$$

(6.1)

where F refers to Σ_l and G to $\Sigma_{m_1 m_2}$. Here a_1, a_2 define the periodicity parallel to the surface—the surface unit mesh—r_l is the position of the atoms within the unit mesh, m_1, m_2 are integers, and $K = k - k_0$ as before. The function $f_l(K)$ is the scattering amplitude of atom l and depends only on $K = |K| = 2k \sin(\vartheta)/(2) = \frac{4\pi}{\lambda} \sin(\vartheta)/(2)$ $(\vartheta = \measuredangle(k, k_0))$.

The exponential factor represents the geometrical phase differences between the incident wave $\exp(ik_0 \cdot r_{lm_1m_2})$ and the waves $\exp(ik \cdot r_{lm_1m_2})/r$ diffracted by the atom in position $r_{lm_1m_2}$.

The intensity of the scattered wave is obtained by summing over the intensities $I = |F|^2 |G|^2$ from all coherence regions. The "structure factor" $|F|^2$ varies only slowly with angle and determines mainly the intensity of the diffracted beams. Because of the neglect of the all-important double scattering in the approximation it is not a meaningful quantity and will not be discussed any further. The "lattice factor" or "interference function" $|G|^2$, however, involves only a summation over the unit meshes which are all completely equivalent from the point of view of scattering theory. Thus it is a generally valid expression. Performing the summation in (6.1) for a periodic structure with the dimensions $M_1 a_1 \times M_2 a_2$ (smaller or equal to the coherence region) gives

$$|G|^2 = \frac{\sin^2\left[\frac{1}{2}(M_1+1)A_1\right]}{\sin^2\left(\frac{1}{2}A_1\right)} \cdot \frac{\sin^2\left[\frac{1}{2}(M_2+1)A_2\right]}{\sin^2\left(\frac{1}{2}A_2\right)} \tag{6.2}$$

with $A_i = K \cdot a_i = K_\parallel \cdot a_i$. This function has strong main maxima whenever $A_i = 2\pi h_i$ (h_i integer). If $K_\parallel$ is expressed in terms of the unit vectors b_1, b_2 of the reciprocal mesh to a_1, a_2: $K_\parallel = 2\pi(n_1 b_1 + n_2 b_2)$, then (6.2) expresses the fact that diffracted waves occur only in certain directions, given by the conditions $k_\parallel - k_{0\parallel} = 2\pi(h_1 b_1 + h_2 b_2)$ and $k = k_0$. Thus diffracted beams and diffraction spots (h_1, h_2) are produced on the screen. The size and shape of these spots is determined according to (6.2) by the dimensions M_1, M_2 of the periodic region, the smallest spot size being that determined by the size of the coherence region. If the periodic region is large in the direction $\pm a_1$ then the diffraction spot is small in the direction $\pm b_1$, and vice versa.

Frequently there are within the coherence region several regions with identical periodicity but displaced relative to each other by amounts $d = d_1 a_1 + d_2 a_2$ which do not belong to the periodicity (d_1, d_2: non-integers). The lattice factor of two such "subdomains" is given by

$$|G[1 + \exp(iK \cdot d)]|^2 = 2|G|^2(1 + \cos K \cdot d). \tag{6.3}$$

Thus the intensity distribution within a diffraction spot (h_1, h_2) [see Eq. (6.2)] is modulated by $(1 + \cos K \cdot d)$. For example the intensity will be zero in the center of the spot when $K \cdot d = (2n+1)\pi$ or $2\pi(h_1 d_1 + h_2 d_2) = (2n+1)\pi$; for a $(1,0)$ spot this condition is fulfilled if $d_1 = (2n+1)/2$, e.g. $d_1 = 1/2$. Such a situation may occur in an adsorption layer which has the same periodicity as the substrate but is displaced relative to it by $a_1/2$. More frequently this spot splitting is seen in adsorption layers which have a larger periodicity than the substrate. A layer with $c(2 \times 2)$ structure on a $\{100\}$ surface of a cubic crystal, i.e. a layer in which the nearest neighbour atoms within the layer are missing, produces, in addition to the integral spots, "extra" half-order spots $[h_i = (2n_i + 1)/2]$. For these spots the condition for splitting is fulfilled also with integral displacements d_1, d_2 of the subdomains. Domain formation is a consequence of the statistical nature of the adsorption process, the limited range of the forces between the atoms and their limited mobility. A detailed analysis therefore requires a statistical treatment.

The location, width and shape of spots can also be analyzed for other forms of surface disorder such as steps, facetting and etch pit formation. The effect of steps is illustrated by a periodic array of steps perpendicular to a unit mesh edge a. If the terraces between the N steps within the coherence diameter are M unit-meshes wide and two steps are separated by the vector s, then

$$G = \sum_{m=0}^{M-1} \exp(im\mathbf{K} \cdot \mathbf{a}) \cdot \sum_{n=0}^{N-1} \exp(in\mathbf{K} \cdot \mathbf{s})$$

so that

$$|G|^2 = \frac{\sin^2\left[\frac{1}{2}(M-1)A\right]}{\sin^2\frac{1}{2}A} \cdot \frac{\sin^2\left[\frac{1}{2}(N-1)S\right]}{\sin^2\frac{1}{2}S}, \tag{6.4}$$

where $S = \mathbf{K} \cdot \mathbf{s}$. The first term determines the spot position for the ideal surface, the second factor the intensity distribution within the spot. Because $S = \mathbf{K} \cdot \mathbf{s}$, and s has a component normal to the surface, the intensity modulation by the second factor is energy dependent: a spot splitting normal to the steps occurs (see Fig. 6.9) which appears and disappears with electron energy. This effect occurs even when the step periodicity is poor [6.11]. For the determination of the absolute or relative spot intensities as function of energy $|G|^2$, of course, has to be multiplied with the dynamical structure factor [6.13]. In facetting surface elements are formed which are inclined to the original surface. They cause additional diffracted beams with directions determined by the conditions $\mathbf{K}_\parallel \cdot \mathbf{a}_i = 2\pi h_i$, $k = k_0$ applied to the unit mesh vectors of these surface elements. These facet beams can easily be recognized because they move in directions different from those expected for the flat surface when the beam energy is changed.

Thus considerable information can be extracted from LEED patterns with the kinematical theory, even about atomic positions (from domain patterns). But because of the importance of double scattering there are also serious limitations. The problem of interpreting spot intensities has already been mentioned, but even the spot positions cannot be uniquely interpreted if different surface meshes are superimposed. This is illustrated in Fig. 6.10: a) represents a LEED pattern from a {100} surface of a cubic crystal (circles) covered with a layer which produces additional spots (crosses) [(100)-(2 × 1) pattern]. If double scattering would not occur the pattern b) plus the same pattern rotated by 90° would explain the diffraction from the layer. However, each diffraction beam from the substrate (circles) can act as a primary beam for the layer (crosses) (and *vice versa*). Thus patterns c) or d) plus the 90° rotated ones as well as many others can also produce pattern a). Therefore

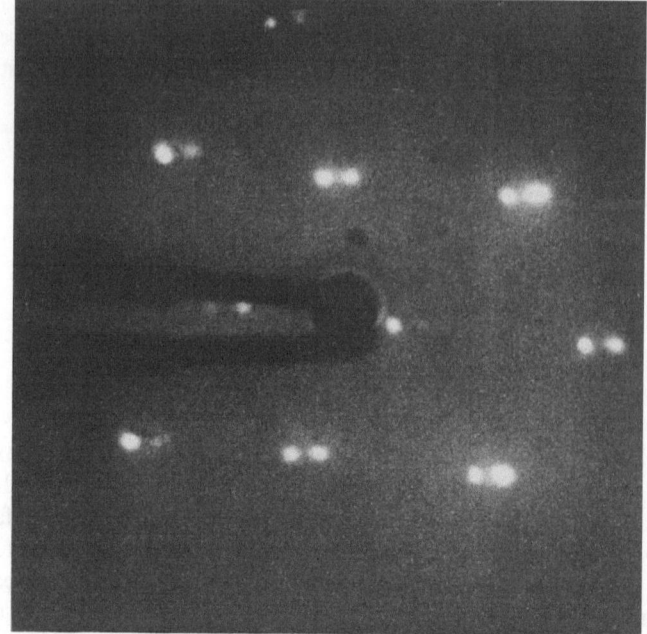

a

b

[1̄1̄0]

[111]

[112]

(001)

Periodicity

Fig. 6.9a and b. LEED pattern from stepped Pt surface and step structure derived from it [6.12]

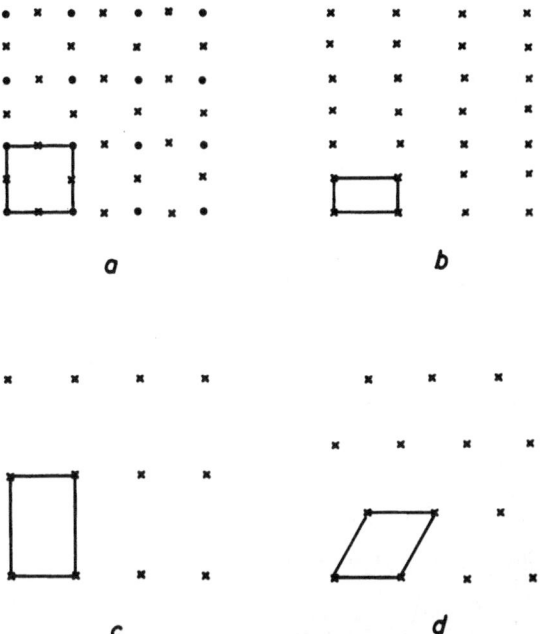

Fig. 6.10a–d. Example for the ambiguity in the determination of the unit cell dimensions due to double scattering. See text for an explanation

additional information is needed for a distinction between these interpretation possibilities which may be obtained with other techniques such as coverage measurements or from an evaluation of the spot intensities.

The evaluation of the spot intensities must take into account all the scattering processes discussed in Section 6.1. There have been attempts to apply standard or modified techniques for the evaluation of X-ray diffraction intensity data to LEED intensity data, such as modifications of the Patterson function method [6.14, 6.15]. An analysis of the Patterson function type assumes implicitly that the wave producing the diffracted wave consists only of the incident plane wave. Applying it to LEED data with the complicated wave field in the specimen should not produce meaningful results unless—as has been suggested to explain the apparent success of the method [6.15]—the transform process used suppresses the contribution of multiple scattering processes and enhances the intensity features produced by the incident wave.

Such suppression and enhancement effects are also used more directly: in LEED data-averaging techniques the observed beam intensities $I_{h_1 h_2}(k, k_0)$ are averaged over a finite energy range [6.16, 6.17]

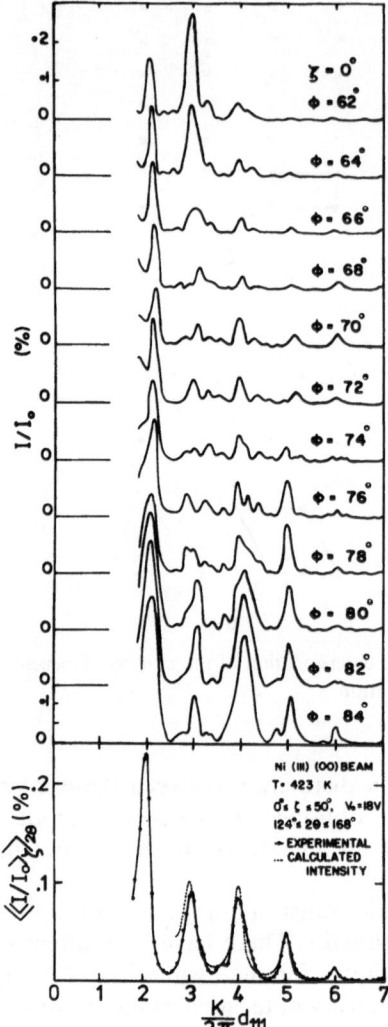

Fig. 6.11. Individual and averaged $I_{00}(V)$ curves of a Ni {111} surface (ξ: azimuth, ϕ: grazing angle of incidence, $2\theta = \vartheta$, $S/S_0 = 2d_{111}\sin\theta/\lambda$) [6.19]

or over many k, k_0 pairs connected by the same $K = k - k_0$ [6.18–6.21] and compared with kinematical model calculations. The rational for the constant K averaging method is the fact that in the kinematical theory $I_{h_1h_2}$ depends only on K so that an average of sufficient data at constant K should give the kinematical intensity. Figure 6.11 shows that at least a kinematical-like curve is obtained on averaging [6.19]. A comparison of the two averaging methods reveals similar accuracy which is given as ± 0.03 Å for the surface layer spacing [6.22].

Dynamical Theory

The dynamical theory takes the re-diffraction of diffracted waves into account and must in the case of LEED include absorption due to the strong inelastic scattering. Also the reduction of the beam intensities due to quasielastic scattering must be included. There is at present no general theory including all these phenomena accurately. Rather absorption is taken into account by an imaginary, energy-dependent effective potential or by an energy-dependent attenuation length λ_{ee} such as calculated in the jellium approximation (Section 6.1) [6.23, 6.24] or choosen as an adjustable parameter [6.25]. Thermal effects are taken into account by multiplying the atomic scattering amplitudes with Debye-Waller factors which are usually assumed to be layer and direction independent [6.26]. The importance of the various simplifications is still a matter of controversy. Some authors consider the accuracy of the ion-core potential (Subsection 6.1.1) to be most critical, some the shape of the complex energy-dependent surface potential (transition region between crystal and vacuum), and others the deviation of the vibrational behavior near the surface from that in the bulk. Although results obtained with various approximations are accumulating (Subsection 6.2.3) it is still premature to say what simplifications are most serious. Very likely all opinions will prove to be correct depending on specimen, temperature and electron energy.

   The basic problem of the dynamical theory can be seen by inspecting the integral equation describing the scattering process

$$\psi(r) = \psi_0(r) + \int G(r, r')\, V(r')\, \psi(r')\, dr'$$
$$\equiv \psi_0(r) + \int G(r, r')\, T(r')\, \psi_0(r')\, dr' , \tag{6.5}$$

where $\psi_0(r)$ is the incident wave, $\psi(r)$ the scattered wave, $G(r, r')$ is the Green's function describing the propagation of the wave scattered in the volume element dr' at position r' to the point r, $V(r)$ is the effective scattering potential taking into account correlation, exchange, thermal scattering and the transition region between crystal and vacuum, and $T(r)$ is the "scattering matrix" of the crystal. Thus the determination of the amplitude of the diffracted wave $\psi(r)$ at a large distance from the crystal $(r \gg r')$ requires the evaluation of $\psi(r)$ within the crystal, of $V(r)$ and of $G(r, r')$. For G the function valid for propagation in jellium has been used in general discussions, and V is usually constructed from band structure muffin tin potentials, i.e. core potentials with spherical symmetry. The scattered wave ψ is evaluated by a variety of techniques. The most frequently used techniques are the Darwin or layer method and the Bethe or "wave function matching" method.

In the Bethe method the Schrödinger equation is solved for the three-dimensionally periodic crystal and the wave functions in the crystal are matched at the surface to those in vacuum. This determines the amplitudes of the vacuum waves. Although this method can be generalized by adding a layer with a layer spacing different from that in the bulk, or with different lateral periodicity at the surface to treat lattice expansion or contraction at the surface, the Darwin method is much better adapted to this situation and to adsorbed layers. In it the crystal is divided into layers parallel to the surface. Within each layer multiple scattering between the various atoms is calculated using various methods, while multiple scattering between layers is taken into account by matching the wave functions at the boundaries between planes. Some of the most successful recent calculation procedures, such as the inelastic-collision model [6.25], the layer Kohn-Korringa-Rostocker (KKR) method [6.27], the RFS (renormalized forward scattering) perturbation theory [6.28] or the t-matrix formalism [6.29] belong to this group of methods. All these calculations require large computers. In cases in which calculations have been made for identical models the various methods now give comparable results.

6.2.3. Results

Clean Surfaces

The lateral periodicity of most surfaces is identical to that in the bulk. However there are a number of exceptions such as the {100} surfaces auf Au, Pt, Ir or various surfaces of Si and Ge. These surfaces exhibit superstructures (Fig. 6.12b) which transform reversibly into the normal periodicity at elevated temperatures, e.g. at 1100 K, 1150 K, and 600 K in the case of the Au {100}, Si {111}, and Ge {111} surfaces, respectively. The superstructures can be easily suppressed or modified by impurities (Fig. 6.12a) but there is no doubt now that they are characteristic of the clean surface. On the {100} surfaces they can be attributed to a slightly distorted hexagonal surface layer, while for the superstructures on Si and Ge surfaces many models have been proposed. At present these structures are still subject of speculation [6.30].

The determination of the periodicity normal to the surface requires an analysis of the beam intensities, which can be done in a meaningful manner only with dynamical calculations, i.e. inclusion of multiple scattering in the calculations or with averaging methods, i.e. elimination of multiple scattering in the measurements. Therefore clean surfaces have been and still are the testing ground of the interpretational procedures. Al, Cu, and Ni surfaces have been most thoroughly studied and illustrate

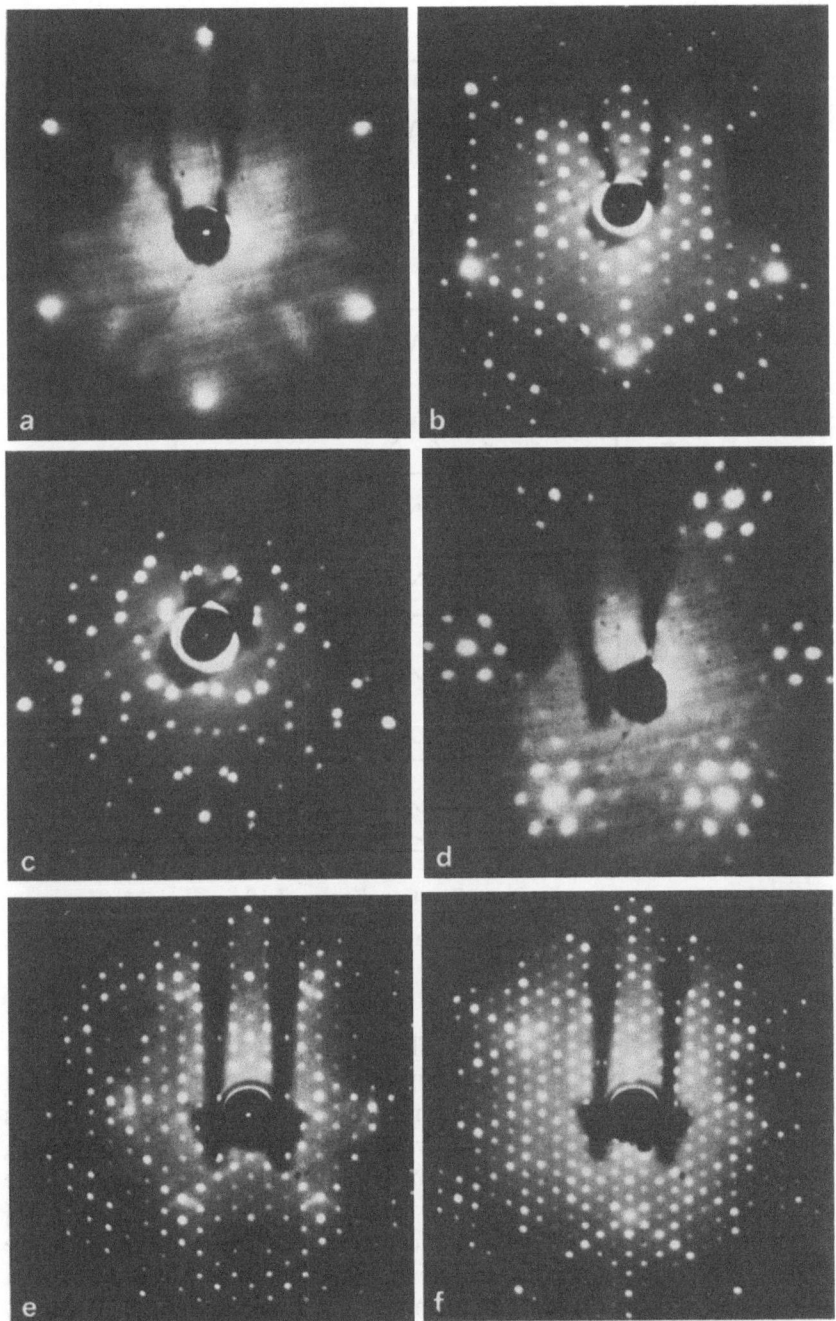

Fig. 6.12a–f. LEED patterns from Si {111} surfaces: (a) impurity stabilized (1 × 1) pattern, (b) (7 × 7) pattern of clean surface, (c) $(\sqrt{19} \times \sqrt{19}) - R(23.5°)$–Ni pattern, (d) (5 × 5)–Cu pattern, (e) (6 × 6)-Au pattern, (f) (7 × 7)-pattern, (a)–(d) 45 eC, (e)–(f) 80 eV

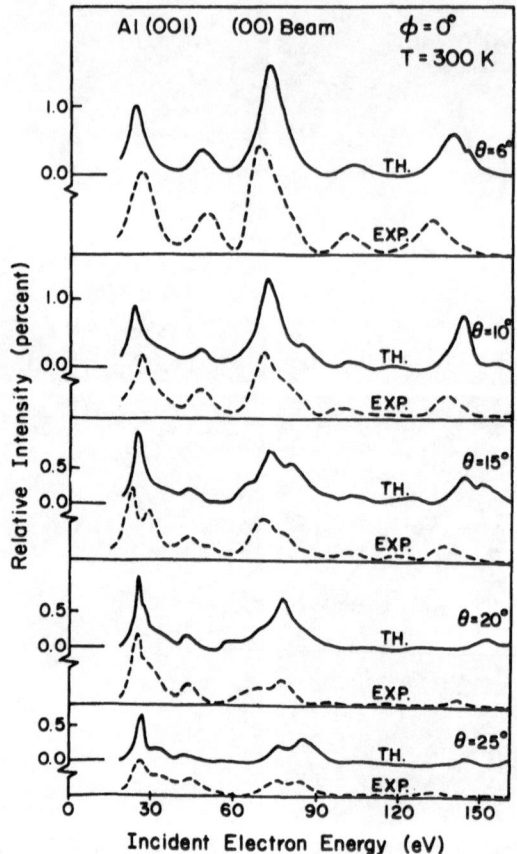

Fig. 6.13. Calculated and measured $I_{00}(V)$ curves from Al{100} surface for various angles of incidence [6.31]

well the present state of the art. The diffraction data most frequently used for comparison are $I_{h_1 h_2}(V)$ curves, i.e. the intensities of diffracted beams $(h_1 h_2)$ as a function of beam energy, obtained at various angles of incidence and temperatures. Criteria for agreement are positions, shapes and widths of the intensity maxima and absolute intensities. The agreement between experiment and recent calculations [6.27, 6.31–6.36] is surprisingly good for {100} and {111} surfaces with bulk lattice spacings normal to the surface (Fig. 6.13), but detailed examination reveals the need for refining the models and calculation procedures. The $I_{h_1 h_2}(V)$ curves for {110} surfaces, as calculated with the same parameters used for {100} and {111} surfaces, differ considerably from the experimental results unless the spacing between the first and second layer is assumed to be 10–15% smaller than that in the bulk [6.31–6.33]. However, even with this surface relaxation agreement is still poor. There are strong indications that steps may, to a large extent, be the cause of the discrepancies [6.12].

Similar results are obtained with data-averaging techniques for Cu, Ni, and Ag [6.18–6.20, 6.22]. The relaxation of the surface interplanar spacing of {100} and {111} surfaces is found to be less than 0.03 Å. The evaluation of the temperature and angle dependence of the intensities indicates that $\langle u^2 \rangle_s \approx 4 \langle u^2 \rangle_B$ in qualitative agreement with theoretical predictions. This suggests again that the layer dependence of the vibrational amplitudes should be included in dynamical calculations. For data-averaging techniques an accuracy of 5% is claimed in the determination of the atomic positions. The results described up to now give some confidence that the evaluation procedures are meaningful and can be applied with caution to more complicated systems.

Adsorption Layers

The most thoroughly studied systems are the $c(2 \times 2)$ structures (Fig. 6.14c) of various adsorbed atoms on the {100} surfaces of Ni and Ag [6.37–6.40]. Models of these adsorption layers are shown in Fig. 6.15. Oxygen on Ni {100} illustrates the results: all authors agree on the fourfold symmetry of the adsorption site so that Model b can be excluded. The normal distances, however, differ considerably: 1.5 ± 0.1 Å (Model a) [6.37], $.9 \pm 0.1$ Å [6.38] and 1.8 ± 0.1 Å (Model c) [6.39], although the same experimental data were used in obtaining the first two values. Figure 6.16 shows how well the calculated curves reproduce the observed $I(V)$ curves [6.39]. Thus there are two problems at present: 1) different methods of analysis lead to different structures for the same experimental data, 2) different authors obtain rather different experimental data. More work, both experimental and theoretical, is therefore necessary before the reliability of the deduced structures can be assessed and before it can be considered as certain that these surfaces are not reconstructed, a question which has been controversial for ten years [6.41]. It is even more premature to judge the reliability of data-averaging and in particular of transform techniques. The energy-averaging technique has been applied to a specific oxygen superstructure on a Rh {100} surface [6.16], the constant momentum transfer(K)-averaging technique to half a monolayer of oxygen on a W {110} surface [6.21] and the Patterson function type analysis to CO on a Pt {100} surface [6.14].

At present the most important application of LEED to adsorption layers is still the determination of the lateral periodicity and of the degree of order (order-disorder transitions, domain size distributions

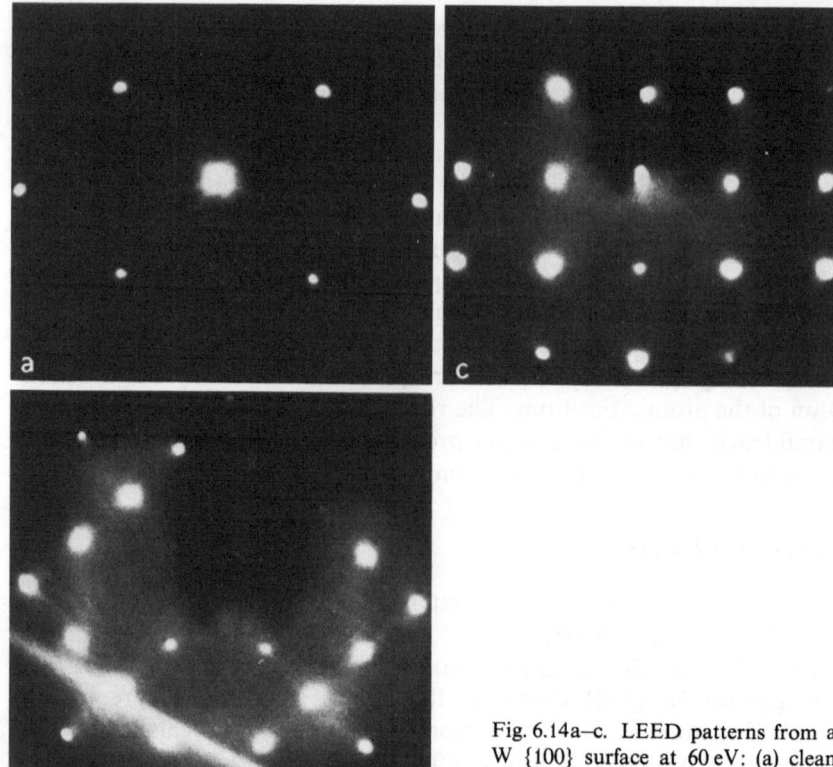

Fig. 6.14a–c. LEED patterns from a W {100} surface at 60 eV: (a) clean (1 × 1) pattern, (b) (2 × 1)-O pattern, (c) c(2 × 2)-Cu pattern

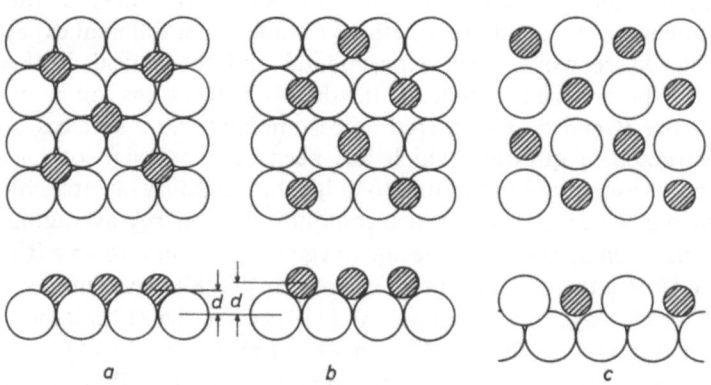

Fig. 6.15a–c. Models of adsorption layers with c(2 × 2) structure on a {100} surface (top and side views)

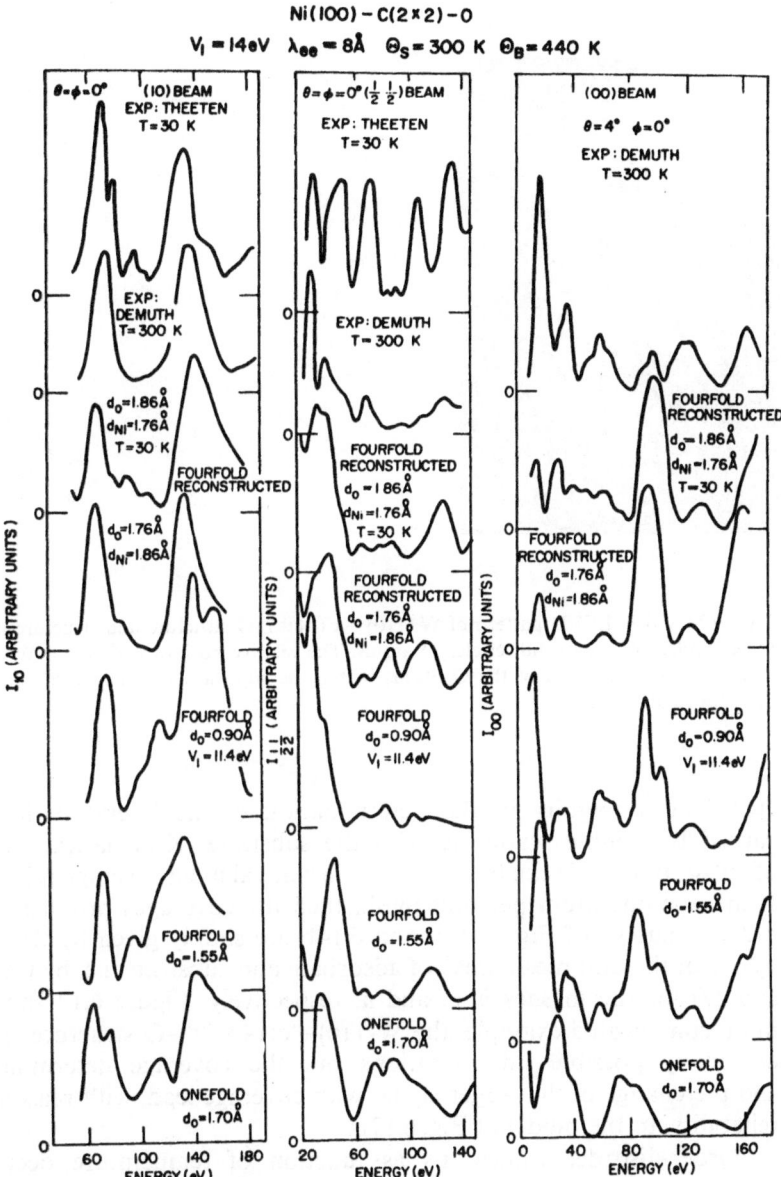

Fig. 6.16. Calculated and measured $I_{h_1 h_2}(V)$ curves for Ni $\{100\} - c(2 \times 2)$-O structure. From left to right: $(h_1 h_2) = (1, 0)$, $(1/2, 1/2)$ and $(0, 0)$. From top to bottom: Experimental curves (Theeten, Demuth), theoretical curves: Model c (fourfold reconstructed), Model a (fourfold) and onefold cordination [6.39]. (V_i: inner potential, θ_S, θ_B: surface and bulk Debye temperatures)

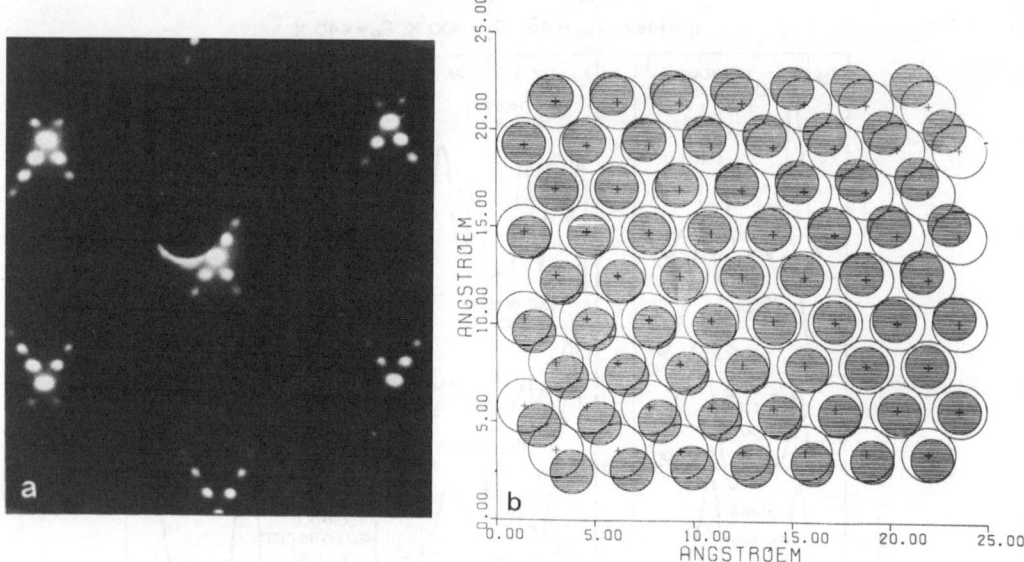

Fig. 6.17a and b. LEED pattern of W{110} – c(14×7)O structure and structure model. Open circles: W atoms, full circles O atoms. The relative positions of O and W atoms, and the precise location of the O atoms within the superstructure unit cell cannot be determined at present

etc.). The ambiguity in the determination of the lateral periodicity caused by double scattering with the substrate (Kinematical Theory of Subsection 6.2.2) can be eliminated in general if the coverage is known. If there is one atom per unit mesh, then the coverages producing the $p(2 \times 1)$ pattern of Figs. 6.10a and 6.14b are simply given by the ratio F_A^*/F_0^* of the unit mesh areas of adsorbate and substrate, i.e. by 0.5, 1.5, and 1.0 for the meshes b, c, and d, respectively. Figure 6.17 shows a more complicated example, the W(110)–"c(14×7)"–O structure. There are many possible interpretations but the coverage determination and the change of the Auger signal with coverage lead, with reasonable reliability, to the model of Fig. 6.17b.

Order-disorder transitions as function of temperature occur in many adsorption systems and can be studied reliably by measuring the intensity of the extra spots due to the adsorbate, taking into account the Debye-Waller factor. AES is very useful in such studies to assure that the adsorbate does not diffuse into the substrate. Some adsorption layers are thermally so stable that they do not disorder before desorption. High coverage oxygen adsorption layers on a W{110} surface as the one shown in Fig. 6.17 are examples. The comparison of the temperature dependence of extra and normal spot intensities gives information to what extent the substrate is involved in the order-disorder process, which gives indirect evidence for the atomic positions (reconstructed versus unreconstructed surface). Similar indirect information on atomic positions (e.g. twofold versus fourfold symmetry of the adsorption site) has been obtained from the spot splitting caused by domain formation. A detailed kinematical analysis of such domain patterns allows conclusions as to the magnitude and range of the lateral interaction forces between adatoms which cause the ordering [6.42, 6.43]. Another type of order-disorder transition is that occurring on compound and alloy surfaces without and with change in composition during the process. An example is the order-disorder process on a Cu_3Au {100} surface in which it was ascertained by AES that no composition change occurred [6.44]. Studies of this type will remain the major application of LEED until the intensity analysis procedures have become more reliable and convenient to use.

6.2.4. Special Topics

Inelastic Low Energy Electron Diffraction (ILEED)

Inelastically scattered electrons can, in general, be observed only in connection with elastic scattering (Subsection 6.1.2) or—in a crystal—with diffraction. Depending upon the sequence of back-diffraction and energy loss one speaks of two processes: diffraction before loss (DL) and loss before diffraction (LD). In view of the magnitude of the elastic back-scattering and inelastic scattering cross-sections (Section 6.1) the LD process is in general more likely. The significance of ILEED lies in the fact that it allows the study of excitation phenomena in the crystal and on its surface which can be studied only or best with slow electrons [6.45]. Such phenomena are, for example, the excitation functions for volume and surface plasmon creation, the surface plasmon dispersion $E_{SP}(k_\parallel)$, or eventually the electron band structure $E(k_\parallel)$ and phonon band structure $\omega(k_\parallel)$ at the surface. General theoretical treatments exist both for electron and phonon excitation [6.46–6.49], but they are to complicated for the interpretation of the experimental results and have to be simplified considerably [6.50].

For the study of electronic excitations (energy loss $\Delta E > 0.5\,eV$) diffractometers, as described in Subsection 6.2.1 with variable retarding potential are sufficient but for vibronic excitations ($\Delta E \approx 10$–$100\,meV$) monochromatized primary beams and high-resolution analyzers are necessary. Because of the experimental difficulties of high resolution systems [6.51] most of the ILEED studies have been concerned with electronic excitations, in particular of surface plasmons, and excitations in adsorbed atoms. An example for the first type of study is the determination of the surface plasmon dispersion relation $E_{SP}(k_\parallel)$ of an Al{111} surface [6.50, 6,62], which is important for the understanding of the charge distribution near the surface. Excitations in adsorbed atoms have been studied for CO on a Ni{100} surface and have been correlated with other data to determine the energy level diagram of the adsorbed atom [6.53]. In loss measurements, in which no k-dependence of the energy loss is sought, angle-averaging detectors such as display type systems can be used. The interpretation of the observed energy losses is frequently not unique. For example, three losses at about 2, 8, and 15 eV have been observed on Si surfaces [6.46, 6.54], which cannot be assigned to direct interband transitions or plasmon creation. The 15 eV loss is probably due to a surface excitation but the lower two losses can be caused either by non-direct interband transitions [6.46] or by surface excitations [6.54].

Polarized Electrons

In all techniques discussed up to now only the charge and energy of the electron is used to obtain information on the specimen. It has been known for some time that the scattering of slow electrons also depends on its spin [6.55, 6.56]. Therefore it is reasonable to expect that slow electrons can be spin polarized by diffraction and that diffraction of slow polarized electrons is spin dependent. Dynamical theories of the diffraction of slow polarized electrons have been developed [6.57, 6.58]. Recent detailed calculations [6.59] for a W{100} surface predict strong polarization effects in the case of the first surface resonance i.e. when the first diffracted beam [the (10) beam] is parallel to the surface. Experiments are in progress or in preparation in some laboratories. Diffraction of slow polarized electrons can be used to obtain information on the electron spin distribution at the surface, e.g. in magnetic superstructures, due to the spin dependence of the scattering amplitudes. They should also allow a distinction between scattering from heavy and light atoms because the scattering by light atoms with paired electrons spins is spin-independent.

Low Energy Electron Diffraction Microscopy

Back scattered or diffracted slow electrons can be used for imaging the surface in a similar manner as emitted or reflected electrons are used in emission and mirror microscopy. To achieve this the specimen has to be made the cathode in an electrical immersion lens. Calculations for a typical electrostatic immersion lens give an optimum resolution of about 20 Å for 50 eV electrons [6.60]. This resolution is better than that of any other imaging method suitable for flat surfaces. Furthermore, imaging and diffraction can be combined in the same system by imaging the primary image plane of the immersion lens with an intermediate lens as it is done in conventional transmission electron microscopy. Thus the lateral distribution of surface features with different structures can be studied. Several systems of this type have been described [6.41, 6.61–6.64], but considerable technological development work still has to be done before the theoretical resolution will be approached.

6.3. Auger Electron Methods

6.3.1. Physical Principles

Free Atoms

Impact of energetic radiation on atoms produces inner shell vacancies (i) which are filled within 10^{-17} to 10^{-12} sec by electrons from outer shells (j). The energy liberated in this process is radiated as characteristic X-ray radiation or it is transferred to another outer shell (k) electron which is then emitted as an "Auger electron". This Auger electron has a kinetic energy characteristic of the atom, which is given by

$$E_{ijk} = E_i(Z) - E_j(Z) - E_k(Z + \Delta Z), \qquad (6.6)$$

where E_i, E_j, and E_k are the binding energies of the electrons in the shells ($K, L, M, ...$) involved. The Auger electron is ejected from a singly charged ion. Therefore its binding energy is somewhat larger than that of the same electron in the neutral atom and can be approximated by that of an atom with somewhat larger nuclear charge $Z (0.5 \lesssim \Delta Z \lesssim 1)$. Auger electron energies calculated with (6.5) from the known atomic energy levels have been tabulated [6.65]. The number of possible Auger transitions increases rapidly with the number of electrons i.e. with Z. The type of coupling between angular momentum and spin of the electrons (*LS-*, *jj-* or intermediate coupling) determines the number of Auger lines possible for a given combination of shells. For example, the *KLL* spectra of very light atoms (*LS-*coupling) consist of 5 lines, those of very heavy atoms (*jj-*coupling) of 6 lines while those of intermediate atoms should have 10 lines. Experiment, however, reveals many more lines, as shown in Fig. 6.18 for Ne which has *LS-*coupling [6.66].

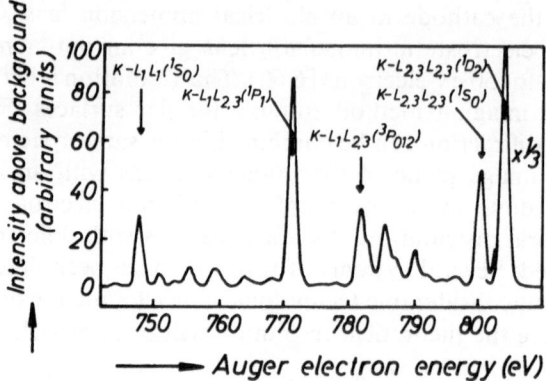

Fig. 6.18. *KLL* Auger spectrum of gaseous Ne excited by
3.2 keV electrons [6.66]

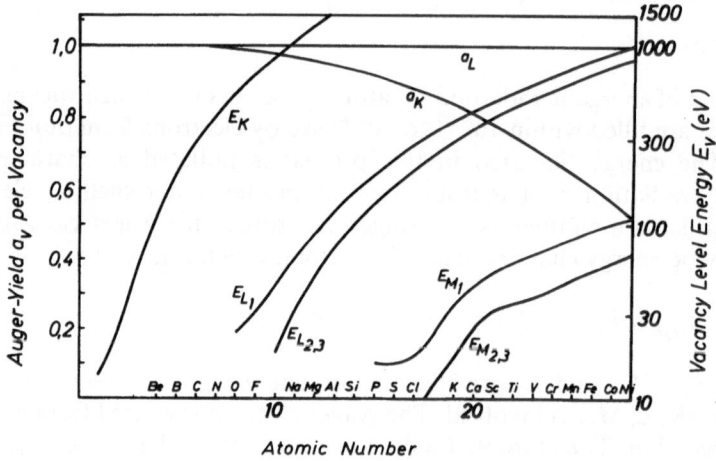

Fig. 6.19. Inner shell ionization energies E_i and transition probabilities a_i for
light atoms

The terms between brackets describe the final state configurations of
the doubly ionized atom. The weaker additional transitions are due to
Auger emission from doubly ionized atoms which produces triply
ionized atoms. Such transitions involving initial double ionization are
quite common at high primary energies. In Auger-electron excitation
by ions even higher degrees of ionization occur giving rise to very
complicated Auger spectra. This is well illustrated by comparing Ne *KLL*
spectra excited by H^+, He^+, or O^{+5} beams with Fig. 6.18 [6.68, 6.69].

Transitions in which an initial vacancy is filled by an electron from the same shell, e.g. a L_1 vacancy by an electron from the L_2 shell are called Coster-Kronig transitions. These transitions contribute considerably to the production of Auger electrons and thus to the complexity of their spectrum. A rough picture of the dependence of the Auger energies upon the atomic number can be obtained by plotting the energies of the initial vacancy versus Z (Fig. 6.19). This is possible because the levels j and k have frequently binding energies which are small compared to E_i and may be neglected to a first approximation. Figure 6.19 shows what inner shells can be used if the energy analyzer has an upper energy limit of 1500 eV: the K shell up to Al or the L shell up to beyond Ni. Higher Auger-electron energies are frequently impractical for intensity reasons.

The intensity of an Auger line is determined by the probability for creation of an inner shell vacancy, i.e. the ionization cross-section Q_i and the probability a_{ijk} that the filling of this vacancy is accompanied by an Auger or Coster-Kronig transition

$$I(E_{ijk}) \sim Q_i \cdot a_{ijk} . \tag{6.7}$$

In the case of ionization by electrons Q_i increases rapidly from threshold – which is determined by the ionization energy E_i – to a maximum at about $3E_i$, and decreases afterwards again. Figure 6.20 shows an example for the agreement between experiment $(+, \bigcirc)$ and various theoretical approximations (curves) [6.70]. The dotted curve represents the simple function

$$Q_i = \frac{a}{E_i^2} \frac{\ln U}{U} \, [\text{Å}^2] \tag{6.8}$$

(E_i in eV, $U = E/E_i$, $a = 960$ in Fig. 6.20) which gives a fair approximation for many ionization processes with a range from 600 to 1100 [6.2]. The function has a maximum

$$Q_i^{\text{max}} \approx 0.38 \, a/E_i^2 \, [\text{Å}^2] \tag{6.9}$$

at $E_{\text{max}} \approx 2.72 E_i$, somewhat lower than observed. This leads to maximum attenuation coefficients $\mu_i^{\text{max}} = Q_i^{\text{max}} \lesssim 1 \cdot 10^{-2} \, \text{Å}^{-1}$ for all materials and ionization energies $E_i \gtrsim 50$ eV, confirming the statement made in Subsection 6.1.2 that attenuation due to inner shell ionization is small compared to attenuation by valence electrons. Equations (6.8) and (6.9) lead to the following rules for AES with electrons:

1) the energy of the incident beam should be at least $2.5 \, E_i$ for efficient ionization,

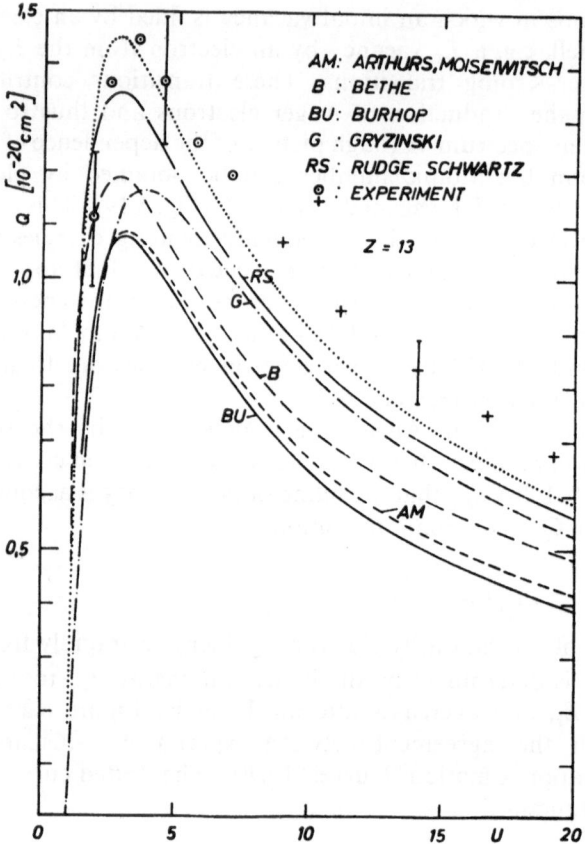

Fig. 6.20. Cross-section for ionization of the Al K shell by electrons according to theory and experiment [6.70]

2) E_i should be smaller than about 1500 eV if an ionization cross-section exceeding 10^{-20} cm^2 is to be obtained.

The second factor in (6.7), the Auger or Coster-Kronig yield a_{ijk} is the probability that a vacancy i is filled by an Auger transition involving levels j and k. It can be obtained reliably only by detailed numerical calculations which have been reviewed recently [6.71]. The results for transitions involving K and L shells are plotted in Fig. 6.19. It is evident that nearly all vacancies are filled by Auger transitions ($a_i = \Sigma\, a_{ijk} \approx 1$) as long as $E_i > 1500$ eV. At higher E_i values X-ray emission becomes increasingly stronger. This fact is a further reason for limiting AES to energies below about 1500 eV.

The Auger-electron line width of free atoms is determined by the lifetime of the vacancy i which ranges from fractions of an eV for small

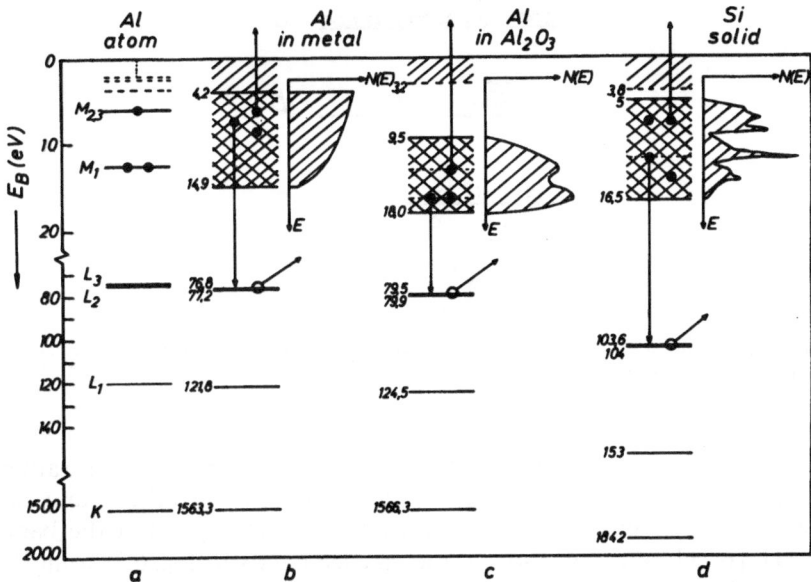

Fig. 6.21a–d. Energy-level schemes and densities of states of Al, in various environments, and of Si

E_i to about 10 eV for high E_i values ($\approx$ 1500 eV), another reason for the use of shells with low ionization energies. The angular distribution of Auger electrons is in general isotropic or nearly isotropic (see *Details of the Auger Emission Process* in Subsection 6.3.4).

Condensed Matter

Figure 6.21 illustrates some of the differences between free atoms and atoms in condensed matter. The major effect of the atomic environment is the broadening of the valence levels into bands with considerable width and variations of the electronic density of state $N(E)$. A second effect is the shift of the energy levels which is, however, similar for all levels and thus has little influence on the energy $E_i - E_j$ available for the Auger electrons. The broadening of the valence levels causes a corresponding broadening of the Auger lines involving these levels. If both states j, k are located in the valence band the Auger line width is in principle twice the valence band width. Because of the variations of $N(E)$ the Auger line is more or less structured. By deconvolution of the Auger line [6.72, 6.73] it is therefore possible in principle to obtain $N(E)$. However, the complications due to the details of the Auger process in solids (*Details of the Auger Emission Process* in Subsection

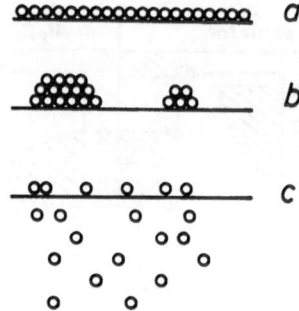

Fig. 6.22a–c. Atomic distributions in the surface region giving rise to different Auger signals in spite of same concentration

6.3.4) and due to inelastic scattering make a reliable analysis rather difficult. The same is true for the derivation of the orbital angular momentum (l) characteristic of electrons in various parts of the band [6.74], [6.75]. A comparison of the valence bands in Fig. 6.21 a and b clearly shows that the chemical environment (metal or oxide) can cause considerable "chemical" shifts of the Auger lines; they may be as large as 15 eV in strongly bound oxides.

The most important effect which the environment exerts on the Auger process in condensed matter is the attenuation of the Auger electrons created at a distance x from the surface on their way because of inelastic scattering (Subsection 6.1.2). Inelastic scattering not only reduces the intensity of the main features of an Auger line, but also redistributes the intensity on the low energy side of the strong features in a manner determined by the energy loss distribution which can be rather complicated. Well defined energy losses such as the volume and surface plasmon losses in Al cause clearly recognizable loss peaks. Energy loss spectra with little structure cause mainly smearing out of features on the low energy side of the main peak.

In principle the attenuation effects can be taken into account rigorously in the evaluations, but the mathematical expressions contain so many poorly known quantities that for practical purposes only highly simplified formulas can be used [6.45]. The consequence is that Auger electron intensities are a good measure for the number of the emitting atoms only if these are distributed in a two-dimensional manner, as indicated in Fig. 6.22a. Other distributions give quite different intensities for low-energy Auger electrons ($\lambda_{ee} \approx 3$–5 Å). If Auger electrons with widely varying energies are emitted (e.g. 50 eV and 1000 eV), the differences in mean-free path cause different effective sampling depths and thus give information on the distribution. Without knowledge of this distribution quantitative AES is not possible. For two-dimensional distributions quantitative measurements can be made easily after calibration with other techniques such as quartz oscillator measurements (mass), radio tracer studies (nucleus) or ion beam deposition (charge). Use of such calibrations for surfaces with different roughness, temperature or surface composition requires considerable caution because the calibrated atoms may be distributed in a different way or the backscattering from the substrate may differ.

6.3.2. Experimental Methods

AES requires energy analysis of the electrons emitted from the surface to be tested with proper subtraction or suppression of the background on which the weak Auger lines are superimposed. The background is small in X-ray and ion bombardment excitation, but large in electron excitation. It is usually suppressed by differentiating—this emphasizes features which vary rapidly with energy—or by pulse counting techniques. There are many types of electron spectrometers suitable for AES, but only the most frequently used ones will be briefly discussed: the retarding field, the cylindrical mirror and the hemi-spherical type. Figure 6.8a suggests immediately the use of the display type LEED system for electron spectroscopy in the retarding field mode: in principle only the retarding potential has to be changed so that the "pass" energy E_a of the electrons varies. In such a "high-pass filter" electrons with energies $E > E_a$ are detected on the collector (fluorescent screen): $I_{a0} = \int_{E_a}^{E_0} I(E)\, dE$.

In order to obtain the energy distribution $I(E)$ the signal has to be differentiated which is done by superimposing an ac voltage with frequency $v\,(10^2–10^5\ \text{Hz})$ on the retarding voltage and by detecting only the ac signal on the detector with lock-in techniques. The resulting ac signal is only due to the electrons in the "energy window" determined by the amplitude of the modulation voltage. By tuning the lock-in detector to frequency v the first derivative of I_{a0}, i.e. $I(E)$, is obtained. In general, however, the signal with frequency $2v$ which is proportional to $dI(E)/dE$ is measured in order to suppress the background. Thus not Auger lines but their derivatives are observed. The amplitudes of the derivatives ("Auger amplitudes") are frequently a good measure for the intensity of the Auger line provided the modulation amplitude is small enough so that no averaging occurs. The maximum modulation amplitude permissible obviously depends on the shape and width of the Auger line and can vary from 1 to 10 V peak to peak. The advantage of this type of energy analyzer is its simplicity, its combination with LEED and its insensitivity to primary beam size and location on the

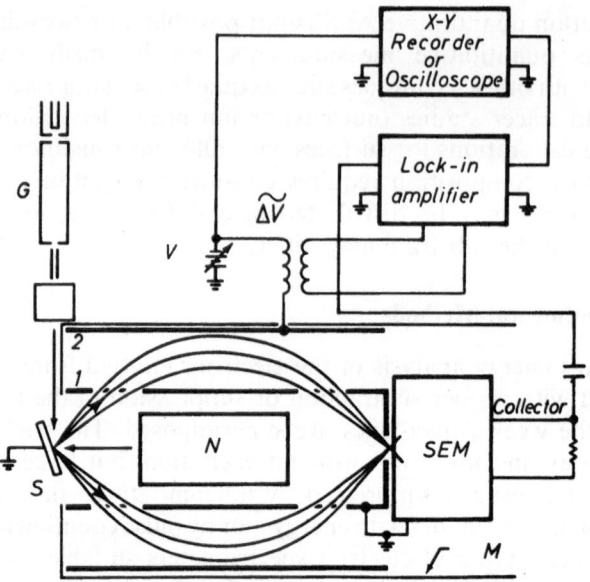

Fig. 6.23. Schematic of cylindrical mirror analyzer. Explanation see text

specimen. It is also very useful if the energy distribution is to be averaged over a large solid angle. Its disadvantage is the high noise $N \sim \sqrt{I_{a0}}$ produced by the electrons with $E_a < E < E_0$. However, by proper averaging and curvefitting techniques this disadvantage can be considerably reduced [6.76].

The high noise of the retarding field analyzers can be avoided if a band-pass analyzer is used which transmits only electrons with the energy of interest so that $N \sim \sqrt{I(E)}$ instead of $N \sim \sqrt{I_{a0}}$. Both the cylindrical mirror analyzer (CMA) shown schematically in Fig. 6.23 and the hemi-spherical analyzer belong to this class of instruments. Energy analysis in a CMA occurs in the radial electrostatic field between the coaxial cylinders 1, 2 in Fig. 6.23, in a hemi-spherical analyzer in the radial field between two concentric hemispheres. The primary electron beam is produced by a normal (N) or grazing (G) incidence electron gun. The Auger signal is very sensitive to beam size and location on the specimen. It is usual to differentiate also in this analyzer by superimposing a modulation voltage $\widetilde{\Delta V}$ on the analyzer voltage V and to use lock-in detection. Figure 6.24 shows an example of $N(E)$ and of $dN(E)/dE$ obtained from the same sample with a CMA. For comparable investment in electronics the S/N ratio of a CMA is higher by a factor of 10^3 than that of a retarding field analyzer. Typical analyzer characteristics are:

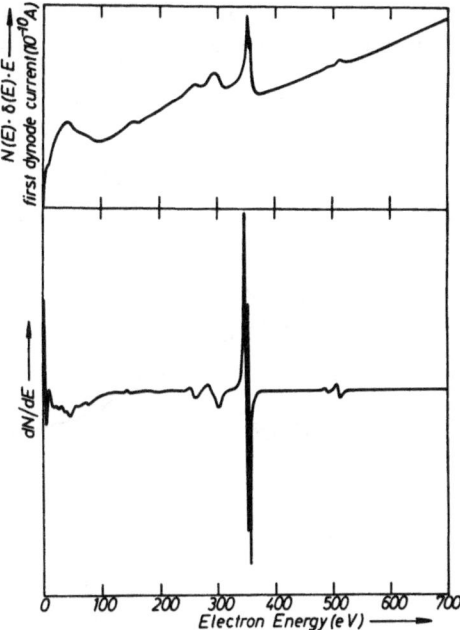

Fig. 6.24. Energy distribution $N(E)$ and its derivative $dN(E)/dE$ of Ag excited by 3 keV electrons as obtained with a CMA. $\delta(E)$ secondary emission coefficient of first dynode (Courtesy of P.E.I.)

energy resolution: 1 %, transmission: 10 % for standard diameter ($\approx$ 10 cm) systems. By scaling up the diameter ($\approx$ 1 m) [6.77] or operating two analyzers in series much higher resolution (< 0.1 %) has been obtained.

Major experimental problems are electron beam induced specimen changes such as dissociation, carbon contamination buildup, electron stimulated desorption (ESD) or electron stimulated adsorption (ESA). Therefore, low beam currents e.g. $< 10^{-6}$ Å must frequently be used, increasing the demands on the detection and data processing system, if high resolution is required. Several procedures for optimization of detection and data processing have been discussed [6.76, 6.78, 6.79]. If higher beam currents can be tolerated the large S/N ratio allows very short time constants in the detection system and thus rapid scan times so that rapidly changing surface processes can be studied.

6.3.3. Results

The determination of surface composition, including trace element analysis by AES, has contributed greatly to the understanding of surfaces. For example, the Si {111}-surface with (7 × 7) structure (Fig. 6.12b)

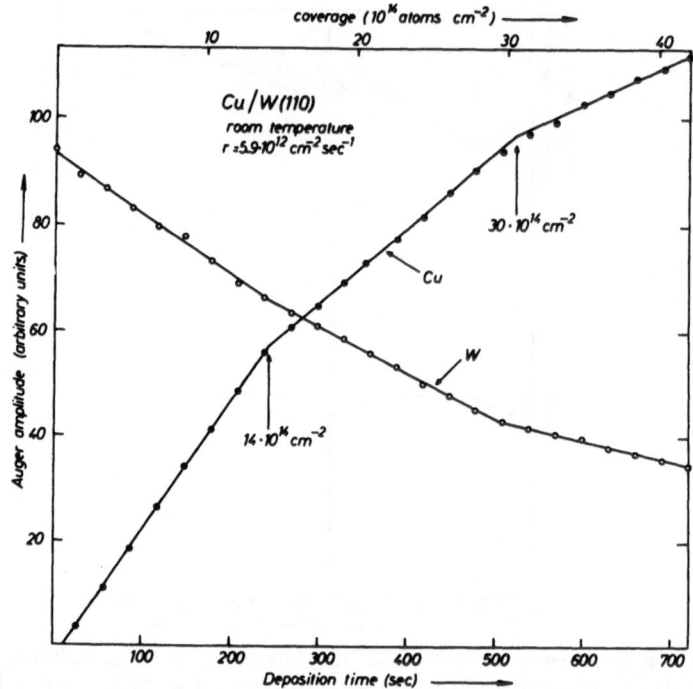

Fig. 6.25. Cu and W Auger amplitudes as functions of Cu coverage on a clean W {110} surface [6.82]

was shown to be clean, while the same surface with $(\sqrt{19} \times \sqrt{19})$ $- R(23.5°)$, (5×5) and (6×6) structures (Fig. 6.12c–e) was found to contain Ni, Cu, and Au, respectively. Similarly, the {100} surfaces of Au, Pt, and Ir with "(5×1)" or "(5×20)" structures are clean by present AES standards. On the other hand, many impurities were revealed by AES on surfaces previously considered to be clean on the basis of their good (1×1) LEED patterns. Some atoms, when located on top of the surface, can be detected if present in less than 1% of a monolayer; in general several percent are sufficient. The surface composition of alloys after or during heating, ion bombardment or chemical reaction has been studied considerably with AES. Information on the distribution of the components normal to the surface could be obtained by using emitted electrons with different characteristic energies from the same specimen. In alloys consisting of a component with low and a component with high specific surface energy, such as Pt–Sn [6.80] or Au–Sn [6.81] alloys, the surface is enriched with the low surface-energy component after equilibration, as expected from thermodynamics. Ion bombardment or chemical reaction can lead to quite different surface compositions depending upon relative sputtering yields and chemical reactivities. The observation of impurity segregation on free surfaces and internal surfaces, as revealed by fracturing, is another important result of AES studies. However, it must be kept in mind that quantitative analysis is difficult when the concentration varies with depth.

The situation is much simpler when the atoms to be studied are distributed only two-dimensionally as is the case in many adsorption systems. An example is shown in Fig. 6.25. Auger amplitudes of Cu and W from a W{110} surface are plotted as a function of the amount of Cu adsorbed on the crystal. Calibration was done with a quartz oscillator microbalance and by mass spectrometry. The Cu Auger signal shows two breaks as does the W signal, caused by the layerwise adsorption of Cu and the short mean-free path of the 60 eV Cu and 170 eV W Auger electrons. The slope ratio allows an accurate determination of the mean-free paths in this system. Of more interest is the fact that the second break does not occur at twice the coverage N_1 of the first break but at $(2.15 \pm 0.03) N_1$ in agreement with the LEED patterns corresponding to the two coverages [6.82]. Another example is the determination of the oxygen coverage producing the LEED pattern of Fig. 6.17 which made an important contribution to the interpretation of this pattern. The fact the AES can be done also at high temperature and gas pressures up to 10^{-5} Torr allows the study of adsorption under quasi-equilibrium conditions [6.83, 6.84]. Caution is necessary in such work because of the temperature dependence of the Auger signal.

Usually undesirable, but potentially useful electron beam effects are the dissociation of oxides (electron beam masking), desorption and adsorption of gases, as well as carbon build-up in the presence of a carbon-containing gaseous environment. Examples are the beam reduction of SiO_2 which leads to Si enrichment in the beam area [6.85], the electron stimulated desorption of C from BeO in the presence of oxygen gas [6.86] or the electron beam induced adsorption of O_2 on Si [6.87]. In spite of these beam effects oxidation studies can be done – and many have been done – by intermittent AES with low beam currents. The highest ESD cross-sections of oxygen are of the order 10^{-18} cm^2 [6.88]. At low current density $i = 1 \cdot 10^{-2}$ A cm$^{-2} = 1.6 \cdot 10^{17}$ El./cm^2 sec so that the need for careful experimentation is evident.

6.3.4. Special Topics

Details of the Auger Emission Process

In Subsection 6.3.1 and 6.3.2 a highly simplified picture of the Auger emission process was presented. The Auger emitter was essentially considered to be undisturbed in the emission process except for allowing a somewhat different force acting on the Auger electron, as expressed by the ΔZ term in $E_k(Z + \Delta Z)$. For Auger processes in solids it was assumed that the valence electrons could be described by the three-dimensional band structure $E(k)$ and density of states $N(E)$, which are meaningful only for nonlocalized processes. With increasing number of precise high resolution measurements it has become evident that the Auger process causes a strong localized disturbance of the emitter, so that k is not a good quantum number. This is true both for free-electron like metals and for d-band metals. Thus the density of states of Al obtained by deconvolution of the $L_{2,3}VV$ Auger line clearly disagrees with the theoretical $N(E)$ [6.89]. The same is true for the $M_{4,5}VV$ spectrum of Ag, which can be explained in terms of atomic-like final states with multiplet splitting [6.89]. Another example is the Cu $L_{2,3}MM$ spectrum which shows strong indications of spin splitting and lines much narrower than expected on the basis of the valence band width [6.90]. Thus conclusions concerning the undisturbed specimen are meaningful only if the localization of the Auger process is taken into account. It cannot be expected that the density of states obtained by deconvolution of Auger spectra should agree with that of three-dimensional band structure calculations. Rather it will be a local density of states, modified by selection rules.

The disturbance [6.91–6.93] caused in the atom from which the Auger electron is emitted may be divided into several parts [6.91, 6.92]: "dynamic" relaxation and "static" relaxation which in turn split into atomic and extra-atomic relaxations. In addition, the multiplet coupling in the final state, which causes the splitting mentioned above, has to be taken into account. Dynamic relaxation describes the rearrangement of the atom accompanying ionization. The dynamic relaxation energies of electrons are automatically included in (6.6) for $\Delta Z = 0$ if experimental binding energies are used. Static relaxation describes the rearrangement of the atom connected with the transition of the j electron into the i hole. Static atomic relaxation is the process which occurs in the free atom, static extra-atomic relaxation represents the response of the atomic environment in the solid to the formation of the j hole. In metals the positive charge on the ionized atom is very likely screened completely by conduction electrons, so that a localized picture is also appropriate for extra-atomic relaxation. The contributions of the various processes to the Auger energy have been determined for the Cu $L_{2,3}M_{4,5}M_{4,5}$ Auger spectrum to be ~ 25 eV, 10 eV and 20 eV for the multiplet, the static atomic and the static extra-atomic relaxation contributions, respectively [6.92]. The Auger energies calculated with (6.6) for $\Delta Z = 0$ but with inclusion of these terms agree within 2–3 eV with experiment.

The magnitude of the extra-atomic relaxation energy which obviously should depend considerably on the immediate environment of the emitting atom suggests that "chemical" shifts have to be analyzed in much more detail than has been done in the past. Another effect of the environment is its influence on the lifetime of the core hole states which determines the i state contribution to the linewidth. This lifetime is found to depend on the valence electron density in the emitting atom and its immediate environment, as determined by the interionic distance [6.94].

The strong localization of the Auger process makes it plausible why another much discussed phenomenon, the plasmon gain of Auger electrons, is very unlikely to occur, at least in d-band metals. Plasmon gains were originally invoked to explain the high energy satellites of the main Auger peaks in various metals [6.95–6.98]. Such a process is expected according to theory, provided that extra-atomic relaxation can be treated in a nonlocalized picture and that the core hole life-time is sufficiently short. Then the plasmons created during the core hole production have not already decayed by the time the Auger electron is emitted and the Auger electron can absorb the energy of one of the plasmons destroyed in its emission process [6.99]. At least the second condition is not fulfilled for the example calculated (the Al $L_{2,3}VV$ Auger process), because the lifetime derived from transition rate calculations [6.100] is $2 \cdot 10^{-13}$ sec, two orders of magnitude larger than required. Alternately the "plasmon gain" peaks have been ascribed to double ionization of core states [6.101–6.104], to internal photoemission [6.102] and to the density of states [6.105]. There is strong evidence that in Li [6.106], Na, Mg, Al, and Si [6.107, 6.108], the first explanation is correct and that the second process is negligable as compared to the first one, even at higher Auger energies [6.109]. The last process appears unlikely because of the localization of the Auger process discussed before. The only material in which plasmon gain cannot be excluded at present seems to be Be in which peaks expected for plasmon gain and double ionization are well separated and observed with the expected intensities [6.110].

After the emission process which apparently has to be treated in a localized picture the Auger electron has to propagate through the solid, which modifies the energy and angular distributions. This has not only negative effects. For example, if the energy losses are very distinct such as in free electron-like metals the plasmon satellites on the low-energy side of the main Auger peaks may be used to determine the depth distribution of the emitting atom by measuring the satellite intensity as a function of exit angle [6.111]. In single crystals the (quasi) elastic scattering of the Auger electrons on their way to the surface modifies the

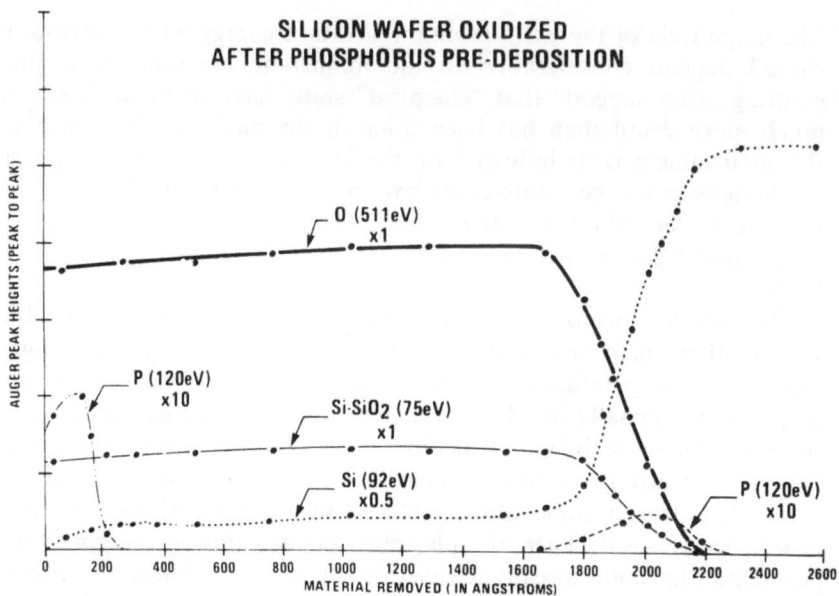

Fig. 6.26. In-depth analysis of an oxydized phosphorus doped silicon surface (Courtesy of P.E.I.)

angular distribution in a manner known from Kikuchi patterns in electron diffraction [6.112–6.114]. This is to be expected because the Auger emission process produces essentially a spherical wave similar to the waves causing Kikuchi patterns. The Auger electrons from atoms located on top of the surface should not show a Kikuchi-like anisotropy, which allows qualitative conclusions on the depth distribution of impurities [6.114]. It should, however, be noted that the angular distribution of Auger electrons from isolated atoms is already somewhat anisotropic if the initial vacancy has a total angular momentum $j > 1/2$ and if the final two vacancy state has a quantum number $J > 1/2$ [6.115].

In-Depth Auger Analysis and Auger Electron Microscopy

The small escape depth of Auger electrons make them an ideal tool for the study of the atomic distribution normal to the surface by successive removal of surface layers. This can be done in several ways but the most convenient method is ion-bombardment sputtering which is possible while the Auger spectrum is being measured [6.116, 6.117]. Typical sputtering conditions are: Ar gas pressure $1-5 \cdot 10^{-5}$ Torr, Ar^+ beam energy 0.5–2 keV, Ar^+ beam current density 0.2–20 $\mu A/cm^2$ and beam diameters on the specimen of several mm to about 1 cm. If the ultimate depth resolution as determined by the escape depth of the Auger electron (≈ 10 Å) is to be achieved, then the ion beam current density must be constant over an area larger than the diameter of the primary electron beam. Otherwise the area from which the Auger electrons are emitted is not flat and the Auger signal represents an average over a range of depths. At sufficiently low sputtering rates (1–10 Å/min depending on concentration gradients) several Auger lines can be monitored sequentially. Figure 6.26 shows an example for such a multiline analysis. Four peaks, the oxygen, phosphorus, silicon 92 eV and the silicon 75 eV line characteristic of Si in SiO_2 are measured while the oxide layer is sputtered away. The availability of convenient commercial data acquisition systems ("multiplexers") has led to a rapid development of in-depth AES.

There are, of course, limitations to this technique: the sputtering cross-sections vary from atom to atom and depend upon its chemical environment. An unusually drastic example is the removal of equal amounts of Mo deposited on various surfaces [6.118]: Mo forms on W a uniform film with a normal sputtering profile while on Cu, Au, and Al it agglomerates and can be observed over a large depth range. Also considerable surface roughening occurs during sputtering, in particular in impure materials. Another disturbing effect is the ion bombardment induced movement of ions in insulating films [6.119] leading to an accumulation of such ions on the boundaries of the film. The problems occurring in ion etching of insulators are also well illustrated by the example of mica [6.120].

The fact that Auger electrons are quite effectively produced by energetic electrons suggests using an Auger analyzer as detector in scanning microscopes in order to produce a high resolusion image of the lateral distribution of the chemical surface composition [6.121–6.123]. Although such an arrangement is, in principle, capable of high resolution due to the small size of the scanning beam on the specimen, the S/N ratio is low. Another approach is to start from a CMA with its high S/N ratio and to incorporate an electron gun which produces a very small spot on the specimen [6.122]. Such a system with a beam diameter of 5–10 µm has recently been developed and some first results are shown in Fig. 6.27, illustrating its capabilities. The specimen is a Si integrated circuit of which an area containing a transistor has been imaged with Si, O, P, and Au Auger electrons, respectively, in the top four panels; the lower half of the picture compares the Au image with the Mo image obtained after removing enough material by ion bombardment so that the Mo interface between the Au leads and the Si substrate becomes visible. Efforts are being made in several laboratories to improve the resolution so that Auger electron microscopy may become an even

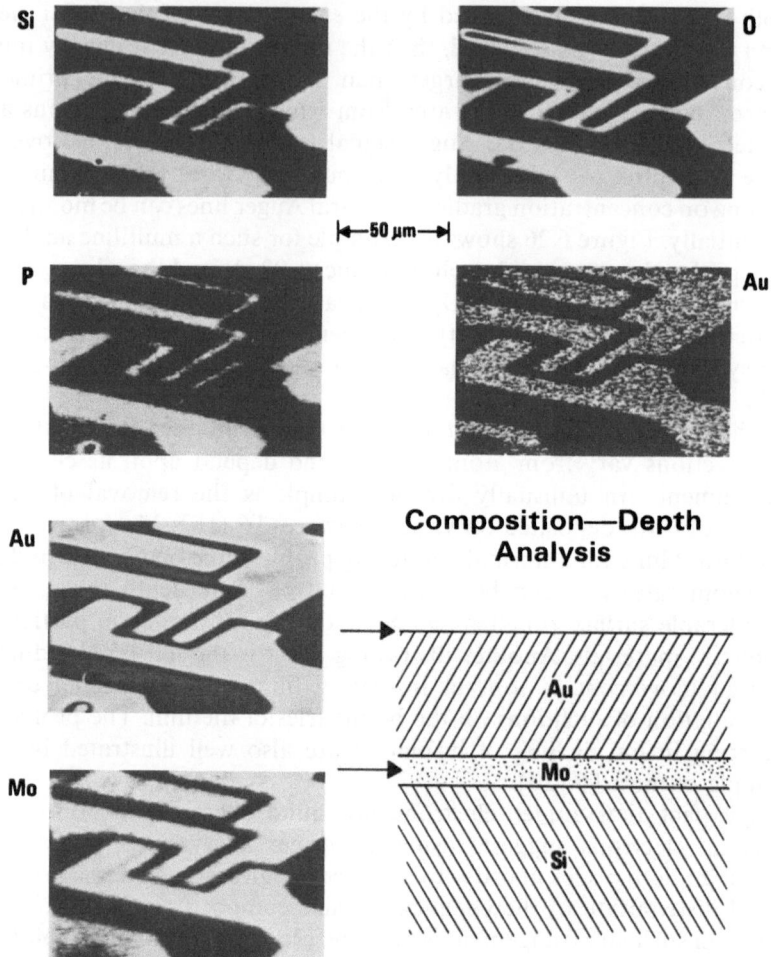

Fig. 6.27. Auger electron images from a transistor on a silicon chip. See text for an explanation (Courtesy of P.E.I.)

more powerful tool in surface and thin film analysis. The significance of these developments for modern technology is well illustrated in a recent review of the industrial applications of AES [6.124].

6.4. LEED Structure Nomenclature and Superstructures

We give here a brief account of commonly used nomenclature. A threefold periodic ideal crystal has a twofold periodic ideal surface lattice. The unit mesh of this two-dimensional lattice is defined by the two unit

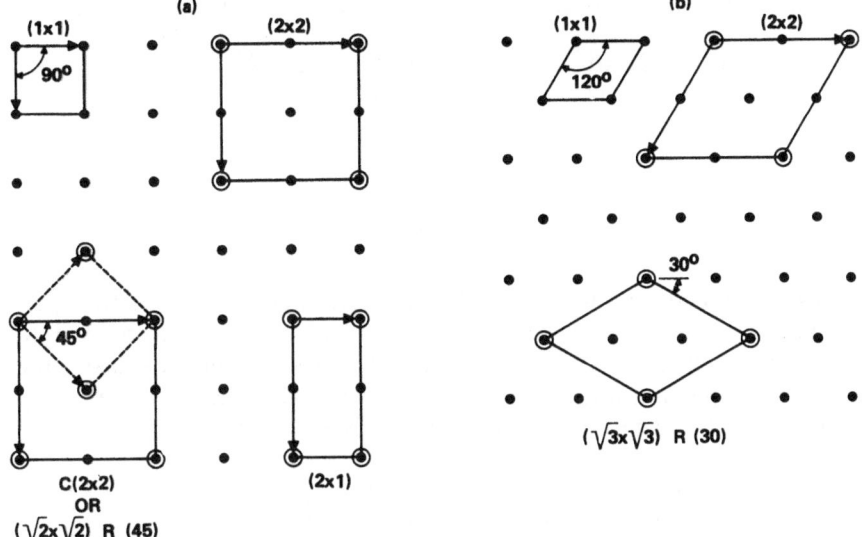

Fig. 6.28a and b. Wood's notation for simple superstructures. Small full circles substrate atoms, large open circles overlayer atoms. (a) Substrate with fourfold symmetry, (b) substrate with sixfold symmetry. The c stands for centered, the (2×2) structure is frequently written as $p(2 \times 2)$ with p standing for primitive to express the absence of an atom in the center

mesh vectors a_1, a_2. A surface structure with these unit mesh vectors is called a (1×1) structure. Frequently the surface periodicity, however, is different from the ideal one due to distortions, reconstruction or adsorbed atoms, for example. For such superstructures two notations are usual, Wood's notation and the matrix notation.

In Wood's notation the length of the unit mesh vectors a_1^s, a_2^s of the superstructures are given as multiples of the lengths of the vectors a_i and their directions are specified by the angle of rotation α^s of the superstructure unit mesh relative to the ideal unit mesh: $\left(\dfrac{a_1^s}{a_1} \times \dfrac{a_2^s}{a_2} \right) R(\alpha^s)$.

Some simple examples for this notation are given in Fig. 6.28. By definition this notation is only applicable if both axes are rotated by the same amount but it can be easily generalized by specifying two angles α_1^s, a_2^s, if this is not the case.

In the matrix notation the vectors a_1^s, a_2^s are expressed by their components a_{11}^s, a_{12}^s, a_{21}^s, a_{22}^s with respect to the ideal unit mesh vectors a_1, a_2

$$\begin{pmatrix} a_1^s \\ a_2^s \end{pmatrix} = \begin{pmatrix} a_{11}^s & a_{12}^s \\ a_{21}^s & a_{22}^s \end{pmatrix} \begin{pmatrix} a_1 \\ a_2 \end{pmatrix} = A^s \begin{pmatrix} a_1 \\ a_2^s \end{pmatrix}. \tag{6.10}$$

For example, the matrices of the superstructures shown in Fig. 6.28 are

$$\begin{pmatrix} 1 & 0 \\ 0 & 1 \end{pmatrix}, \begin{pmatrix} 2 & 0 \\ 0 & 2 \end{pmatrix}, \begin{pmatrix} 1 & -1 \\ 1 & 1 \end{pmatrix}, \begin{pmatrix} 2 & 0 \\ 0 & 1 \end{pmatrix}, \begin{pmatrix} 1 & 0 \\ 0 & 1 \end{pmatrix}, \begin{pmatrix} 2 & 0 \\ 0 & 2 \end{pmatrix},$$

and

$$\begin{pmatrix} 1 & -1 \\ 1 & 2 \end{pmatrix}.$$

This nomenclature allows for different angles between ideal and super-structure unit mesh vectors and makes the transition from real to reciprocal lattice – of which LEED provides an image – easy: the corresponding unit mesh vectors in the reciprocal lattice b_i, b_i^s are simply related by the reciprocal transposed matrix $B^s = (\tilde{A}^s)^{-1}$

$$\begin{pmatrix} b_1^s \\ b_2^s \end{pmatrix} = B^s \begin{pmatrix} b_1 \\ b_2 \end{pmatrix} = \frac{1}{|A^s|} \begin{pmatrix} a_{22}^s & -a_{12}^s \\ -a_{21}^s & a_{11}^s \end{pmatrix} \begin{pmatrix} b_1 \\ b_2 \end{pmatrix}. \tag{6.11}$$

Using the matrix notation the interpretation of superstructures can be made purely mathematically without geometrical constructions. It is, however, often convenient to make a graphical analysis based on the physical process which is frequently producing complicated super-structures, i.e., double or multiple scattering: each beam scattered by the surface layer acts as a primary beam for the substrate and *vice versa*. If, therefore, each substrate spot is used as the origin of the spot pattern of the chosen overlayer model the observed LEED pattern should be obtained. The choice of the overlayer model is considerably aided by the knowledge of the absolute coverage associated with the super-structure pattern: the area of the overlayer reciprocal unit mesh in cm^{-2} gives just the coverage in atoms/cm^2 if there is one atom per real space unit mesh. Another criterion is frequently, but not always, the intensity of the superstructure spots: the most intense spots are the most likely candidates for defining the vectors b_i^s. With such aids a relatively reliable interpretation of complex LEED patterns such as that shown in Fig. 6.17 becomes possible in spite of the fundamental ambiguity of the geometry of the LEED pattern.

6.5. Appendix:
Recent Reviews of LEED and Auger Phenomena

1. M. LAZNIČKA: *LEED-Surface Structures of Solids* (Union of Czechoslovak Mathematicians and Physicists, Prague, 1972) (LEED).
2. M. B. WEBB, M. G. LAGALLY: Solid State Phys. **28**, 302 (1973) (LEED).
3. C. B. DUKE: Advan. Chem. Phys. **27**, 1 (1974) (LEED).

4. J.A.Strozier: In *Surface Physics of Crystalline Solids*, ed. by J.M.BLAKELY (Academic Press, New York 1974 (LEED)).
5. J.B.PENDRY: *Low Energy Electron Diffraction* (Academic Press London 1974) (LEED).
6. G.ERTL, J.KÜPPERS: *Low Energy Electrons and Surface Chemistry* (Verlag Chemie, Weinheim 1974) (LEED, AES).
7. C.C.CHANG: In *Characterization of Solid Surfaces*, ed. by P.F.KANE, G.B.LARRABE (Plenum Press, New York, 1974) (AES).

References

6.1. E.BAUER: J. Vac. Sci. Technol. **7**, 3 (1970).
6.2. E.BAUER: In 14ème Cours AVCP, Verbier (1972).
6.3. H.N.BROWNE, E.BAUER: Unpublished Repts. NASA Contr. No. R-05-030-001 (1965–1969).
6.4. J.C.TRACY: J. Vac. Sci. Technol. **11**, 280 (1974); Solid State Commun. (to be published).
6.5. P.J.FEIBELMAN: Surface Sci. **36**, 558 (1973).
6.6. C.R.BRUNDLE: J. Vac. Sci. Technol. **11**, 212 (1974).
6.7. C.J.POWELL: To be published.
6.8. P.J.FEIBELMAN, C.B.DUKE, A.BAGCHI: Phys. Rev. B**5**, 2436 (1972).
6.9. H.RAETHER: In Springer Tracts Mod. Phys. **38**, 84 (1965).
6.10. R.F.ALLEN, G.P.ALLDREDGE, F.W.DE WETTE: Phys. Rev. B**4**, 1661 (1971).
6.11. J.E.HOUSTON, R.L.PARK: Surface Sci. **26**, 269 (1971).
6.12. B.LANG, R.W.JOYNER, G.A.SOMORJAI: Surface Sci. **30**, 440 (1972).
6.13. G.E.LARAMORE, J.E.HOUSTON, R.L.PARK: J. Vac. Sci. Technol. **10**, 196 (1973); Surface Sci. **34**, 477 (1973).
6.14. T.A.CLARKE, R.MASON, M.TESCARI: Surface Sci. **30**, 553 (1972), **40**, 1 (1973); Proc. Roy. Soc. (London) A**331**, 321 (1972).
6.15. U.LANDMAN, D.L.ADAMS: J. Vac. Sci. Technol. **11**, 195 (1974).
6.16. C.W.TUCKER, C.B.DUKE: Surface Sci. **23**, 411 (1970); **24**, 31 (1971); **29**, 237 (1972).
6.17. C.B.DUKE, D.L.SMITH: Phys. Rev. B**5**, 4730 (1972).
6.18. M.G.LAGALLY, T.C.NGOC, M.B.WEBB: Phys. Rev. Letters **26**, 1557 (1971); J. Vac. Sci. Technol. **9**, 645 (1972).
6.19. T.C.NGOC, M.G.LAGALLY, M.B.WEBB: Surface Sci. **35**, 117 (1973).
6.20. W.N.UNERTL, M.B.WEBB: J. Vac. Sci. Technol. **11**, 193 (1974); Surface Sci. (to be published).
6.21. J.C.BUCHHOLZ, M.G.LAGALLY: J. Vac. Sci. Technol. **11**, 194 (1974); Surface Sci. (to be published).
6.22. J.M.BURKSTRAND, G.G.KLEIMAN: J. Vac. Sci. Technol. **11**, 192 (1972); Phys. Rev. (to be published).
6.23. E.BAUER, H.N.BROWNE: 1st LEED Theory Seminar, Brooklyn, 1967 (unpublished).
6.24. J.A.STROZIER, JR., R.O.JONES: Phys. Rev. B**3**, 3228 (1971).
6.25. C.B.DUKE, C.W.TUCKER, JR.: Surface Sci. **15**, 231 (1969).
6.26. C.B.DUKE, G.E.LARAMORE: Phys. Rev. B**2**, 4765, 4783 (1970).
6.27. D.W.JEPSEN, P.M.MARCUS, F.JONA: Phys. Rev. B**5**, 3933 (1972).
6.28. J.B.PENDRY: J. Phys. C**4**, 3095 (1971).
6.29. R.H.TAIT, S.Y.TONG, T.N.RHODIN: Phys. Rev. Letters **28**, 553 (1972).
6.30. J.C.PHILLIPS: Surface Sci. **40**, 459 (1973).
6.31. S.Y.TONG, T.N.RHODIN, R.H.TAIT: Phys. Rev. B**8**, 430 (1973).
6.32. C.B.DUKE, N.O.LIPARI, U.LANDMAN: Phys. Rev. B**8**, 2454 (1973).
6.33. M.R.MARTIN, G.A.SOMORJAI: Phys. Rev. B**7**, 3607 (1973).

6.34. G.E.LARAMORE: Phys. Rev. B. **9**, 1204 (1974).

6.35. S.ANDERSSON, J.B.PENDRY: J. Phys. C**6**, 601 (1973).

6.36. G.E.LARAMORE: Phys. Rev. B**8**, 515 (1973).

6.37. S.ANDERSSON, B.KASEMO, J.B.PENDRY, M.A.VAN HOVE: Phys. Rev. Letters **31**, 595 (1973).

6.38. J.E.DEMUTH, D.W.JEPSEN, P.M.MARCUS: Phys. Rev. Letters **31**, 540 (1973); Solid State Commun. **13**, 1311 (1973); J. Phys. C**6**, 307 (1973).

6.39. C.B.DUKE, N.O.LIPARI, G.E.LARAMORE, J.B.THEETEN: Solid State Commun. **13**, 579 (1973); J. Vac. Sci. Technol. **11**, 180 (1974).

6.40. A.IGNATIEV, F.JONA, D.W.JEPSEN, P.M.MARCUS: Surface Sci. **40**, 439 (1973).

6.41. E.BAUER: In *Adsorption et Croissance Cristalline* (CNRS, Paris 1965), p. 21.

6.42. R.HECKINGBOTTOM: Surface Sci. **27**, 370 (1971).

6.43. C.E.CARROLL: Surface Sci. **32**, 119 (1972).

6.44. V.S.SUNDARAM, B.FARRELL, R.S.ALBEN, W.D.ROBERTSON: Phys. Rev. Letters **31**, 1136 (1973).

6.45. E.BAUER: Vacuum **22**, 539 (1972).

6.46. E.BAUER: Z. Physik **224**, 19 (1969).

6.47. J.I.GERSTEN: Phys. Rev. **188**, 774 (1969; B**2**, 3457 (1970).

6.48. C.B.DUKE, U.LANDMAN: Phys. Rev. B**6**, 2956, 2968 (1972).

6.49. V.ROUNDY, D.W.MILLS: Phys. Rev. B**5**, 1347 (1972); E.EVANS, D.L.MILLS: Phys. Rev. B**5**, 4126 (1972); B**7**, 853 (1973).

6.50. C.B.DUKE, U.LANDMAN: Phys. Rev. B**7**, 1368 (1973); B**8**, 505 (1973).

6.51. H.IBACH: J. Vac. Sci. Technol. **9**, 713 (1972).

6.52. J.O.PORTEUS, W.N.FAITH: J. Vac. Sci. Technol. **9**, 1062 (1972); Phys. Rev. B**8**, 491 (1973).

6.53. J.KÜPPERS: Surface Sci. **36**, 53 (1973).

6.54. J.E.ROWE, H.IBACH: Phys. Rev. Letters **31**, 102 (1973).

6.55. E.BAUER: In *Techniques of Metals Research*, Vol. II, Part 2, ed. by R.F.BUNSHAH (Interscience, New York, 1969), p. 624.

6.56. J.KESSLER: Rev. Mod. Phys. **41**, 3 (1969).

6.57. P.J.JENNINGS: Surface Sci. **20**, 18 (1970); **33**, 1 (1972).

6.58. R.FEDER: phys. stat. solidi (b) **49**, 699 (1972; **56**, K43 (1973).

6.59. R.FEDER: phys. stat. solidi (b) **62**, 135 (1974).

6.60. D.R.CRUISE, E.BAUER: J. Appl. Phys. **35**, 3080 (1964).

6.61. G.TURNER, E.BAUER: J. Appl. Phys. **35**, 3080 (1964).

6.62. V.DRAHÖS: In *Proc. 5th European Congr. Electron Microscopy* (Institute of Physics, London, 1972), p. 34.

6.63. W.KOCH, B.BISCHOFF, E.BAUER: In *Proc. 5th European Congr. Electron Microscopy* (Institute of Physics, London, 1972), p. 58.

6.64. L.LAYDEVANT, C.GUITTARD, R.BERNARD: In *Proc. 5th European Congr. Electron Microscopy* (Institute of Physics, London, 1972), p. 662.

6.65. W.A.COGHLAN, R.E.CLAUSING: Atomic Data **5**, 318 (1973).

6.66. M.O.KRAUSE, F.A.STEVIE, L.J.LEVIS, T.A.CARLSON, W.E.MODDEMAN: Phys. Letters **31**A, 81 (1970).

6.67. J.D.GARCIA, R.J.FORTNER, R.M.KAVANAGH: Rev. Mod. Phys. **45**, 111 (1973).

6.68. N.STOLTERFOHT, H.GABLER, U.LEITHÄUSER: Phys. Letters **45**A, 351 (1973).

6.69. D.L.MATTHEWS, B.M.JOHNSON, J.J.MACKEY, C.F.MOORE: Phys. Letters **45**A, 447 (1973); Phys. Rev. Letters **31**, 1331 (1973).

6.70. W.HINK, A.ZIEGLER: Z. Physik **226**, 222 (1969).

6.71. W.BAMBYNEK, B.CRASEMANN, R.W.FINK, H.-U.FREUND, H.MARK, C.D.SWIFT, R.E.PRICE, P.VENUGOPALA RAO: Rev. Mod. Phys. **44**, 716 (1972).

6.72. W.M.MULAIRE, W.T.PERIA: Surface Sci. **26**, 125 (1971).

6.73. E. N. Sickafus: Surface Sci. **36**, 472 (1973); Phys. Rev. B**7**, 5100 (1973); J. Vac. Sci. Technol. **12**, 43 (1973).
6.74. R. G. Musket, R. J. Fortner: Phys. Rev. Letters **26**, 80 (1971).
6.75. E. J. LeJeune, Jr., R. D. Dixon: J. Appl. Phys. **43**, 1998 (1972).
6.76. F. Fiermans, J. Vennick: Surface Sci. **38**, 237 (1973).
6.77. P. H. Citrin, R. W. Shaw, Jr., T. D. Thomas: In *Electron Spectroscopy*, ed. by D. A. Shirley (North-Holland Publ. Co., Amsterdam, 1972), p. 105.
6.78. J. E. Houston: Surface Sci. **38**, 283 (1973), Appl. Phys. Letters **24**, 42 (1974); Rev. Sci. Instrum. (to be published).
6.79. J. T. Grant, T. W. Haas, J. E. Houston: Phys. Letters **45**A, 309 (1973); Surface Sci. **42**, 1 (1974).
6.80. R. Bouwman, R. Biloen: Surface Sci. **41**, 348 (1974).
6.81. S. Thomas: Appl. Phys. Letters **24**, 1 (1974).
6.82. E. Bauer, H. Poppa, G. Todd, F. Bonczek: J. Appl. Phys. **45**, 5164 (1974).
6.83. A. E. Dabiri, V. S. Aramati, R. E. Stickney: Surface Sci. **40**, 205 (1973).
6.84. E. B. Bas, U. Bänninger: Surface Sci. **41**, 1 (1974).
6.85. S. Thomas: J. Appl. Phys. **45**, 161 (1974).
6.86. B. Goldstein: Surface Sci. **39**, 261 (1973).
6.87. R. E. Kirby, D. Lichtman, J. W. Dieball: Surface Sci. **41**, 447, 467 (1974).
6.88. T. Madey, J. Yates: J. Vac. Sci. Technol. **8**, 525 (1971).
6.89. J. C. Powell: Phys. Rev. Letters **30**, 1179 (1973).
6.90. G. Schön: J. Electron Spectrosc. **1**, 377 (1972/73); Phys. Letters **42**A, 381 (1973).
6.91. D. A. Shirley: Chem. Phys. Letters **16**, 220 (1972); **17**, 312 (1972); Phys. Rev. A**7**, 1520 (1973).
6.92. S. P. Kowalczyk, R. A. Pollak, F. R. McFeely, L. Ley, D. A. Shirley: Phys. Rev. B**8**, 2387 (1973).
6.93. J. A. D. Matthew: Surface Sci. **40**, 451 (1973).
6.94. P. H. Citrin: Phys. Rev. Letters **31**, 1164 (1973).
6.95. L. H. Jenkins, M. F. Chung: Surface Sci. **26**, 151, 649 (1971); **28**, 409 (1971); **33**, 159 (1972).
6.96. M. Suleman, E. B. Pattinson: J. Phys. F**1**, L21 (1971).
6.97. H. G. Maguire, P. D. Augustus: J. Phys. C**4**, L174 (1971).
6.98. B. D. Powell, D. P. Woodruff: Surface Sci. **33**, 437 (1972).
6.99. C. M. K. Watts: J. Phys. F**2**, 574 (1972).
6.100. D. L. Walters, C. P. Bhalla: Phys. Rev. A**4**, 2164 (1971).
6.101. J. T. Grant, T. W. Haas: Surface Sci. **23**, 347 (1970).
6.102. C. J. Powell: Appl. Phys. Letters **20**, 335 (1972), Solid State Comm. **10**, 1161 (1972).
6.103. J. E. Rowe, S. B. Christman: J. Vac. Sci. Technol. **10**, 276 (1973).
6.104. H. Löfgren, L. Walldén: Solid State Comm. **12**, 19 (1973).
6.105. M. Salmerón: Surface Sci. **41**, 584 (1974).
6.106. D. M. Zehner, R. E. Clausing, G. E. McGuire, L. H. Jenkins: Solid State Comm. **13**, 681 (1973).
6.107. W. F. Hanson, E. T. Arakawa: Z. Physik **251**, 271 (1972).
6.108. M. Salmerón, A. M. Baró, J. M. Rojo: Surface Sci. **41**, 11 (1974).
6.109. J. A. D. Matthew: Solid State Comm. **13**, 1203 (1973).
6.110. L. H. Jenkins, D. M. Zehner, M. F. Chung: Surface Sci. **38**, 327 (1973).
6.111. P. J. Feibelman: Phys. Rev. B**7**, 2305 (1973).
6.112. L. McDonnell, D. P. Woodruff: Vacuum **22**, 477 (1972).
6.113. B. W. Holland, L. McDonnell, D. P. Woodruff: Solid State Comm. **11**, 991 (1972).
6.114. T. W. Rusch, J. P. Bertino, W. P. Ellis: Appl. Phys. Letters **23**, 359 (1973).

6.115. B. CLEFF, W. MEHLHORN: Phys. Letters A 37, 3 (1971).

6.116. P. W. PALMBERG: J. Vac. Sci. Technol. 9, 160 (1972); 10, 274 (1973).

6.117. J. M. MORABITO: J. Vac. Sci. Technol. 10, 278 (1973).

6.118. M. L. TARNG, G. K. WEHNER: J. Appl. Phys. 43, 2268 (1972); 44, 1534 (1973).

6.119. D. V. McCAUGHAN, R. A. KUSHNER, V. T. MURPHY: Phys. Rev. Letters 30, 614 (1973).

6.120. P. STAIB: Radiation Effects 18, 217 (1973).

6.121. N. C. MacDONALD: Appl. Phys. Letters 16, 76 (1970); 19, 315 (1971).

6.122. L. A. HARRIS: J. Vac. Sci. Technol. 11, 23 (1974).

6.123. K. HAYAKAWA, H. OKANO, S. KOWASE, S. YAMAMOTO: J. Appl. Phys. 44, 2575 (1973).

6.124. J. M. MORABITO: Thin Solid Films 19, 21 (1973).

7. Concepts in Heterogeneous Catalysis

M. Boudart

7.1. Definitions

A single catalytic reaction consists of a closed sequence of elementary processes or steps. Summation of these steps, multiplied each by an appropriate *stoichiometric number* reproduces the stoichiometric equation for the reaction. In the first step, a species called the catalyst enters as a reactant whereas it appears as a product in the last step of the sequence.

The catalyst may be an enzyme, a complex in a liquid solution, a gaseous molecule or a grouping of atoms at the surface of a solid called the *active site* and denoted by an asterisk *. In the latter case, catalysis is called heterogeneous. It is only a special case of a general phenomenon. The great technological advantage of heterogeneous catalysis is the easy separation between the solid catalyst and the fluid reaction medium.

In this chapter, I shall discuss only heterogeneous catalysis, or catalysis for short, remembering that the active site is only a member of a large population of homogeneous chemical analogs. The wealth of chemical catalysis is also illustrated by the number of solids used in catalysis: metals, semiconductors or insulators, X-ray crystalline or amorphous materials, clusters of a few atoms or large crystals.

Table 7.1. Dissociative and associative mechanisms of catalysis: ammonia synthesis: value of stoichiometric number σ

Dissocative	σ	Associative	σ
$2* + N_2 \rightleftarrows 2*N$	1	$* + N_2 \rightleftarrows *N_2$	1
$2* + H_2 \rightleftarrows 2*H$	3	$*N_2 + H_2 \rightleftarrows *N_2H_2$	1
$*N + *H \rightleftarrows *NH + *$	2	$*N_2H_2 + H_2 \rightleftarrows *N_2H_4$	1
$*NH + *H \rightleftarrows *NH_2 + *$	2	$*N_2H_4 + H_2 \rightleftarrows 2NH_3 + *$	1
$*NH_2 + *H \rightleftarrows *NH_3 + *$	2		
$*NH_3 \rightleftarrows NH_3 + *$	2	$N_2 + 3H_2 \rightleftarrows 2NH_3$	
$N_2 + 3H_2 \rightleftarrows 2NH_3$			

The two basic mechanisms of catalysis are *dissociative* and *associative*. As an example, let us postulate two possible sequences of steps in ammonia synthesis (Table 7.1). Qualitative considerations suggest that the steps in a dissociative mechanism will have a higher probability (entropy factor) but also higher energy barrier (energy factor) than their associative counterparts [7.1]. Hence a dissociative mechanism may be favored at high temperatures while at low temperatures associate mechanisms might prevail. For instance, in the case of isotope equilibration reactions involving H_2 or O_2:

$$H_2 + D_2 \rightleftharpoons 2HD$$

$$^{32}O_2 + {}^{36}O_2 \rightleftharpoons 2\,^{34}O_2$$

it is very remarkable that they proceed at a fast rate on a variety of solids at very low temperatures at which most chemical processes are imperceptibly slow. This suggests associative mechanisms.

Thus, in the case of the H_2–D_2 equilibration, two possible dissociative and associative mechanisms can be considered. They are normally associated with the names of *Bonhoeffer-Farkas* (or generally *Langmuir-Hinshelwood*) on the one hand and those of *Eley-Rideal* on the other hand [7.2] (Table 7.2). In an Eley-Rideal mechanism, reaction takes place between a chemisorbed species and a non-chemisorbed species. Since the latter may, however, be physisorbed or held molecularly in a precursor state, the distinction between a Langmuir-Hinshelwood and a Eley-Rideal process is unambiguous only if, in the latter case, the non-chemisorbed species strikes the chemisorbed one directly from the fluid phase.

The first quantitative measure of the catalytic act is the rate of the single reaction called the *activity* of the catalyst. The best measure of the rate, which lends itself to comparison of the activity of various catalysts is the *turnover number* defined as the rate per mole of site, that rate itself being defined as the rate of change of the extent of reaction (also in moles) with time. In most cases, only an average or nominal number may be

Table 7.2. Langmuir-Hinshelwood and Eley-Rideal mechanisms [7.2]: H_2–D_2 equilibration: $H_2 + D_2 \rightleftharpoons 2HD$

Langmuir-Hinshelwood (or Bonhoeffer-Farkas)	Eley-Rideal
$2* + H_2 \rightleftharpoons 2*H$	$*H + D_2 \rightleftharpoons *D + HD$
$2* + D_2 \rightleftharpoons 2*D$	$*D + H_2 \rightleftharpoons *H + HD$
$*H + *D \rightleftharpoons HD + 2*$	

obtained, as the number and types of sites may not be known. The method of determination of the number of sites must be specified as well as the conditions under which the rate is measured. Frequently, turnover numbers of heterogeneous catalytic reactions are between 10^{-2} and 10^2 per second [7.3].

The activity of a catalyst is rarely its most important characteristic. *Selectivity*, which is defined as a ratio of rates, that of a desired reaction over the sum of rates of other undesirable side reactions, is often more important to achieve and more subtle to understand. At any rate, no discussion of catalysis is worthwhile without a quantitative assesment of rates.

The habit of catalytic materials is dictated by technological usage and by the frequent need for a very large specific surface area, of the order of $10^2 \, m^2 \, g^{-1}$. Thus porous materials with average pore dimensions around 10 nm are often used as the catalyst itself or as a *support* or *carrier* for the catalytic material. The use of the word *substrate* to denote the support or the catalytic surface is to be avoided as it has been pre-empted in catalysis to denote the reacting molecules transformes *by* the catalyst. Clearly, the *texture* of a porous catalyst, i.e. the shape of a replica of the material is of great importance in determining activity and selectivity because of the unavoidable gradients of temperature and concentration in the porous medium. The interplay between chemical reactivity and the physical phenomena of heat and mass transfer will not be discussed further in this chapter. It is covered in a number of textbooks [7.4, 5].

Quantitative measures of the texture of catalytic materials, namely the total specific surface area and the pore size distribution, can be obtained from isotherms of physisorbed nitrogen, according to standard methods which have received universal acceptance. The *dispersion* of a supported material, defined as the fraction of atoms of this material which is exposed at the surface, can be determined by selective chemisorption which titrates the surface atoms of interest. Although no general method is available to measure dispersion, reliable methods have been developed in a number of cases [7.6].

7.2. Affinity, Reactivity and Catalytic Activity [7.7]

Consider the reaction $A_1 + A_2 \rightleftarrows B_1 + B_2$ taking place through the simplest two-step catalytic sequence

1) $S_1 + A_1 \rightleftarrows B_1 + S_2$

2) $S_2 + A_2 \rightleftarrows B_2 + S_1$

over a constant number of identical sites denoted by S_1 if empty and S_2 if occupied. From the constancy of number of sites and the equality of rates of Steps 1) and 2) at the steady-state, the turnover number is

$$N = \frac{k_1 k_2 (A_1)(A_2) - k_{-1} k_{-2} (B_1)(B_2)}{k_1(A_1) + k_{-2}(B_2) + k_2(A_2) + k_{-1}(B_1)}, \tag{7.1}$$

where the k's denote rate constants with appropriate subscripts for forward and reverse processes, and quantities between parentheses are number densities.

If we now consider different collections of sites, we may ask how N will vary from one collection to the next. We assume that sites differ in their affinity for the reactants and products. Let us call $\underset{\sim}{A}_1$ the standard affinity for adsorption of A_1 divided by RT. Similarly $\underset{\sim}{A}_2$ will be the standard affinity for adsorption of B_2, divided by RT. These affinities are not independent: their differences is equal to the standard affinity for the overall reaction, divided by RT which does not vary with the nature of the sites and therefore may be regarded as a constant

$$\underset{\sim}{A}_1 - \underset{\sim}{A}_2 = \text{const}. \tag{7.2}$$

Now the reactivity of the sites, as expressed by their rate constants must be related to their thermodynamic affinity. A general relationship between rate constants and affinities is that of BRØNSTED, as found experimentally in acid base catalysis, free radical reactions and chemisorption at solid surfaces [8.8]

$$k = G \exp(\alpha \underset{\sim}{A}), \tag{7.3}$$

where G is a constant, and α is a so-called transfer coefficient

$$0 < \alpha < 1$$

with values frequently in the vicinity of $\frac{1}{2}$. It is easy to use the Brønsted relations (7.3) for the rate constants of our problem, taking into account (7.2) which permits us to retain as sole independent variable the affinity $\underset{\sim}{A}$ without the need for a subscript for adsorption of A_1 divided by RT. We assume that α is the same for both steps and recall that in general the ratio of rate constants for an elementary process is given by $\exp \underset{\sim}{A}$. The resulting relations are

$$k_1 = k_1^0 \exp[\alpha(\underset{\sim}{A} - \underset{\sim}{A}_0)]; \quad k_{-2} = k_{-2}^0 \exp[\alpha(\underset{\sim}{A} - \underset{\sim}{A}_0)]$$

$$k_2 = k_2^0 \exp[(\alpha - 1)(\underset{\sim}{A} - \underset{\sim}{A}_0)]; \quad k_{-1} = k_{-1}^0 \exp[(\alpha - 1)(\underset{\sim}{A} - \underset{\sim}{A}_0)], \tag{7.4}$$

where rate constants with superscripts ^{0} denote values of the rate constants for the sites with the highest value of $\underset{\sim}{A}$, namely $\underset{\sim}{A}_0$.

For the case $\alpha = 1/2$, it is now easy to answer the question: what are the best sites for the catalytic reaction, i.e. the value of $\underset{\sim}{A}$ (namely $\underset{\sim}{A}_{max}$) that maximizes the turnover number. Indeed substitution of (7.4) into (7.2) yields an expression for N with a numerator that does not depend on $\underset{\sim}{A}$. Taking the derivative of the denominator with respect to $\underset{\sim}{A} - \underset{\sim}{A}_0$ and equating it is zero, gives the required condition for maximizing N:

$$k_1(A_1) + k_{-2}(B_2) = k_2(A_2) + k_{-1}(B_1). \tag{7.5}$$

Comparing this result with the steady state condition expressing the equality of rates of both steps shows at once that (7.5) is satisfied when the concentration of empty sites is equal to that of occupied ones. Thus the optimum sites are those with a value of $\underset{\sim}{A}$ corresponding to half coverage. This is an expression of the *principle of Sabatier* according to which the best catalyst is that which has an affinity for reactants which is neither too small nor too large. There are many examples of so-called *volcano-shaped curves* following BALANDIN [7.7] where the rate of a reaction when plotted versus an energy parameter which is a measure of the affinity of the surface for a reactant, first increases, goes through a maximum and then decreases as the energy parameter becomes still higher.

Another consequence of the Brønsted relations (7.4) with $\alpha = 1/2$ is that sites with affinities larger or smaller by the same amount $\Delta \underset{\sim}{A}$ than those of the optimum sites, have the same turnover number. Moreover, it is easy to show [7.8] that that same turnover number is $\exp(f/4)$ times smaller than that corresponding to the optimum sites, where $f = 2\Delta \underset{\sim}{A}$. These are most interesting results. From a *thermodynamic* standpoint alone, sites differing in their adsorption affinity by an amount f would be expected to differ in their occupancy, at least at low coverage, by an amount of $\exp(f)$. From the standpoint of *reactivity*, this amount is reduced to $\exp(f/2)$ by a Brønsted relationship with $\alpha = 2$. But from a *catalytic* standpoint, these sites have the same turnover number, if they have affinities equidistant from the optimum value. This is easy to understand: for adsorption to take place at a certain rate at the steady state of the catalytic reaction, the sites must be reactive but they must also be free. If the affinity is large, the reactivity is high but the number of free sites is low. Conversely, if the affinity is small, the reactivity is low but the number of free sites is high. The optimum catalyst realizes a compromise. But insofar as the fraction of surface cover under optimum conditions may be in the vicinity of 1/2, it appears unlikely that a good catalyst will function under conditions of low surface coverage. Thus, unless active sites are isolated and far removed from each other, a situation which may arise frequently but is inherently far from the optimum, a good catalyst should operate under conditions where interactions between chemisorbed species are likely to be important and cannot be neglected. Besides, active sites may not all be the same. These two situations suggest that a catalytic surface is likely to be *non-ideal*, in the sense that it does not conform to a Langmuir model. In other words, rate constants ought to depend on surface coverage. Two ways to handle this problem will be reviewed briefly in turn.

7.3. Non-Ideal Catalytic Surfaces: the Ammonia Synthesis

If there is any reason to believe that more than one type of site participates in the catalytic reaction, the reasoning that led to the expression for the turnover number (7.1) must be modified, either by a summation or by an integration if the need arises. In particular, the non-uniformity of the surface can be expressed by a distribution function

$$ds = \underline{a} \exp(-\gamma \underline{A}) \, d(\underline{A}), \tag{7.6}$$

where ds is the number of sites per cm^2 with a value of $\underline{A}$ between $\underline{A}$ and $d\underline{A}$, γ is a parameter characteristic of the distribution, and $\underline{a}$ a constant that can be determined by a normalization condition.

Then, with the same Brønsted conditions (7.4) as before, it can be shown, following Temkin, that for the two step catalytic reaction discussed earlier, the integrated turnover number becomes

$$N = \eta \; \frac{k_1^0 k_2^0 (A_1)(A_2) - k_{-1}^0 k_{-2}^0 (B_1)(B_2)}{[k_1^0(A_1) + k_2^0(B_2)]^m \, [k_2^0(A_2) + k_{-1}^0(B_1)]^{1-m}}, \tag{7.7}$$

where η is a numerical sonstant and $m = \alpha - \gamma$. Details are given in [7.8].

The physical meaning of the distribution function (7.6) lies in the fact that chemisorption isotherms on real surfaces frequently obey the FREUNDLICH form [7.4]

$$\theta \propto p^\gamma \tag{7.8}$$

or the Frumkin-Temkin form [7.4]

$$\theta \propto \ln p, \tag{7.9}$$

where θ is the fraction of surface covered by a gas at pressure p in equilibrium with the surface at temperature T. The isotherms (7.8) and (7.9) can be derived from the distribution function (7.6) for an arbitrary value of γ, $0 < \gamma = 1$, and for $\gamma = 0$, respectively. Besides, the rate of chemisorption of a gas on a non-uniform surface with a distribution function (7.7), in the case $\gamma = 0$, can be shown [7.8] to be

$$r = r_0 \exp(-\beta\theta) \tag{7.10}$$

which is a very frequently found empirical expression associated with the name of several workers, especially that of ELOVICH [7.9].

Thus, Temkin's phenomenological approach to kinetics on non-ideal surfaces finds its justification in thermodynamics and kinetic studies of surface chemistry. Besides, a simplified form of (7.7) is the famous equation of TEMKIN and PYZHEV [7.10] which has been used successfully since 1939 to represent kinetic data of ammonia synthesis and decomposition at low and high pressures. A slight variation thereof is used today in the design of ammonia catalytic converters. The Temkin-Pyzhev equation is obtained from a simplified mechanism for ammonia synthesis in which it is postulated that nitrogen chemisorption is the rate determining step and that chemisorbed nitrogen is the most abundant surface intermediate [7.4]. These assumptions lead to a two-step mechanism

1) $* + N_2 \rightleftarrows\!\wedge\!\rightleftarrows *N_2$

2) $*N_2 + 3H_2 \longleftrightarrow 2NH_3 + *$

in which the symbol $\wedge$ means rate determining and $\leftrightarrow$ stands for equilibrium. Noting that terms in the denominator of (7.7) involving k_1^0 and k_{-1}^0 can be neglected in front of those involving k_2^0 and k_{-2}^0 since the latter parameters refer to a fast equilibrated step, we get for the synthesis reaction sufficiently far from equilibrium [7.11]:

$$N = \eta\, k_1^0 [K_2^0]^m (N_2) \left[\frac{(H_2)^3}{(NH_3)^2} \right]^m , \tag{7.11}$$

where $K = k_2^0 / k_{-2}^0$ is the equilibrium constant for the second step. The success of (7.11) in fitting data is not, of course, a sufficient justification of the theory behind it. But there are many other arguments in favor of the latter [7.12]. One of the quantitative ones rests on a comparison of the rates of ammonia synthesis on the same catalyst with H_2 and with D_2. Under identical conditions, it is clear from (7.11) that the ratio of turnover numbers should be

$$N_D/N_H = (K_{2,D}^0/K_{2,H}^0)^m \tag{7.12}$$

with subscripts H and D pertaining to the reactions with H_2 and D_2 respectively. However, the right hand side of (7.12) is K^m, where K is the equilibrium constant of the reaction

$$3D_2 + 2NH_3 \longleftrightarrow 3H_2 + 2ND_3$$

which involves only gas phase species and can therefore be obtained from thermodynamic tables. Thus the isotope effect in ammonia synthesis with H_2 and D_2 can be predicted ahead of time of the mechanism behind the equation of TEMKIN and PYZHEV is correct [7.10]. Experiments veryfying these predictions have now been performed over an extended temperature range and the results with $m = 1/2$ are in excellent agreement with expectations [7.13]. It must be stressed that the value of $m = 1/2$ was used in many previous investigations and was therefore not adjusted so as to bring about agreement between theory and experiment in the study of the isotope effect.

In spite of this quantitative success, the nature of the most abundant surface species in ammonia synthesis is not known as yet. Thus an equation identical to that of TEMKIN and PYZHEV is obtained if it is assumed that N_2 dissociates into atoms in the rate determining step and that N atoms are the most abundant surface intermediate. Hence, a choice between associative and dissociative mechanisms of ammonia synthesis (Table 7.1) will have to rely on future direct indentification of the adsorbed intermediates during the reaction by an adequate spectroscopic technique. This need is general in catalysis [7.14]. In particular, even more sophisticated kinetic methods based on the use of tracers are not able to decide between the alternatives presented in Table 7.1. Thus, according to HORIUTI the stoichiometric number of the rate determining process, as determined by rate measurements at equilibrium (by means of tracers) and away from equilibrium, should be unity both for the dissociative and the associative mechanisms if the chemisorption of N_2 is the rate determining process [7.7]. In fact, measurements of the stoichiometric number of the rate determining process in ammonia synthesis do yield a value of unity which confirms the generally accepted view that chemisorption of N_2 is indeed the rate determining process if the mechanism is dissociative. But if the mechanism is associative, the obtained value of the stoichiometric number of the rate determining process is clearly ambiguous. A useful concept in catalysis which can be applied quantitatively in the case of ammonia synthesis is that virtual pressure of a reactant. Since, on an iron catalyst, adsorbed nitrogen is in equilibrium with gas phase H_2 and NH_3 but not with gas phase N_2, it can be considered to be in virtual equilibrium in the gas phase with H_2 and NH_3 at a virtual pressure defined by the equilibrium constant K of the reaction

$$2NH_3 \xleftrightarrow{K} N_2 + 3H_2 \,.$$

At 673 K, the value of K is 6×10^3 atm^2. Thus, when ammonia decomposes on a catalyst at 673 K in a mixture of NH_3 and H_2 both at 1 atm, the virtual pressure of N_2 over the catalyst surface is 6000 atm. Thus it is possible to control the virtual pressure of a reactant over a catalyst surface over a very wide range of values not readily reached in reality.

In a study of ammonia synthesis, the virtual pressure of N_2 was varied over 4 orders of magnitude while the fraction of the surface of the iron catalyst covered with nitrogen, as determined by microgravimetry, varied only between 0.45 and 0.53 [7.15]. This result is another clear indication of a broadly non-uniform or non-ideal surface with a logarithmic adsorption isotherm of the form (7.9), as determined experimentally in a separate experiment. It also illustrates the quantitative expression of the Sabatier principle according to which the fraction of surface covered is in the vicinity of 1/2 for an optimum catalyst.

7.4. Non-Ideal Catalytic Surfaces: the Water Gas Shift Reaction

There is another way to take non-ideality into account. It rests on the measurement of the *thermodynamic activity* of an element which is transferred from one molecule to the next in the catalytic reaction. This phenomenological approach has been developed by WAGNER [7.16] and his school. A particularly clear example is that of the water gas shift reaction over a foil of wustite FeO

$$CO + H_2O \rightleftharpoons CO_2 + H_2 \tag{7.13}$$

taking place by oxygen atom transfer between CO and CO_2 on the one hand and H_2 and H_2O on the other hand, in two elementary processes

$$\begin{aligned} &1) \quad * + CO_2 \rightleftharpoons CO + *O \\ &2) \quad *O + H_2 \rightleftharpoons H_2O + * \,. \end{aligned}$$

The rate r_1 of the first step is written as usual with the notation $k_1/k_{-1} = K_1$

$$r_1 = k_1(*) (CO_2) - k_{-1}(CO) (O*)$$

$$= k_1(*) (CO_2)\left[1 - \frac{1}{K_1} \frac{(*O) (CO)}{(*) (CO_2)}\right]. \tag{7.14}$$

In these expressions, concentrations have been used instead of thermodynamic activities with the penalty that the rate "constants" may then depend on surface composition. The non-ideality of the surface is handled by defining the *thermodynamic activity* of atomic oxygen at the surface as

$$a_0 = \frac{(*O)}{(*)}. \tag{7.15}$$

But at equilibrium

$$a_0 = \frac{(*O)}{(*)} = K_1 \frac{(CO_2)}{(CO)}. \tag{7.16}$$

The standard state is chosen so that $a_0 = 1$ in an equimolar mixture of (CO_2) and (CO), which is equvalent to writing $K_1 = 1$ in (7.16). Thus finally, the thermodynamic activity of oxygen is

$$a_0 = \frac{(CO_2)}{(CO)}$$

Substitution of (7.15) into (7.14) yields

$$r_1 = k_1'(CO_2)\left[1 - \frac{a_0}{(CO_2)/(CO)}\right], \tag{7.17}$$

where $k_1' = k_1(*)$, clearly not a constant but a function of surface composition. Now, the rate r_1 can be measured by passing a mixture of CO and CO_2 over the surface at a temperature at which a_0 is the same in the bulk and at the surface, so that a bulk physical measurement such as electrical conductivity of the catalyst foil which is determined by the non-stoichiometry of the oxide which in turn depends on a_0, can yield a value of a_0. After equilibrium is reached, the ratio of CO_2 to CO is changed slightly in stepwise manner and the rate r_1 can be determined by following the *relaxation* of the system to a new equilibrium, by means of conductivity measurements [7.16]. By such measurements, Stotz

[7.17] could not only verify the form of (7.17) but find the dependence of k'_1 on surface composition

$$k'_1 \propto a_0^{-0.6}. \tag{7.18}$$

It can be shown that an expression of this kind is expected also from the treatment of TEMKIN presented in Section 7.3. Thus, both phenomenological approaches, that of WAGNER and that of TEMKIN are two different manners by which the non-ideality of catalytic surfaces can be handled in kinetics. It is somewhat surprising that so much in surface chemistry and catalysis is done routinely without considering explicity the problem of non-ideality of surfaces.

In the case of the water gas shift reaction on FeO, STOTZ [7.17] carried out similar measurements of r_2. He then found that the experimental value of a_0 at which both rates r_1 and r_2 were equal, was the same as that measured during the steady state water gas shift reaction taking place at the same rate $r = r_1 = r_2$. Again, as was the case for ammonia synthesis and the Temkin approach, this result is a remarkable quantitative check of Wagner's approach. But here also other methods are required before more can be said about the nature of the sites or the nature of oxygen bound to these sites.

7.5. Structural or Geometric Factors

The most distinctive feature of a crystalline solid is anisotropy. At the surface, this is manifested by surface atoms with different values of their coordination number C_i, where C_i denotes an atom with i nearest neighbors. Also, especially on metal surfaces, it is easy to recognize groupings of atoms with a symmetry that matches that of a chemisorbed molecule. That grouping is sometimes called a *multiplet* after BALANDIN or an *ensemble* after KOBOZEV [7.5]. A typical multiplet is that found on the (111) faces of a *fcc* metal which matches the hexagonal symmetry of a benzene molecule. I shall refer to these *geometric* factors which depend on atoms with specified values of C_i as *structural* factors. Chemical intuition dictates that the rate of a catalytic reaction should depend on anisotropy or structure. We would normally expect that a catalytic reaction be *structure sensitive* even though the difference in rate between atoms with different C_i may not be as large as anticipated on the basis of differences in affinity alone, for reasons discussed in Section 7.2.

One way to check whether a catalytic reaction is structure sensitive or not, is to measure its rate on different crystallographic plane. Although such experiments have been carried out by various workers in the past, only recent results taking advantage of modern methods of surface analysis can be retained as reliable. One typical result is that of McAllister and Hansen who showed that the rate of decomposition of ammonia is ca. 10 times larger on the (111) face than on the (100) and (110) faces of *bcc* tungsten [7.18]. A much more surprising result is that of Ertl and Koch who found no appreciable difference in the rate of oxidation of carbon monoxide on the three low index faces of *fcc* palladium as well as on a polycrystalline palladium wire [7.19]. That the decomposition of ammonia on tungsten is structure sensitive is not surprising. That the oxidation of carbon monoxide on palladium is structure insensitive is an unexpected observation.

But it is not an isolated phenomenon. Several reactions have been found to be structure insensitive over supported Group VIII metals. In the case of supported metals, the experiment consists in preparing a series of catalysts with increasing particle size between 1 and 10 nm. Qualitative and quantitative considerations indicate that the relative fractions of atoms with a given C_i changes appreciably with particle size in that critical range. Thus, as particles of a *bcc* structure grow from 1 to 5 nm, the relative amount of C_7 surface atoms increases by almost an order of magnitude [7.20]. Therefore, if the turnover number for a given reaction does not change over a series of catalysts with metal particles from 1 to 10 nm corresponding to metal dispersion between 1 and 0.1, the evidence is that the reaction is structure insensitive. The reverse is not necessarily true: if the turnover number changes with dispersion, the reaction can be considered to be structure sensitive only after other possible effects such as electronic interaction between metal and support can be ruled out.

An early example of a structure insensitive reaction on supported platinum was the hydrogenation of cyclopropane to propane [7.21]. Striking confirmation of this finding is provided by a recent study of the same reaction on a stepped surface of a platinum single crystal: the turnover number was found to be almost the same as that found in the earlier work on supported platinum particles about 1.5 nm in average size [7.22].

Another example of a structure insensitive reaction is that of the hydrogenation of cyclohexene on supported platinum in the liquid phase [7.23] and in the gas phase [7.24]. Other examples of structure insensitive reactions include hydrogenation of benzene and the H_2-D_2 equilibration [7.25]. By contrast, the hydrogenolysis of ethane [7.26] and of n-hexane [7.27] on rhodium and platinum, respectively, appear to be structure sensitive reactions.

The different behavior of reactions involving C–C bonds on the one hand and C–H or H–H bonds on the other hand is illustrated by a study

Table 7.3. Hydrogenation of cyclohexene on supported platinum at 295 K (g) and 307 K (l): turnover number N on catalysts with degree of Pt dispersion D [7.23]; (g) and (l) stand for gas phase and liquid phase respectively

Support	%-wt. Pt	D	N/s^{-1} (g)	N/s^{-1} (l)
SiO_2	1.5	1.00	2.73	8.90
SiO_2	0.38	1.00	2.64	8.77
γ-Al_2O_3	0.6	0.70	—	8.37
SiO_2	2.3	0.62	2.75	8.43
SiO_2	0.8	0.34	—	8.65
η-Al_2O_3	1.96	0.23	—	7.98
SiO_2	3.7	0.14	2.53	8.37

of the hydrogenolysis (cracking) of cyclopropane running in parallel with its hydrogenation on two forms of chromium oxide: a crystalline form and an X-ray amorphous form [7.28]. The cracking reaction was found to be structure sensitive while the hydrogenation reaction was structure insensitive. Structure here is defined by the difference between short-range and long-range order of the same material. Another example is the reaction between H_2 and O_2 on supported platinum catalysts [7.29]. In excess oxygen, the reaction appears to be structure insensitive while in excess hydrogen, the same reaction is clearly structure sensitive on the same samples under otherwise identical experimental conditions.

The following speculations may explain these latter results. If the interaction between reactants and surface is strong, a reconstruction of the surface with formation of a two-dimensional layer forming a coincidence lattice with the subjacent layer is possible, as indicated by the work of BÉNARD and coworkers [7.30]. If this happens, structure insensitivity becomes understandable [7.31]. On the other hand, if the interaction between reactants and surface is weak, structure sensitivity is expected to be observable. Thus in the case of the reaction between H_2 and O_2 on platinum, the surface may be reconstructed in excess oxygen but not in excess hydrogen because of the strong and weak interactions of these adsorbated with the surface. Thus structure insensitivity may be expected in the first case but not in the other.

To explain the structure sensitivity of hydrogenolysis reactions may require a different concept. Although molecular details are still missing, it is generally agreed that hydrogenolysis of hydrocarbons on metals is accomplished by extensive dissociation of the chemisorbed hydrocarbon [7.11]. This should require a number of sites. i.e. an ensemble of sites, thus a particular surface structure.

These speculations are of great interest in connection with recent results on alloy catalysts. Many studies have been devoted to the activity

Table 7.4. Reaction between H_2 and O_2 on Pt/SiO_2 at 273 K: rate constant k on catalysts with degree of Pt dispersion D [7.29]

%-wt Pt	D	$k \times 10^3/cm\ s^{-1}$
a) Excess O_2		
3.7	0.14	2.5
2.3	0.62	3.5
0.53	0.625	4.6
0.38	1.00	2.0
b) Excess H_2		
3.7	0.14	3.1
2.3	0.62	10.7
0.53	0.625	10.5
0.38	1.00	20.0

of alloys as catalysts but it is only very recently that *selectivity* has been emphasized. A typical discovery is that of SINFELT et al. [7.32]. When copper is added to nickel the turnover number of the alloy for dehydrogenation of cyclohexane (a structure insensitive reaction) hardly changes until the alloy contains more than 80 mol-% copper. By contrast, the turnover number for hydrogenolysis of ethane (a structure sensitive reaction) goes down by more than three orders of magnitude after addition of less than 10 mol-% copper. Whatever the final interpretation of these results will be, the difference in catalytic behavior from one reaction to the next, i.e. selectivity, is best explained at the present time in terms of geometric or structural or ensemble effects.

7.6. Electronic or Ligand Factors

A recurring theme in surface catalysis has been the vague idea of chemical unsaturation of the sites. With progress in coordination and organometallic chemistry of soluble complexes, the idea has received more attention with the idea that a site exhibits *coordinative unsaturation* by the loss at the surface of one or more ligands [7.33]. Even though this refinement has remained qualitative, for instance in the treatment by BOND of metallic surface orbitals [7.34], it has clarified the old broad concept of *electronic factors* in catalysis. Indeed, with metals and alloys especially, it appears preferable to talk about *ligand effects* rather than electronic factors to designate the influence on a given site of nearest neighbor sites [7.35].

When it comes to compare the activity of different metals for a given reaction from the viewpoint of electronic factors, the only quantitative index that has been used very frequently since first proposed in 1950 [7.36] has been the percentage d bond character from Pauling's theory of metals [7.37]. The relative success of these correlations in bringing some order in *reactivity patterns* especially among Group VIII and Group Ib metals [7.38] suggests that even highly approximate theories of the metallic bond are greatly useful in catalysis. This is because of the very large differences in turnover number, differing by as much as ten orders of magnitude, exhibited by different metals.

A much more puzzling problem is that the *specificity* of a given catalyst for a certain reaction can become so important that only one, or perhaps a very few catalysts are capable of carrying out that reaction with a high selectivity. A good example is the specificity of metallic silver in carrying out the oxidation of ethylene to ethylene oxide [7.39]. A reasonable Eley-Rideal mechanism for the reaction postulates that selective oxidation of ethylene takes place when the molecule hits chemisorbed O_2 bound linearly (end-on) to a silver atom. By contrast, if ethylene reacts with O adatoms, complete oxidation of the molecule to carbon dioxide and water place. If this mechanism is correct, the reason for the specificity of silver lies in the unique reactivity of the binding state of O_2 at its surface.

Another example of specificity is the unique ability of supported iridium to catalyze the decomposition of hydrazine in small rocket motors used in space navigation. In spite of intensive research motivated by the scarcity of iridium, no other metal has been found satisfactory. In this case, no mechanism is known.

A third example deals with the reactions of 2,2-dimethylpropane which was found to be readily hydrogenolyzed to smaller fragments on all Group VIII metals as well as on copper and gold but is isomerized to 2 methylbutane *only* on iridium, platinum and gold [7.40]. For the latter reaction, the bond shift mechanism proposed by ANDERSON and AVERY [7.41] may require a shift in *surface valence* so that the specificity of iridium, platinum and gold in this isomerization may well be related to the multiplicity and ready interconversions of multiple valences suggested for these three metals, but not for others, by RHODIN and coworkers [7.42] on the basis of binding energies of foreign metal atoms on a tungsten FIM tip as well as from LEED observations of hexagonal overlayers on cubic faces of Ir, Pt, and Au.

The need for an improved chemical theory of metals is also felt in the interpretation of the many results obtained on alloy catalysts. The challenge for an electronic or ligand interpretation is particularly strong in those cases where large changes in catalytic activity are obtained upon alloying and where also surface composition is relatively well known. The latter is of course not equal in general to bulk composition [7.43].

An alloy which has been studied many times as a catalyst is copper-nickel. In these alloys copper enrichment of the surface takes place as a result of the thermodynamic drive for lowering surface free energy by accumulation at the surface of the component with the lower sublimation energy. There is now general agreement on the fact that Cu addition to Ni yields copper rich surfaces of a composition that changes rapidly as copper is first introduced into nickel and then more slowly after subsequent addition of copper. In a study of nickel-copper alloy catalysts for reactions of n-hexane and hydrogen mixtures, PONEC and SACHTLER reported that with increasing additions of Cu to Ni, up to 23 mol-%, the rate of hydrogenolysis decreases markedly with addition of Cu but the mode of cracking of n-hexane remains that characteristic of pure Ni [7.44]. These observations are explained in terms of structural or geometric or ensemble factors. It must be remembered that, as noted earlier, the hydrogenolysis of n-hexane on platinum has been shown to be a structure sensitive reaction [7,27]. However, as more Cu is alloyed into Ni, the selectivity for the reactions of n-hexane changes markedly although rates are not affected substantially: the selectivity toward isomerization of n-hexane increases sharply while the cracking pattern also changes, both modes of reactivity becoming those observed with platinum rather than with nickel. These observations with copper rich nickel alloys are interpreted in terms of a ligand or electronic effect. A definitive explanation of these phenomena remains to be provided but at a phenomenological level, the contrast between geometric and electronic factors for a given alloy is noteworthy.

Another aspect of the electronic factor in catalysis has been traditionally linked with semiconductors in an *electronic theory of semiconductor catalysis* particularly associated with the name of VOL'KENSHTEIN [7.45]. This is a formal theory which has led over a period of 25 years to an abundant experimental literature. It would be unfair to attempt to review it briefly and equally unfair to ignore it altogether. It will be simply introduced by means of a fairly typical example, the oxidation of carbon monoxide on zinc oxide.

In such a reaction on an oxide surface, it is natural to expect that an oxygen ad-atom will be an intermediate. But what should its charge be? The oxygen anion O^{2-} should be quite unreactive. Thus, naturally, the choice narrows down to O and O^-. The relative population at the surface of these two intermediates must be dictated by an equilibrium

$$O + e \rightleftharpoons O^-, \tag{7.19}$$

the position of which will be determined by the chemical potential of the electrons e or the Fermi level of the semiconductor. The equilibrium can be shifted by doping, by light or by external electric fields. Hence the possibility of controlling catalytic activity by introduction of impurities. Alternatively photocatalytic or electrocatalytic phenomena are conceivable. Indeed they all have been observed on zinc oxide as a catalyst.

For the reaction under discussion, attempts to shift the equilibrium (7.19) by doping zinc oxide with foreign ions (lithium and indium, or gallium) were made by CHON and PRATER on the one hand [7.46] and by AMIGUES and TEICHNER [7.47] on the other hand. The first group of workers found an effect of doping on catalytic activity and a correlation between catalytic rates and the concentration of electronic carriers as measured by the Hall effect. They concluded that the charged surface species O^- was the intermediate in the reaction. By contrast, the second group of workers found no effect of doping on catalytic activity and concluded that the uncharged surface species was the reaction intermediate. But the conditions used by the two groups were different [7.48]. The first group used relatively high temperatures, low pressures and large crystallite sizes. The second group used relatively low temperatures, high pressures and small crystallite sizes. The latter conditions, which are fairly typical of conventional catalysis, may indeed shift the equilibrium (7.19) to the left with the result that the electron concentration is buffered and insensitive to doping. The conditions of the first group of workers are, however, typical of the kind of catalysis practiced by WAGNER and his school [7.16], who pioneered the ideas subsequently developed by VOL'KENSHTEIN and others. Although these conditions are typical of conventional catalysis, they are those encountered in the catalytic treatment of automotive exhaust, a problem of current interest where the electronic theory of semiconductor catalysis may yet find applications.

7.7. Promoters and Poisons

Conceptually, promoters and poisons must be considered together: they consist of additives introduced with the reactants or during the catalyst preparation and they affect catalyst activity or selectivity. Some of the possible mechanisms will be considered in turn.

A first kind of promoter, said to be *textural*, prevents loss of surface area of the catalyst. A typical example is alumina Al_2O_3 introduced in small quantities (ca. 3%) during the preparation of ammonia synthesis iron catalysts. After reduction, the catalyst consists of metallic iron particles with a mean diameter of 35 nm. The alumina Al_2O_3 is unreduced and covers about half of the iron surface, preventing sintering of the metallic particles [7.49]. Another possible mechanism of textural promotion of iron by alumina has been suggested as a result of the examination of Mössbauer spectra[1] of reduced promoted catalysts [7.50]. At least part of the alumina may remain inside the iron crystallites in the form of very small (1.5 nm) inclusions. The latter would contribute to the elastic stress of the slightly strained metallic particles and a stable particle size would result as a balance between elastic stress and surface free energy.

Another kind of promoter is *structural* or *chemical*. An example is chlorine in the selective oxidation of ethylene to ethylene oxide on silver catalysts (see Section 7.6). It appears that chemisorbed chlorine inhibits the activated or non-activated dissociative chemisorption of oxygen on the metal. Since oxygen ad-atoms are responsible for the non-selective oxidation of ethylene, the role of the promoter is to enhance selectivity. Since the promoter occupies part of the surface, its role may also be conceived as that of a *selective poison*.

As an illustration of selective poisoning, and especially of its use in the assignment of sites responsible for one kind of reaction, γ-alumina which has been pretreated in vacuo at 800 K, catalyzes the isomerization of 1-butene to 2-butene at room temperature as well as the H_2–D_2 equilibration [7.51]. The latter reaction is suppressed by the adsorption of 1.2×10^{13} molecules of CO_2 per cm^2 of alumina surface. Under these conditions, the isomerization reaction proceeds at about the same rate as before selective poisoning by CO_2. Clearly, the sites responsible for the two reactions are different. The number density of sites capable of carrying out the isotopic equilibration is less than or equal to 1.2×10^{13} cm^{-2} as CO_2 may have blocked other sites besides the ones responsible for the suppressed reaction. The sites for H_2–D_2 equilibration may be associated with coordinatively unsaturated surface Al^{3+} ions.

There are as many types of poisons as there are types of sites (see Section 7.8). Some are reversibly held, i.e. they can be removed under reaction conditions as is the case for surface oxygen during ammonia synthesis. Some are irreversibly held as are most types of carbonaceous residues accumulating on catalytic surfaces during reactions involving hydrocarbons. These residues must be removed in a special step of catalyst *regeneration*. In fact, catalyst *deactivation* during use is the rule rather than the exception: it is the central problem in the transfer of any catalyst from the laboratory to the plant [7.52].

Reversible poisons are called also *inhibitors* especially when they are participants in the reaction. Thus reaction products are often inhibitors as is the case for ammonia in ammonia synthesis on iron according to (7.11). The reason for this is clear from the postulated equilibrium

$$*N_2 + 3H_2 \longleftrightarrow 2NH_3 + *$$

[1] See Topics in Applied Physics, Vol. 5.

which can and has been studied seperately. Similar equilibria involving H_2O and H_2S are

$$*O + H_2 \longleftrightarrow H_2O + *$$

$$*S + H_2 \longleftrightarrow H_2S + *.$$

They explain inhibition of reactions by H_2O or H_2S, respectively. But they also provide a convenient means of covering the surface of a catalyst with small and controlled quantities of oxygen or sulfur adatoms [7.53, 54] by regulating the virtual pressure of oxygen or sulfur (see Section 7.3).

7.8. Active Centers

The fraction of sites which is active in a given reaction depends on the catalyst and on the reaction, as recognized fifty years ago by TAYLOR who called *active centers* the site or group of sites taking part in the reaction [7.55]. The identification of the active centers and the structure of their complexes with the reactive intermediates under reaction conditions is the central goal of research in catalysis. The following examples have been chosen among many to illustrate some possibilities.

As an example of *electronic defects*, paramagnetic V-centers detected and counted by electron spin resonance spectroscopy have been found to be responsible for the H_2–D_2 equilibration at 78 K on magnesium oxide powders [7.56]. The concentration of these centers could be varied over four orders of magnitude by pre-treatment of the samples in vacuo at temperatures between 800 and 1100 K. The catalytic activity could be correlated with the concentration of the V-centers. These are believed to consist of three O^- surface ions in a triangular array corresponding to (111) planes of magnesium oxide. The catalytic site involves two of these O^- defects plus a neighboring OH^- group. A deuterium molecule is pictured as approaching two O^- sites to form a triangular transition state with the neighboring proton. The transition state then dissociates with release of HD. The reaction thus appears to be of the Eley-Rideal type.

Similar electronic defect centers can be produced by X-ray pre-irradiation of silica gel containing aluminum impurities and associated with these impurities, as shown by hyperfine splitting of the electron spin resonance signal [7.57]. These centers were also found to be responsible for the H_2–D_2 equilibration at 78 K [7.58].

The same chemical system, silica-alumina, is an active catalyst for reactions involving carbonium ion intermediates as a result of the *Brønsted acidity* associated with the *protons* required for charge compensation as an Al^{3+} ion is replacing a Si^{4+} ion in the structure [7.59]. These acid sites have been studied in great detail in *zeolites*, especially synthetic faujasite or Y-zeolites which are crystalline porous aluminosilicates, as opposed to silica-alumina which consists of X-ray amorphous gels [7.60]. Many examples of *Lewis acid* sites are also known, for instance, Al^{3+} surface ions with unsaturated coordination on alumina, as already discussed in Section 7.6. Sites consisting of *Lewis bases* are also known, for example, $O^=$ ions which are sufficiently strong electron donors, at corners of magnesium oxide cubic crystallites to catalyze reactions taking place via carbanion intermediates [7.61]. In the latter case, the concentration of the strongly basic sites was determined by electron spin resonance spectroscopy of the radial anion found upon adsorption of tetracyanoethylene on these sites, and the turnover number for the double bond isomerization of 1-butene taking place through carbanion intermediates was found to be almost constant when the turnover number was calculated on the basis of the site concentration measured by electron spin resonance and when the concentration of these sites was varied over a modest range by thermal pretreatment of the samples in vacuo.

In all the examples cited thus far, the concentration of active centers is only a small — sometimes very small — fraction of the total number of sites per unit surface area. Thus in the case of the V-centers at the surface of pure magnesium oxide, the surface density of centers was measured to be only $10^9 \, cm^{-2}$ for the samples with the highest catalytic activity. It follows that the reactions studied on these centers are structure sensitive. For example in the case of magnesium oxide, the sites responsible for the H_2–D_2 equilibration at 78 K are associated with metastable (111) planes of the solid. These are present only when the solid is formed from another phase such as magnesium hydroxide or magnesium hydroxycarbonates. Samples that are recrystallized to form cubic crystals of magnesium oxide which do not expose (111) planes are devoid of any catalytic activity, irreversibly so.

This is just one of the many examples suggesting that electronic or ionic defects associated with small particles of a metastable phase are frequently responsible for catalytic activity. It must be noted that, for the cases discussed thus far, the identification of the active centers as electronic defects or acidic and basic sites is quite conclusive but the mechanism of the reactions on these active centers is not known as yet. For instance, in the case of hydrocarbon cracking reactions proceeding on acid sites through carbonium ion intermediates, inadequate knowledge of the reaction mechanism is illustrated by the lack of explanation for the reported vast difference in catalytic activity for cracking of isooctane on silica alumina gels and decationated X-zeolites containing only protons as charge compensating cations [7.62]. In both cases, Brønsted acidic sites differ relatively little in their strength or concentration as measured by adsorption of ammonia [7.63], but the catalytic activity of the X-ray amorphous and crystalline materials of similar chemical composition, differs by almost four order of magnitude under comparable conditions.

Another type of active centers consists of coordinatively unsaturated transition metal cations, sometimes in an unusual oxidation state. For example, W^{3+} surface ions found on the edges of WS_2 crystallites and of WS_2 crystallites promoted with Ni^{2+} ions [7.64–66]. The concentration of W^{3+} ions was determined by their electron spin resonance spectra. The role of the Ni^{2+} ions which are inserted between close-packed sulfur ion planes of a layer lattice is to increase the concentration of W^{3+} ions which must be formed from W^{4+} to maintain electrical neutrality. The rate of hydrogenation of benzene was found to correlate with the intensity of the electron spin resonance signal as the concentration of the W^{3+} ions responsible for the signal and the catalytic activity increased by three orders of magnitude. Again, crystalline anisotropy effects are expected to be very important: the hydrogenation of benzene over these WS_2 catalysts cannot take place on the basal planes and is thus structure sensitive since it is energetically prohibitive to remove S^{2-} ions from the basal planes to achieve coordinative unsaturation of the tungsten ions, as calculated by ARLMAN [7.67] for the similar case of layer structures of $TiCl_3$ active in ZIEGLER-NATTA *stereospecific* polymerization of alkenes [7.68]. In the latter case, coordinatively unsaturated Ti^{3+} ions at the edges of the crystallites are part of the active centers responsible for the most selective form of catalysis achieved on an industrial scale by means of solid catalysts in very small particle size.

Ultimately, the identification of the active centers is complete only if the structure of the complexes they form with the reactive inter-mediates is also determined. This goal has now been achieved in a remarkable series of studies by KOKES and his coworkers [7.69]. The system investigated was the hydrogenation of ethylene in zinc oxide, proceeding through the most durable catalytic mechanism first pro-posed for this reaction by POLANYI and HORIUTI fourty years ago [7.70].

According to this mechanism, the alkene R is chemisorbed asso-ciatively on a catalytic site, H_2 ia chemisorbed dissociatively on two sites, the chemisorbed alkene reacts step-wise with two chemisorbed H atoms to produce first a chemisorbed alkyl radical RH, the so-called halfhydrogenated state, then finally the alkane RH_2

$$* + R \rightleftharpoons *R$$

$$2* + H_2 \rightleftharpoons 2*H$$

$$*R + *H \rightleftharpoons *RH + *$$

$$*RH + *H \longrightarrow RH_2 + 2*.$$

Whereas the last step is clearly irreversible at the temperatures at which hydrogenation of alkenes is studied, the first three steps may be reversible. If so, when the alkene is hydrogenated with dideuterium, multiply exchanged alkenes and alkanes can appear. The study of the patterns of exchange for alkane-D_2 and alkene-D_2 over many catalysts has led to a wealth of interesting conclusions and predictions concerning the nature and stereochemistry of chemisorbed intermediates, over metallic [7.71] and non-metallic catalysts [7.72].

In addition, in the case of hydrogenation of ethylene on zinc oxide, possibly because the catalyst has a rather low activity so that the concentration of the active intermediates is relatively large, it has been possible to observe the intermediates directly by infra-red absorption spectrophotometry. In particular, KOKES and coworkers have established that *R is a π-complex between ethylene and Zn^{2+} ions, that *H intermediates are bound to either Zn^{2+} or O^{2-} ions and that the half-hydrogenated state is an ethyl radical bound to the zinc half of the active center. The molecular picture may not be complete but it is much more than a sketch and it rests on direct spectroscopic observations during reaction.

These examples of active centers indicate the progress currently made in the study of heterogeneous catalysis on conventional high surface area catalysts, as well as the great diversity in their chemistry. While catalysis by metals and alloys continues to attract a lot of attention as discussed in Section 7.5 and 7.6, knowledge of catalysis by semiconductors and insulators also progresses rapidly. In fact one of the most interesting forms of industrial catalysis involves both metallic and non-metallic active centers: it is called *bifunctional catalysis* and a good example is the isomerization of n-pentane to i-pentane in the presence of platinum catalysts supported on acidic alumina. The mechanism of the reaction is as follows: n-pentane is first dehydrogenated to n-pentene on platinum, the n-alkene is then isomerized to i-alkene on the acidic sites of the alumina and finally the i-pentene is rehydrogenated on the platinum sites to the final product, i-pentane. The concept of bi- or multifunctional catalysis has many applications in nature and industry [7.74]. Of particular interest is the fact that the two catalytic functions must be sufficiently near to each other to avoid diffusional limitations as intermediates have to be transported from one function to the other. These transport phenomena dictate the size of the particles of both catalytic phases. Thus bifunctional catalysis is another example and for different reason of a catalytic phenomenon which can be observed only with sufficiently small particles and not on large crystals.

Acknowledgement. Partial support of this work by Exxon Research and Engineering Company is gratefully acknowledged.

References

7.1. G. K. BORESKOV: In *The Second Japan-Soviet Catalysis Seminar, New Approach to Catalysis* (Catalysis Society of Japan, Tokyo 1973), p. 114.

7.2. E. K. RIDEAL: *Concepts in Catalysis*, (Academic Press, New York, 1968), p. 113.

7.3. R. L. BURWELL, JR., M. BOUDART: In *Investigation of Rates and Mechanisms of Reactions*, Part I, ed. E. S. LEWIS (John Wiley & Sons, New York, 1974), Chapt. 12.

7.4. J. M. THOMAS, W. J. THOMAS: *Introduction to Principles of Heterogeneous Catalysis* (Academic Press, New York, 1967).

7.5. A. CLARK: *Theory of Adsorption and Catalysis* (Academic Press, New York, 1970).

7.6. S. J. GREGG, K. S. W. SINGH: *Adsorption, Surface Area and Porosity* (Academic Press, New York, 1967).

7.7. M. BOUDART: *Kinetics of Chemical Processes* (Prentice Hall, Engelwood Cliffs, N. Y., 1968), Chapter 9.

7.8. M. BOUDART: *Physical Chemistry*, eds. H. EYRING, D. HENDERSON and W. JOST, Vol. 7, Chapt. 7, *Heterogeneous Catalysis* (Academic Press, New York 1975).

7.9. M. J. D. LOW: Chem. Rev. **60**, 267 (1960).

7.10. A. OZAKI, H. TAYLOR, M. BOUDART: Proc. Roy. Soc. (London) A **258**, 47 (1960).

7.11. M. BOUDART: AIChE Journal **18**, 465 (1972).

7.12. A. OZAKI: In *Fixation of Dinitrogen*, Vol. 1, ed. W. F. HARDY (John Wiley, New York, 1975), Chapt. 4.

7.13. E. I. SHAPATINA, V. L. KUCHAEV, M. I. TEMKIN: Kinet. Katal. **12**, 1476 (1971).

7.14. K. TAMARU: Adv. Catal. Relat. Subj. **15**, 65 (1964).

7.15. J. H. de BOER (ed.): *Mechanism of Heterogeneous Catalysis* (North-Holland Publishing Co., Amsterdam, 1960).

7.16. C. WAGNER: Adv. Catal. Relat. Subj. **21**, 323 (1970).

7.17. S. STOTZ: Ber. Bunsenges. **70**, 37 (1966).

7.18. J. McALLISTER, R. S. HANSEN: J. Chem. Phys. **59**, 414 (1973).

7.19. G. L. ERTL, J. KOCH: In *Proceedings of Vth Intern. Congress Catalysis*, ed. by J. W. HIGHTOWER (North-Holland Publishing Co., Amsterdam, 1973), p. 969.

7.20. R. VAN HARDEVELD, F. HARTOG: Surface Sci. **15**, 189 (1969).

7.21. M. BOUDART, A. W. ALDAG, J. E. BENSON, N. A. DOUGHARTY, C. G. HARKINS: J. Catal. **6**, 92 (1966).

7.22. D. R. KAHN, E. E. PETERSEN, G. A. SOMORJAI: J. Catal. **34**, 294 (1974).

7.23. M. BOUDART, R. J. MADON: AIChE Journal (to be published), also R. J. MADON, Ph. D. Thesis, Stanford 1974.

7.24. M. BOUDART, E. SEGAL: J. Catal. (to be published), also: R. J. MADON, Ph. D. Thesis, Stanford 1974.

7.25. M. BOUDART: Accounts Chem. Res. (to be published), also: R. J. MADON, Ph. D. Thesis, Stanford 1974.

7.26. D. J. C. YATES, J. H. SINFELT: J. Catal. **8**, 348 (1967).

7.27. J. R. ANDERSON, Y. SHIMOYAMA: In *Proceedings of Vth Int. Congress Catalysis*, ed. by J. W. HIGHTOWER (North-Holland Publishing Co., Amsterdam, 1973), p. 695.

7.28. S. R. DYNE, J. B. BUTT, G. L. HALLER: J. Catal. **25**, 391 (1972).

7.29. F. V. HANSON, M. BOUDART: J. Catal. (to be published); also F. V. HANSON: Ph. D. Thesis, Stanford, 1975.

7.30. J. BÉNARD: Catal. Rev. **3**, 93 (1970).

7.31. M. BOUDART: J. Vac. Sci. Technol. **12**, 329 (1975).

7.32. J. H. SINFELT, J. L. CARTER, D. J. C. YATES: J. Catal. **24**, 283 (1972).

7.33. R. L. BURWELL, JR., G. L. HALLER, K. C. TAYLOR, J. F. READ: Advan. Catal. **20**, 1 (1969).

7.34. G. C. BOND: Disc. Faraday Soc. **41**, 200 (1966).

7.35. Y. SOMA-NOTO, W. M. H. SACHTLER: J. Catal. **32**, 315 (1974).

7.36. M. Boudart: J. Am. Chem. Soc. **70**, 1040 (1950).

7.37. L. Pauling: Proc. Roy. Soc. (London) A **196**, 343 (1959).

7.38. J. H. Sinfelt: Catal. Rev. **9**, 147 (1974).

7.39. P. A. Kilty, W. M. H. Sachtler: Catal. Rev. **10**, 1 (1974).

7.40. L. D. Ptak, M. Boudart: J. Catal. **16**, 90 (1970).

7.41. J. R. Anderson, N. R. Avery: J. Catal. **5**, 446 (1966).

7.42. T. N. Rhodin, P. W. Palmberg, E. W. Plummer: In *The Structure and Chemistry of Solid Surfaces*, ed by G. A. Somorjai (John Wiley and Sons, Inc., New York, 1969), paper No. 22.

7.43. F. Williams, M. Boudart: J. Catal. **30**, 438 (1973).

7.44. V. Ponec, W. M. H. Sachtler: *Proceedings of Vth Intern. Congress Catalysis* (North-Holland Publishing Co., Amsterdam, 1973), p. 645.

7.45. F. F. Vol'kenshtein: *Fiziko-khimya poverkhnosti poluprovodnikov* (Nauka Publishers, Moscow, 1973), Chapt. 5.

7.46. H. Chon, C. D. Prater: Disc. Faraday Soc. **41**, 380 (1966).

7.47. P. Amigues, S. J. Teichner: ibid. p. 362.

7.48. M. Boudart: Proc. of the Robert A. Welch Foundation Conferences on Chem. Res. XIV: Solid State Chemistry, Houston, Tex. (1970), p. 299.

7.49. P. H. Emmett: In: *Structure and Properties of Solid Surfaces*, ed. by R. Gomer C. S. Smith (The University of Chicago Press, Chicago, 1953), p. 414.

7.50. H. Topsøe, J. A. Dumesic, M. Boudart: J. Catal. **28**, 447 (1973).

7.51. J. W. Hightower: Accounts Chem. Res. 1975 (to be published).

7.52. J. B. Butt: Advan. Chem. Series **109**, 259 (1972).

7.53. O. D. Gonzales, G. Parravano: J. Am. Chem. Soc. **78**, 4533 (1956).

7.54. J. Oudar: Compt. Rend. **249**, 91 (1959).

7.55. H. S. Taylor: Proc. Roy. Soc. (London) A **108**, 105 (1925).

7.56. M. Boudart, A. Delbouille, E. G. Derouane, V. Indovina, A. B. Walters: J. Am. Chem. Soc. **94**, 6622 (1972).

7.57. Yu. A. Mishchenko, B. K. Boreskov: *Kinet. Katal.* **6**, 842, (1965).

7.58. H. W. Kohn: J. Catal. **2**, 208 (1968).

7.59. K. Tanabe: *Solid Acids and Bases* (Academic Press, New York, 1970).

7.60. P. B. Weisz: Ann. Rev. Phys. Chem. **21**, 175 (1970).

7.61. M. J. Baird, J. H. Lunsford: J. Catal. **26**, 440 (1972).

7.62. J. N. Miale, N. Y. Chen, P. B. Weisz: J. Catal. **5**, 278 (1966).

7.63. J. E. Benson, K. Uchiba, M. Boudart: J. Catal. **9**, 91 (1967).

7.64. R. J. H. Voorhoeve, J. C. M. Stuiver: J. Catal. **23**, 228 (1971).

7.65. R. J. H. Voorhoeve: J. Catal. **23**, 236 (1971).

7.66. R. J. H. Voorhoeve, J. C. M. Stuiver: J. Catal. **23**, 243 (1971).

7.67. E. J. Arlman: Rec. Trav. Chim. Pays-Bas **87**, 1217 (1968).

7.68. T. Keii: "Kinetics of Ziegler-Natta Polymerization", Kodansha, Tokyo 1972.

7.69. R. J. Kokes: Accounts Chem. Res. **6**, 226 (1973).

7.70. M. Polanyi, J. Horiuti: Trans. Faraday Soc. **30**, 1164 (1934).

7.71. R. L. Burwell, Jr.: Accounts Chem. Res. **2**, 289 (1969).

7.72. C. Kemball: Annals New York Acad. Sci. **213**, 90 (1973).

7.73. P. B. Weisz: Advan. Catal. **13**, 137 (1967).

Additional References with Titles

Chapter 2

B. BELL, A. MADHUKAR: A Theory of Chemisorption on Metallic Surfaces: The Role of Intra-Adsorbate Coulomb Correlation and Surface Structure (to be published).

W. BRENIG, K. SCHÖNHAMMER: On the theory of chemisorption. Z. Physik **267**, 201 (1974).

T. B. GRIMLEY, C. PISANI: Chemisorption theory in the Hartree-Fock approximation. J. Phys. C **7**, 2831 (1974).

Chapter 4

4.1. Thermal Desorption

M. BALOOCH, M. J. CARDILLO, D. R. MILLER, R. E. STICKNEY: Molecular beam study of the apparent activation barrier associated with adsorption of hydrogen on copper. Surface Sci. **46**, 358 (1974).

R. CHEN: On the analysis of thermal desorption curves. Surface Sci. **43**, 657 (1974).

C. T. FOXON, M. R. BOUDRY, B. A. JOYCE: Evalution of surface kinetic data by the transform analysis of modulated molecular beam measurements. Surface Sci. **44**, 69 (1974).

D. A. KING: Thermal desorption from metal surfaces: A review. Surface Sci. **47**, 384 (1975).

D. A. KING, M. G. WELLS: Reaction mechanisms in chemisorption kinetics: Nitrogen on W(100). Proc. Roy. Soc. **339**, 245 (1974).

E. V. KORNELSEN, D. H. O'HARA: Analysis of thermal desorption spectra using a computer graphics system. J. Vac. Sci. Technol. **11**, 885 (1974).

F. M. LORD, J. S. KITTELBERGER: On the determination of activation energies in thermal desorption experiments. Surface Sci. **43**, 173 (1974).

M. R. SHANABARGER: Clarification of the kinetics observed in isothermal and programmed thermal desorption measurements. Surface Sci. **44**, 297 (1974).

M. SMUTEK: Unified and generalized treatment of thermal desorption data. Vacuum **24**, 173 (1974).

4.2. Electron Impact Desorption and

4.3. Photodesorption

J. L. GERSTEN, R. JANOW, N. TZOAR: Theory of photodesorption. Phys. Rev. B **11**, 1267 (1975).

D. LICHTMAN: Electron- and photon-induced desorption. J. Nucl. Mat. **53**, 285 (1974).

V. K. RYABCHUK, L. L. BASOV, A. A. LISACHENKO, F. I. VILESOV: Time-of-flight determination of the kinetic energy of photodesorption products (the NO/Al_2O_3-system). Sov. Phys. Techn. Phys. **18**, 1349 (1974).

YA. P. ZINGERMAN: Electron-stimulated desorption of oxygen on a single-crystal surface of tungsten. Fiz. Tved. Tela **16**, 1795 (1974) (Sov. Phys. Solid State **16**, 1168 (1974).

4.4. Ion Impact Desorption

S. M. LIU, W. E. RODGERS, E. L. KNUTH: Interaction of hyperthermal atomic beams with solid surfaces. J. Chem. Phys. **61**, 902 (1974).

Z. SROUBEK: Theoretical and experimental study of the ionization processes during the low energy ion sputtering. Surface Sci. **44**, 47 (1974).

4.5. Field Desorption

J. A. PANITZ: The crystallographic distribution of field-desorbed species. J. Vac. Sci. Technol. **11**, 206 (1974).

W. A. SCHMIDT, O. FRANK, J. H. BLOCK: Investigations of field desorption products from silver surfaces by mass spectrometry. Surface Sci. **44**, 185 (1974).

Chapter 6

General

Proc. 2nd Internat. Conf. on Solid Surfaces, 1974, Japan; J. Appl. Phys. Suppl. 2, Pt. 2, 1974; in particular p. 607, p. 795 (LEED, AES).

The Solid-Vacuum Interface (Proc. 3rd Symp. on Surface Physics, 1974); Surface Sci. **47**, Nr. 1 (1975) (AES, LEED).

S. Y. TONG: *Progress in Surface Science* (Pergamon Press, Oxford 1975).

To Section 6.2

Dynamical Calculations

J. E. DEMUTH, D. W. JEPSEN, P. M. MARKUS: Phys. Rev. Letters **32**, 1182 (1974); – Surface Sci. **45**, 733 (1974); – J. Phys. C. (Solid State Phys.) **8**, L 25 (1975).

J. E. DEMUTH, P. M. MARKUS, D. W. JEPSEN: Phys. Rev. B **11**, 1460 (1975).

C. B. DUKE, N. O. LIPARI, G. E. LARAMORE: J. Vac. Sci. Technol. **12**, 222 (1975).

M. VAN HOVE, S. Y. TONG: J. Vac. Sci. Technol. **12**, 230 (1975).

S. Y. TONG: Solid State Commun. **16**, 91 (1975).

Averaging and Transform Techniques

D. L. ADAMS, U. LANDMAN: Phys. Rev. Letters **33**, 585 (1974).

J. M. BURKSTRAND, G. G. KLEIMAN, F. ARLINGHAUS: Surface Sci. **46**, 43 (1974).

M. G. LAGALLY, J. C. BUCHHOLZ, G.-C. WANG: J. Vac. Sci. Technol. **12**, 213 (1975).

L. McDONNELL, D. P. WOODRUFF, K. A. R. MITCHELL: Surface Sci. **45**, 1 (1974).

K. A. R. MITCHELL, D. P. WOODRUFF, G. W. VERNON: Surface Sci. **46**, 418 (1974).

Imperfect Surfaces

C. B. DUKE, A. LIEBSCH: Phys. Rev. B **9**, 1126, 1150 (1974).

W. P. ELLIS: Surface Sci. **45**, 569 (1974).

G. ERTL, M. PLANCHER: Surface Sci. **48**, 364 (1975).

To Section 6.3

S. AKSELA, J. VÄYRYNEN, H. AKSELA: Phys. Rev. Letters **33**, 999 (1974).

G. BETZ, G. K. WEHNER, L. TOTH, A. JOSHI: J. Appl. Phys. **45**, 5312 (1974).

C. R. BRUNDLE: J. Vac. Sci. Technol. **11**, 212 (1974).

M. A. CHESTERS, B. J. HOPKINS, A. R. JONES, R. NATHAN: Surface Sci. **45**, 740 (1974); – J. Phys. C (Solid State Phys.) **7**, 4486 (1974).

J. W. GADZUK: Phys. Rev. B **9**, 1978 (1974).

J. T. GRANT, M. P. HOOKER, R. W. SPRINGER, T. W. HAAS: J. Vac. Sci. Technol. **12**, 481 (1975).

K. O. GROENEVELD, R. SPOHR: Vakuum-Technik **23**, 225 (1974).

T. W. HAAS, D. J. POCKER: J. Vac. Sci. Technol. **11**, 1087 (1974).

L. A. HARRIS: J. Vac. Sci. Technol. **11**, 23 (1974).

L. C. ISETT, J. M. BLAKELY: Rev. Sci. Instr. **45**, 1382 (1974).

A. P. JANSSEN, R. C. SCHOONMAKER, A. CHAMBERS, M. PRUTTON: Surface Sci. **45**, 45 (1974).

S. P. KOWALCZYK, L. LEY, F. R. MCFEELY, R. A. POLLAK, D. A. SHIRLEY: Phys. Rev. B **9**, 381 (1974).

T. NARUSAWA, S. KOMIYA: J. Vac. Sci. Technol. **11**, 312 (1974).

D. A. SHIRLEY: Phys. Rev. A **9**, 1549 (1974).

F. J. SZALKOWSKI, G. A. SOMARJAI: J. Chem. Phys. **61**, 2064 (1974).

Author Index

Adams, D. L.   74–76, 83, 115
Alferieff, M.   148, 157, 158
Alldrege, G. P.   13, 28, 37
Allyn, C.   219
Amenomiya, Y.   108
Amigues, P.   291
Anderson, J.   78, 79, 81, 82, 87, 97
Anderson, J. R.   289
Anderson, O. K.   200
Anderson, P. W.   43, 53
Appelbaum, J. A.   12, 25, 26, 35, 151
Arlman, E. J.   295
Armand, G.   122
Ashcroft, N. W.   11, 12, 166, 187
Avery, N. R.   289

Bagchi, A.   158, 162, 163
Bagus, P. S.   178
Baker, J. M.   87, 209
Balandin, A. A.   279, 285
Bardeen, J.   151
Bauer, E.   21
Beck, D. E.   29
Becker, G. E.   170
Bell, A. E.   147, 200, 205
Bendow, B.   124
Bethe, H. A.   165
Blakeley, J. M.   21
Bond, G. C.   288
Bradley, T. L.   120
Brenig, W.   134
Brinkman, W. F.   151
Brundle, C. R.   144
Burkstrand, J. M.   28, 29, 31, 36

Caroli, C.   157, 160
Carter, G.   109
Celli, V.   29
Chon, H.   291
Christensen, N. E.   193, 198
Clavenna, L. R.   87, 114

Cooper, J. W.   166, 170
Crouser, L. E.   199
Cutler, P. H.   152, 153, 155–157
Cvetanovic, R. J.   108
Cyrot-Lackmann, F.   32
Czyzewski, J. J.   133

Davenport, J. W.   33
Davison, S. G.   24
Dawson, P. T.   114
Degras, D. A.   120
Delchar, T. A.   72
Demuth, J. E.   214, 215
Dionne, N. J.   200
Duke, C. B.   148, 151, 157, 158, 187

Earnshaw, J. W.   111
Eastman, D. E.   87, 149, 166, 200, 209, 214, 215
Egelhoff, W. E.   205
Ehrlich, G.   70, 72, 88, 108, 111
Einstein, T. B.   57, 219
Endriz, J. G.   183, 184
Engel, T.   71
Engelhardt, H. A.   105
Ertl, G.   94, 285
Estrup, P. J.   78, 79, 81, 82, 87, 94, 97, 206
Eyring, H.   121, 123

Fauchier, J.   151, 158, 187
Feder, R.   195
Feibelman, P. J.   31, 36
Ferrell, R. A.   29
Feuerbacher, B.   193, 198, 201, 205
Fitton, B.   205
Flood, D. J.   161
Fowler, R. H.   150
Frenkel, J.   106

Gadzuk, J. W.   147, 155, 162, 172–174, 189
Gay, J. G.   36

Gerlach, R. L.   117
Germer, L. H.   21, 74–76
Gomer, R.   23, 71, 85, 88, 120, 124, 131, 136, 137, 144, 147, 211, 212, 219
Goodman, F. O.   122, 124
Goymour, C. G.   115, 122
Griffin, A.   31
Grimley   44, 57
Grobman, W. D.   201
Gurney, R. W.   15
Gustafsson, T.   219

Hagstrum, H. D.   170
Hamann, D. R.   12, 25, 26, 35
Hansen, R. S.   286
Harris, J.   31
Harrison, W. A.   151
Haydock, R.   33
Heine, V.   24
Hertz, J.   57
Hobson, J. P.   111
Hohenberg, P.   6
Horiuti, J.   282, 295

Ibach, H.   198, 199
Inghram, M. G.   136
Inglesfeld, J. E.   29, 30
Ishchuk, V. A.   134
Itskovitch, F. I.   151

Jaklevic, R. C.   160
Jelend, W.   105
Jennings, P. J.   21
Johnson, K. H.   1
Jones, R. O.   25
Jones, W.   7
Juenker, D.   149

Kalkstein, D.   34
Kanamori, J.   53
Kane, E. O.   199
Kaplan, I. G.   173
Keck, J. C.   123
King, D. A.   87, 88, 115, 120, 122
Kjeldish, L. V.   157
Kleinman, L.   13, 28, 36
Kobozev, N. I.   285
Koch, J.   286
Kohn, W.   5, 6, 9–11, 13, 60, 61
Kohrt, C.   88, 120, 212
Kokes, R. J.   295
Kramers, H. A.   123
Kuznietz, M.   166

Lambe, J.   160
Lang, N. D.   1, 9–11, 13, 14, 34
Langreth, D. C.   11, 12
Lee, M. J. G.   162, 163
Levine, J. D.   24
Leung, C.   212
Liebsch, A.   172, 189, 191
Lundquist, B. I.   162
Lyo, S. K.   23

McAllister, J.   286
McCarroll, B.   111
McMillan, W. L.   53
McRae, E. G.   21
Madey, T. E.   114, 169, 205
Mahan, G. D.   187
March, N. H.   7
Markin, A. P.   173
Matysik, K. J.   78
Menzel, D.   88, 105, 124, 131, 144
Mitchell, K.   183–188
Modinos, A.   160, 200
Moore, G. E.   124
Müller, E. W.   136, 145

Natta, G.   295
Neumann, H.   162
Newns, D. M.   43
Nordheim, L.   150

Onffroy, J.   5

Pagni, P. J.   123
Pandey, K. C.   27, 28, 33, 35
Peng, Y. K.   114
Penn, D.   51, 151, 156–158
Peria, W. T.   200
Petermann, L. A.   105
Pettifor, D. G.   34
Phillips, J. C.   27, 28, 33, 35
Piper, T. C.   206
Pisani, C.   113
Plummer, E. W.   22, 36, 87, 147, 154, 155, 158, 169, 191, 195, 196, 200, 205, 208, 212
Polanyi, M.   295, 296
Politzer, B. A.   152, 153, 155–157
Ponec, V.   290
Prater, C. D.   291
Price, W. C.   166
Probst, F. M.   204
Pyzhev, V.   281, 282

Redhead, P. A.   85, 108–110, 124, 131
Rhodin, T. N.   117, 200, 289
Ritchie, R. H.   29
Rivière, J. C.   6
Rowe, J. W.   198, 199
Rubloff, G. W.   205

Sachtler, W. M. H.   290
Salpeter, E. E.   165
Schacch, W. L.   166, 187
Schmidt, L. D.   79, 81, 83, 87, 97, 114, 117
Schrieffer, J. R.   57, 61
Scofield, J. H.   168
Sham, L. J.   9
Shepherd, W. B.   200
Sinfelt, J. H.   219, 288
Slater, J. C.   168
Slater, N. B.   122
Smith, D. P.   200
Smith, J. R.   181
Soven, P.   157, 219
Spicer, W. E.   189
Stern, E. A.   29
Stotz, S.   285
Stratton, R.   151
Sturm, K.   195
Suhl, H.   123
Swanson, L. W.   147, 199

Tamm, I.   23
Tamm, P. W.   79, 81, 83, 97
Tarng, M. L.   149
Taylor, T. N.   94
Tecchner, S. J.   291
Temkin, M. I.   281, 282, 285
Toya, T.   115
Tracy, J. C.   21, 36, 94, 97, 98, 149
Tully, J. C.   173

Vol'kenshtein, F. F.   290, 291

Waelawski, B. J.   169, 190, 195, 196, 201
Wagner, C.   283, 285, 291
Wallis, R. F.   1
Weinberg, W. H.   161
Wikborg, E.   29, 30

Yates, J. T.   88, 114, 210, 214
Ying, S. C.   12, 29, 36, 124
Young, P. L.   158, 211, 212
Young, R. D.   154, 155, 158

Ziegler, K.   295
Zingerman, Ya. P.   134

Subject Index

Active sites   275
Adsorbate-adsorbate interaction   57
— —, attractive   89, 91
— —, repulsive   91, 92, 115
Adsorbate characterization   68
Adsorbates, binding states of   77, 90
—, ordered structure of   91
Adsorption, crystallographic anisotropy
   72
—, dissociative   104
— energy   49, 90 (table) (see also desorp-
   tion energies)
—, kinetics of   70
— of ... see substance in question
— on Cu   92
— — fcc metals   94, 97, 98
— — Mo   95–97
— — Ni   92, 214–216
— — Pt   247
— — Rh 247
— — Ta   95–97
— — W   13, 21, 22, 25, 72–84, 84–90,
   95–97, 107, 112, 115, 119, 159, 196, 200,
   205, 216–219, 247, 250
—, order-disorder transformation in   250
Ag desorption from W   107 (figure)
Alkali adsorption   14
Alkene hydrogenation   295
Arrhenius plot   113
Attenuation of electrons   149 (figure), 230,
   234
Auger emission   253–268
— —, line width in   254
Auger microscopy   266
Auger spectroscopy   71, 253–268
— —, experimental methods in   259–261
— —, line width in   256
— —, relaxation effects in   264, 265
— —, theory of   263–266

Binding states of some adsorbates   77, 90
   (table)

Bonding-antibonding levels   42, 47
Bonhoeffer-Farkas mechanism   276
Brønsted relations   278, 279

Catalysis, active centers in   293–296
—, bi-functional   296
—, electronic factors in   288–291
—, electronic theory of   290
—, reactivity patterns in   289
—, structural factors in   285–288
Catalyst, activity   276
—, deactivation   292
—, dispersion   277
—, poisons   292, 293
—, promoters   291–293
—, selectivity   277
—, specificity   289
—, texture   277
C_2H_4 adsorption on Ni   214–216
— — on W   216–219
Chemical potential   6, 7
Chemisorption energy   49
— — see also adsorption, desorption
CO adsorption on fcc metals   94, 97, 98
— — — Cu   92
— — — Mo   95–97
— — — Ni   92
— — — Pt   247
— — — Ta   95
— — — W   84–89, 90 (table), 95–97,
   115
— — — W, binding states in   85–88,
   115, 206–213
CO from W, desorption of   106, 112, 114
CO on W, dissociation of   87, 210
Compensation effect   120, 123
Complete basis   51
Continuum states   54
Correlation   7, 42, 229
Coster-Kronig transitions   255
Coulomb repulsion   42

Cu-Ni catalyst   290
Cyclohexane hydrogenation   287

Dangling bonds   27
Debye-Waller factor   233
Density function   6
Density functional   16
Density of states
—, joint   186
—, local   42, 45, 52
—, one-dimensional   157, 193, 194
—, total   42, 49
Dielectric response   16
Diffusion, surface   70
Dipole barrier   10
Desorption, electron impact, see electron impact desorption
— energies   19, 116 (table), 117 (table), 118–119 (table)
— flash   108–115
—, isothermal   105
— step   109
—, thermal, see thermal desorption
Dwell time   106

Electron energy analyzers   259, 260
Electron impact desorption   124–135
— — —, cross sections of   128 (table)
— — —, determination of binding states by   129
— — —, desorption probability in   128, 133
— — —, energy distribution of products   132
— — —, isotope effects in   131, 133
— — — of CO from W   128, 129
— — — of hydrogen from W   128
— — — of oxygen from W   128
— — —, practical importance of   134
— — —, theory of   131–134
— — —, threshold energy of   131
Electronic defects   293
Electronic structure   1
Electrons, number density of   6
Electrostatic barrier   7, 15
Eley-Rideal mechanism   276
Energy distributions in field emission, see field emission energy distribution
— — in photoemission, see photo-emission energy distribution
Energy of adsorption, see adsorption energy

Energy of desorption, see desorption energy
Escape depth   148
Exchange energy   7

Facetting   239
Field desorption   136–138
Field emission   145–148, 150–163
— —, energy distributions in, see field emission energy distributions
— —, Fowler-Nordheim equation for   150
— — from adsorbate-covered surfaces   157–162
— — from clean surfaces   151–157
— —, photo-assisted   162, 163
— —, spin polarization in   155–157
— —, transmission probability in   152, 153
— —, W.K.B. description of   153, 154, 156
Field emission energy distributions from Ge   200
— — — — from Mo   200
— — — — from hydrogen covered W   196, 200, 205
— — — — from CO covered W   211–213
— — — — from oxygen covered W   159
— — — —, theory of   151–162
Field ionization   136–138
Flash desorption   108–115
Fowler-Nordheim equation   150
Freundlich isotherm   280
Frumkin-Temkin isotherm   280

Green's function   43
Ground state energy   6
Group orbital   50

Hartree-Fock approximation   41, 42
— —, magnetic and nonmagnetic cases of   47, 48
Heats of adsorption, desorption see ad-desorption energies
H_2-O_2 reaction   228
Hydrocarbon decomposition   214–219
Hydrogen adsorption   18
— —, electronic potential in   21
— — on Mo   95–97
— — on Ta   95–97

— — on W   13, 21, 22, 55, 77–84, 90, 95–97, 112, 119, 196, 200, 201–205
Hydrogenation of alkenes   295
— of cyclohexene   287

Image interaction   42
Image plane   12
Inelastic tunneling   160–162
Ion impact desorption   136
Ion neutralization spectroscopy   144

Jellium   7, 17, 231

Koopman's theorem   166, 178–180

LCAO-MO method   41
LEED   69
—, coherence region in   237, 238
—, dynamical theory of   243, 244
—, experimental methods of   234–236
— from adsorbed layers   247–251
— from clean surfaces   244–247
— from hydrogen on W   21, 78–84
— from CO on W   87
—, inelastic   251, 252
—, kinematical theory of   236–242
—, microscopy   252, 253
—, nomenclature in   268–270
—, spot intensity in   238, 239, 241
—, spot position in   239
Linear response   55
Linearization   17
Local density of states   4, 32, 158
Local orbitals   2
Local spin susceptibility   59
Localized states   46, 53
Longmuir-Hinshelwood mechanism   276
Low energy electron diffraction see LEED

Magnetic, nonmagnetic solutions   47, 48

Newns-Anderson model   43
NH_3 synthesis   275, 280–283
Nitrogen adsorption on Mo   95–97
— — on Ta   95–97
— — on W   72–76, 95–97

Oxygen adsorption on Rh   247
— — on W   159, 247, 250

Phase transitions, 2-dimensional   93–95
— —, order-disorder   250

Phonons, surface   1
Photodesorption   135, 136
Photoelectric threshold   5
Photoemission   148, 150, 163–193
—, angular dependence of   171–174
—, angular resolved   189–193
—, cross sections in   166–171
—, dipole approximation in   165
—, direct   184
—, experimental results of   193–219
—, fingerprint technique in   214
— from CO on W   206–211, 213
— from EuS   166, 167
— from hydrogen on W   22, 196, 201–205
— from oxygen on W   190
— from Si   199
—, relaxation effects in   174–182
—, surface shift in   180–182
—, theory of energy distributions in   151–162
—, study of hydrocarbon decomposition by   214–219
—, surface   182–189
Photoexcitation   163–166
Plasmon gain   265
Plasmons   230, 232, 251
—, bulk   28
—, dispersion relations of   31
—, surface   28
Polarized electrons   252
Pseudopotential   11, 13

Random phase approximation   30
Reconstruction   269
Resonances   53, 57
Response function   17, 29

Sabatier principle   279
Scattering of electrons, elastic   227–229
— — —, inelastic   230–233
— — —, quasielastic   233
Screening charge   12, 16, 21
Screening density   17
Screening length   9
Self-consistency   9, 10, 16, 30, 33
SIMS   136
Step desorption   109
Stepped surfaces   74
Structure factor   237
Superstructures   244
Surface barrier   2
Surface diffusion   70

Surface energy   1
Surface potential   2
— — of Si   26
Surface resonance   195
Surface states   2, 23
Surface structure   64
— — by LEED   244–247
— — of bcc metals   65
— — of fcc metals   67
— — of semiconductors   67
— — of insulators   68
Surface valence   289

Temkin-Pyzhev equation   281, 282
Thermal desorption   102
— —, activation energies of   19, 116, 118, 119
— —, angular distribution of products in 120
— —, isothermal   105
— —, kinetics of   70
— —, measurement of   105–115
— — of electronegative adsorbates   118, 119

— — of metals   116
— — of weakly bound adsorbates   117
— —, order of   103, 104, 113, 116, 117
— —, theory of   120–124
— —, transition state theory of   121–123
Transfer Hamiltonian   155, 156
Tunneling resonance   148, 157
Turnover number   276

Unit mesh   269

Virtual level   22, 23
Virtual state   47
Valence bond method   58

Wannier functions   5
Watergas shift reaction   283–285
Work function   4, 5
— — of hydrogen on W   80
— — of nitrogen on W   76

Zeolite catalyst   293

Applied Physics

A monthly journal

Board of Editors
A. Benninghoven, Münster · **R. Gomer,** Chicago, Ill.
F. Kneubühl, Zürich · **H. K. V. Lotsch,** Heidelberg
H. J. Queisser, Stuttgart · **F. P. Schäfer,** Göttingen
A. Seeger, Stuttgart · **K. Shimoda,** Tokyo
T. Tamir, Brooklyn, N.Y. · **H. P. J. Wijn,** Eindhoven
H. Wolter, Marburg

Coverage
application-oriented experimental and theoretical physics:

Solid-State Physics   Quantum Electronics
Surface Physics   Coherent Optics
Infrared Physics   Integrated Optics
Microwave Acoustics   Electrophysics

Special Features
rapid publication (3-4 months)
no page charges for **concise** reports

Languages
Mostly English; with some German

Articles
review and/or tutorial papers
original reports, and short communications
abstracts of forthcoming papers

Manuscripts
to Springer-Verlag (Attn. H. Lotsch), P.O. Box 105 280
D-69 Heidelberg 1, F.R. Germany

Distributor for North-America:
Springer-Verlag New York Inc., 175 Fifth Avenue, New York. N.Y. 100 10, USA

Springer-Verlag
Berlin Heidelberg New York

Volume 66
30 figures. III, 173 pages. 1973
ISBN 3-540-06189-4 Cloth DM 78,–
ISBN 0-387-06189-4
(North America) Cloth $33.60

Quantum Statistics

in Optics and Solid-State Physics

R.Graham: Statistical Theory of Instabilities in Stationary Nonequilibrium Systems with Applications to Lasers and Nonlinear Optics.
F. Haake: Statistical Treatment of Open Systems by Generalized Master Equations.

Volume 67
III, 69 pages. 1973
ISBN 3-540-06216-5 Cloth DM 38,–
ISBN 0-387-06216-5
(North America) Cloth $16.40

S. Ferrara, R. Gatto, A. F. Grillo:

Conformal Algebra in Space-Time

and Operator Product Expansion

Introduction to the Conformal Group in Space-Time. Broken Conformal Symmetry. Restrictions from Conformal Covariance on Equal-Time Commutators. Manifestly Conformal Covariant Structure of Space-Time. Conformal Invariant Vacuum Expectation Values. Operator Products and Conformal Invariance on the Light-Cone. Consequences of Exact Conformal Symmetry on Operator Product Expansions. Conclusions and Outlook.

Volume 68
77 figures. 48 tables. III, 205 pages. 1973
ISBN 3-540-06341-2 Cloth DM 88,–
ISBN 0-387-06341-2
(North America) Cloth $37.90

Solid-State Physics

D. Schmid: Nuclear Magnetic Double Resonance — Principles and Applications in Solid-State Physics.
D.Bäuerle: Vibrational Spectra of Electron and Hydrogen Centers in Ionic Crystals.
J. Behringer: Factor Group Analysis Revisited and Unified.

Volume 69
13 figures. III, 121 pages. 1973
ISBN 3-540-06376-5 Cloth DM 78,–
ISBN 0-387-06376-5
(North America) Cloth $33.60

Astrophysics

G. Börner: On the Properties of Matter in Neutron Stars.
J. Stewart, M. Walker: Black Holes: the Outside Story.

Prices are subject to change without notice
■ Prospectus with Classified Index of Authors and Titles
Volumes 36—74 on request

Volume 70
II, 135 pages. 1974
ISBN 3-540-06630-6 Cloth DM 77,–
ISBN 0-387-06630-6
(North America) Cloth $33.20

Quantum Optics

G. S. Agarwal: Quantum Statistical Theories of Spontaneous Emission and their Relation to Other Approaches.

Volume 71
116 figures. III, 245 pages. 1974
ISBN 3-540-06641-1 Cloth DM 98,–
ISBN 0-387-06641-1
(North America) Cloth $42.20

Nuclear Physics

H. Überall: Study of Nuclear Structure by Muon Capture.
P. Singer: Emission of Particles Following Muon Capture in Intermediate and Heavy Nuclei.
J. S. Levinger: The Two and Three Body Problem.

Volume 72
32 figures. II, 145 pages. 1974
ISBN 3-540-06742-6 Cloth DM 78,–
ISBN 0-387-06742-6
(North America) Cloth $33.60

D. Langbein:

Theory of Van der Waals Attraction

Introduction. Pair Interactions. Multiplet Interactions. Macroscopic Particles. Retardation. Retarded Dispersion Energy. Schrödinger Formalism. Electrons and Photons.

Volume 73
110 figures. VI, 303 pages. 1975
ISBN 3-540-06943-7 Cloth DM 97,–
ISBN 0-387-06943-7
(North America) Cloth $41.80

Excitons at High Density

Editors: H. Haken, S. Nikitine
Biexcitons. Electron-Hole Droplets. Biexcitons and Droplets. Special Optical Properties of Excitons at High Density. Laser Action of Excitons. Excitonic Polaritons at Higher Densities.

Volume 74
75 figures. III, 153 pages. 1974
ISBN 3-540-06946-1 Cloth DM 78,–
ISBN 0-387-06946-1
(North America) Cloth $33.60

Solid-State Physics

G. Bauer: Determination of Electron Temperatures and of Hot Electron Distribution Functions in Semiconductors.
G. Borstel, H. J. Falge, A. Otto: Surface and Bulk Phonon-Polaritons Observed by Attenuated Total Reflection.